VOLUME TWO HUNDRED AND TWENTY

ADVANCES IN IMAGING AND ELECTRON PHYSICS

VOLUME TWO HUNDRED AND TWENTY

ADVANCES IN IMAGING AND ELECTRON PHYSICS

The Beginnings of Electron Microscopy - Part 1

Edited by

PETER W. HAWKES and MARTIN HŸTCH

CEMES-CNRS
Toulouse, France

Cover photo credit:
From E. Ruska: The Early Development of Electron Lenses and Electron Microscopy (Hirzel, Stuttgart, 1980). Courtesy S. Hirzel Verlag.

Academic Press is an imprint of Elsevier
125 London Wall, London EC2Y 5AS, United Kingdom
525 B Street, Suite 1650, San Diego, CA 92101, United States
50 Hampshire Street, 5th Floor, Cambridge, MA 02139, United States
The Boulevard, Langford Lane, Kidlington, Oxford OX5 1GB, United Kingdom

Notices

Knowledge and best practice in this field are constantly changing. As new research and experience broaden our understanding, changes in research methods, professional practices, or medical treatment may become necessary.

Practitioners and researchers must always rely on their own experience and knowledge in evaluating and using any information, methods, compounds, or experiments described herein. In using such information or methods they should be mindful of their own safety and the safety of others, including parties for whom they have a professional responsibility.

To the fullest extent of the law, neither the Publisher nor the authors, contributors, or editors, assume any liability for any injury and/or damage to persons or property as a matter of products liability, negligence or otherwise, or from any use or operation of any methods, products, instructions, or ideas contained in the material herein.

ISBN: 978-0-323-91507-6
ISSN: 1076-5670

Publisher: Zoe Kruze
Acquisitions Editor: Jason Mitchell
Developmental Editor: Jhon Michael Peñano
Production Project Manager: James Selvam
Designer: Greg Harris

Typeset by VTeX

Contents

Contributors

T.E. Allibone FRS
Formerly Metropolitan–Vickers, Manchester, United Kingdom
Formerly AEI, Aldermaston, United Kingdom

C. Colliex
Laboratoire de Physique des Solides (LPS), CNRS UMR 8502, Bâtiment 510, Université Paris-Sud, Université Paris-Saclay, Orsay, France

V.E. Cosslett
Formerly Cavendish Laboratory, University of Cambridge, Cambridge, England, United Kingdom

Armin Delong
Formerly Institute of Scientific Instruments of the Czechoslovak Academy of Sciences, Brno, Czechoslovakia

D.G. Drummond
Formerly 45 Albert Drive, West Killara, NSW, Australia

P. Duncumb FRS
Formerly Tube Investments Research Laboratories (TIRL), Hinxton Hall, Cambridge, United Kingdom

Gaston Dupouy
Formerly Laboratoire d'Optique Electronique du Centre National de la Recherche Scientifique, Toulouse, France

Humberto Fernández-Morán
Formerly Cryo-Electron Microscope Laboratories, Research Institutes, University of Chicago, Chicago, IL, United States

P. Grivet
Formerly Institut d'Electronique Fondamentale, Université de Paris-Sud, Orsay, France

Peter Hawkes
CEMES-CNRS, Toulouse, France

Bohumila Lencová
Tescan Brno, Brno, Czech Republic

Tom Mulvey
Formerly University of Aston, Birmingham, United Kingdom

Heinz Niedrig
Institut für Optik und Atomare Physik, Technische Universität Berlin, Berlin, Germany

Dieter Typke
Formerly Munich, Germany

Manfred Von Ardenne
Formerly Forschungsinstitut Manfred von Ardenne, Dresden-Weisser Hirsch, German Democratic Republic

Foreword

With one exception, the chapter by Charles Süsskind on Ladislaus Marton, the contributions to the 1985 edition of 'Beginnings' were all written by the pioneers themselves. This inevitably left many major gaps. Early work on electron microscopy at the AEG Research Institute was entirely absent for Ernst Brüche, Hans Mahl, Carl Ramsauer and Otto Scherzer had already passed away. (Scherzer had agreed to contribute but was finally unable to do so.) Hans Boersch was also too unwell at the time. In this second edition, I have filled in a few of the gaps. Boersch now appears in the form of a biographical article by Hans Niedrig, translated from the original in the house journal of the German Electron Microscopy Society (DGE), *Elektronenmikroskopie*. Raymond Castaing is now present: a full study by his former pupil Christian Colliex is reproduced here, accompanied by a tribute by Peter Duncumb and myself. The Royal Society has generously allowed us to reprint the long Memoir on Dennis Gabor by T.E. Allibone. Joachim Frank, who began his scientific life in Walter Hoppe's laboratory, has composed an account of Hoppe's life and work.

I very much wanted to include the pioneers of electron holography in this new edition and am very pleased that Hannes Lichte has agreed to write an account of his own work in which that of Akira Tonomura is intertwined. This could not be completed in time to appear here but will be published in a regular volume of these Advances. For Gottfried Möllenstedt, we reprint his account of the genesis of the electron biprism, first published in *Advances in Optical and Electron Microscopy* but not hitherto available in electronic form. Another of Möllenstedt's historical articles appears in 'Growth of Electron Microscopy' (volume 96 of these Advances). It is not widely known that two electron microscopes with magnetic lenses were produced by the Bosch company in Stuttgart in the early 1950s, at the instigation of Rudolf Rühle (author of *Das Elektronenmikroskop*, Curt E. Schwab, Stuttgart 1949). The life of Rudolf Rühle and the construction of these instruments are described by Hans Gelderblom and Heinz Schwarz. The last of the new articles is an account of Otto Scherzer's career by Dieter Typke, who prepared his doctoral thesis in Scherzer's laboratory in Darmstadt. Typke originally planned to cover the work performed at the AEG Research Institute in Berlin in the 1930s in this article. It was later decided to publish it as a separate chapter since Scherzer did not spend much of his career at the AEG Institute.

Several other major figures of the early years are still missing but do appear in earlier volumes of these Advances. The widow of Bodo von Borries published a biography of her husband in volume 81 (1991) of *Advances in Electronics and Electron Physics*, though some sentences were found excessive by other survivors of the early years. His cousin Götz von Borries has also published a small book on *Bodo von Borries und das Elektronenmikroskop* (Fouqué Literaturverlag, Egelsbach 2001). In 'Helmut Ruska (1908–1973): His role in the evolution of electron microscopy in the life sciences, and especially virology' in volume 182 (2014), Hans Gelderblom and Detlev Krüger have retrieved Ernst Ruska's brother Helmut from oblivion and shown how important was his role in persuading Siemens to launch their first commercial instrument, the Übermikroskop. The article on Dennis Gabor by Tom Mulvey in volume 91 (1995) nicely complements the Royal Society Memoir printed here. The life of Ernst Ruska, again by Mulvey together with Lotte Lambert, in volume 95 (1996) should be read alongside Ruska's own historical volume. Jan Le Poole was present in the original 'Beginnings' and a fuller picture of the man emerges from the biographical article by Mulvey and D.J.J. van de Laak-Tijsen in volume 115 (2001).

I particularly regret the absence of a full biography of Walter Glaser, who contributed so much to the theory of electron optics. The references in his monumental *Grundlagen der Elektronenoptik* (Springer, Vienna 1952), which are very complete, make it clear that Glaser's own work is at the heart of almost every chapter. The composition of that book is evoked vividly by Glaser's research students Hans Grümm and Peter Schiske in volume 96 (1996).

Johannes Picht should also not be forgotten. His *Einfuhrung in die Theorie der Elektronenoptik* (Barth, Leipzig 1939) went into three editions and one of his doctoral students (M. Leitner) drafted the first full study of the aberrations of cylindrical electron lenses before being killed in World War II.

All the contributors to the 1985 edition of 'Beginnings' have since died. I have therefore added an Afterword to each of their chapters. In a few cases (Armin Delong, David Drummond, Jan Le Poole, Cilly Weichan), these were kindly provided by authors close to the subjects.

Nor have the historians and chroniclers of science been idle. We refer to Chapter 26, in which full bibliographical details are provided of several books and articles on various aspects of the history of the subject. To these we add a major publication by Falk Muller, *Jenseits des Lichts. Siemens, AEG und die Anfänge der Elektronenmikroskopie in Deutschland* (Wallstein Verlag, Göttingen 2021). Royal Society Memoirs are available on V. Ellis Cosslett and Sir Charles Oatley, whose papers are housed in the archives of Churchill College, Cambridge. Obituary notices on many German electron microscopists are to be found in *Physikalische Blätter* and *Elektronenmikroskopie* (accessible via the website of the Deutsche Gesellschaft für Elektronenmikroskopie). Biographical papers on deceased members of the French Académie des Sciences can be found in their *Comptes Rendus*.

Peter W. Hawkes

Foreword to first edition

Electron microscopy is still a young subject and some of those who were active during its formative years, contributing to its development or implantation in their various countries, can still be met in their laboratories, while others, although retired, have not broken all contact with the subject. The idea of gathering recollections of the beginnings of electron optics and electron microscopy had been in my mind for some time but it was the publication of Professor E. Ruska's meticulous historical volume[1] that spurred me into action; that and the macabre reality that time was running out.

My initial approach to scientists in many countries met with a largely enthusiastic response, though the few refusals were particularly disappointing as they left whole areas of activity imperfectly represented. Also, several letters elicited no reply and I have no way of telling whether or not they failed to reach their targets. I mention these points here to explain the absence of a number of names, which might occasion surprise.

One country is, however, not represented for a quite different reason. Interest in electron optics and the associated instruments in Russia dates back to 1935 and the first commercial instrument appeared in 1946. Russian scientists have remained active in the field ever since. Despite innumerable letters and telegrams to members of the Russian electron microscope community, I have failed to obtain any response and I therefore address this more public appeal, to anyone concerned with the beginnings of electron microscopy in the USSR, to get in touch with me with a view to remedying this omission, perhaps in a future volume of these *Advances*.

This is not the only historical gap and I have therefore ventured to include a brief account of the whereabouts of further information on the history of the subject. Thus some of those who were not able to speak for themselves figure in these pages, if only at second hand.

I am sorry to have to record that a few of those who had planned to contribute, at least provisionally, have been prevented from doing so by ill health or by death. Helmut Johannson, who agreed to contribute, died in April, 1982. Otto Scherzer, who made no firm commitment but

[1] E. Ruska: Die Frühe Entwicklung der Elektronenlinsen und der Elektronenmikroskopie, *Acta Hist. Leopoldina* (1979), Nr 12, 136 pp.; translated into English by T. Mulvey as "The Early Development of Electron Lenses and Electron Microscopy" (Hirzel, Stuttgart, 1980).

from whom I still hoped for a short piece, perhaps on the first derivation of "Scherzer's Theorem," also died in 1982. Dr. R. W. G. Wyckoff sent me the short account published here but was prevented by illness from expanding it as he planned. Professor Hans Boersch has been obliged to withdraw for health reasons. Dr. A. C. van Dorsten has been delayed, but his contribution will appear in a future volume of the *Advances*.

It only remains for me to thank all the contributors for the time and hard work that they have devoted to preparing their articles for this book. From the outset, I made it clear that both formal history and more informal reminiscence would be welcome, best of all perhaps being a combination of the two. This explains the differences in length and tone between the various articles, and I trust that the reader will agree that this is not the least attractive feature of this collection. It also explains why the war figures so prominently in the articles by members of various occupied countries, where the first instruments were assembled in the grim conditions recalled here.

Finally, I am particularly grateful to Professor E. Ruska for agreeing to write a preface and for furnishing much invaluable information and numerous documents.

Peter W. Hawkes

Note on the references. The various European and International Conferences on electron microscopy are referred to so frequently that we merely give place and date in the individual lists of references. The full publishing details of these and the international high-voltage electron microscopy meetings are as follows:

Delft, 1949: *Proceedings of the Conference on Electron Microscopy*, Delft, 4–8 July, 1949, A. L. Houwink, J. B. Le Poole, and W. A. Le Rütte, eds. (Hoogland, Delft, 1950).

Paris, 1950: *Comptes Rendus du Premier Congrès International de Microscopie Electronique*, Paris, 14–22 September, 1950 (Editions de la Revue d'Optique Théorique et Instrumentale, Paris, 1953).

London, 1954: *The Proceedings of the Third International Conference on Electron Microscopy*, London, 1954, R. Ross, ed. (Royal Microscopical Society, London, 1956).

Gent, 1954: *Rapport Europees Congrès Toegepaste Electronenmicroscopie*, Gent, 7–10 April, 1954, edited and published by G. Vandermeersche (Uccle-Bruxelles, 1954).

Toulouse, 1955: *Les Techniques Récentes en Microscopie Electronique et Corpusculaire*, Toulouse, 4–8 April, 1955 (C.N.R.S., Paris, 1956).

Stockholm, 1956: *Electron Microscopy. Proceedings of the Stockholm Conference*, September, 1956, F. J. Sjöstrand and J. Rhodin, eds. (Almqvist and Wiksells, Stockholm, 1957).

Tokyo, 1956: *Electron Microscopy. Proceedings of the First Regional Conference in Asia and Oceania,* Tokyo, 1956 (Electrotechnical Laboratory, Tokyo, 1957).

Berlin, 1958: *Vierter Internationaler Kongress für Electronenmikroskopie*, Berlin, 10–17 September, 1958, *Verhandlungen*, W. Bargmann, G. Möllenstedt, H. Niehrs, D. Peters, E. Ruska, and C. Wolpers, eds. (Springer, Berlin, Göttingen, Heidelberg, 1960), 2 Vols.

Delft, 1960: *The Proceedings of the European Regional Conference on Electron Microscopy*, Delft, 1960, A. L. Houwink and B. J. Spit, eds. (Nederlandse Vereniging voor Elektronenmicroscopie, Delft n.d.), 2 Vols.

Philadelphia, 1962; *Electron Microscopy. Fifth International Congress for Electron Microscopy*, Philadelphia, Pennsylvania, 29 August to 5 September, 1962, S. S. Breese, ed. (Academic Press, New York, 1962), 2 Vols.

Prague, 1964: *Electron Microscopy 1964. Proceedings of the Third European Regional Conference*, Prague, M. Titlbach, ed. (Publishing House of the Czechoslovak Academy of Sciences, Prague, 1964), 2 Vols.

Kyoto, 1966: *Electron Microscopy 1966. Sixth International Congress for Electron Microscopy*, Kyoto, R. Uyeda, ed. (Maruzen, Tokyo, 1966), 2 Vols.

Rome, 1968: *Electron Microscopy 1968. Pre-Congress Abstracts of Papers Presented at the Fourth Regional Conference*, Rome, D. S. Bocciarelli, ed. (Tipografia Poliglotta Vaticana, Rome, 1968), 2 Vols.

HVEM Monroeville, 1969: *Current Developments in High Voltage Electron Microscopy (First National Conference)*, Monroeville, 17–19 June, 1969. Proceedings not published but *Micron* **1** (1969), 220–307, contains official reports of the meeting based on the session chairmen's notes.

Grenoble, 1970: *Microscopie Électronique 1970, Résumés des Communications Présentées au Septième Congrès International*, Grenoble, P. Favard, ed. (Société Française de Microscopie Electronique, Paris, 1970), 3 Vols.

HVEM Stockholm, 1971: The Proceedings of the Second International Conference on High-Voltage Electron Microscopy, Stockholm, 14–16 April, 1971; published as *Jernkontorets Annaler* **155** (1971), No. 8.

Manchester, 1972: *Electron Microscopy 1972. Proceedings of the Fifth European Congress on Electron Microscopy*, Manchester (Institute of Physics, London, 1972).

HVEM Oxford, 1973: *High Voltage Electron Microscopy. Proceedings of the Third International Conference*, Oxford, August, 1973, P. R. Swann, C. J. Humphreys, and M. J. Goringe, eds. (Academic Press, London and New York, 1974).

Canberra, 1974: *Electron Microscopy 1974. Abstracts of Papers Presented to the Eighth International Congress on Electron Microscopy*, Canberra, J. V. Sanders and D. J. Goodchild, eds. (Australian Academy of Science, Canberra, 1974), 2 Vols.

HVEM Toulouse, 1975: *Microscopie Electronique à Haute Tension. Textes des Communications Présentées au 4e Congrès International*, Toulouse, 1–4 Septembre, 1975, B. Jouffrey and P. Favard, eds. (SFME Paris, 1976).

Jerusalem, 1976: *Electron Microscopy 1976. Proceedings of the Sixth European Congress on Electron Microscopy*, Jerusalem, D. G. Brandon (Vol. I) and Y. Ben-Shaul (Vol. II), eds. (Tal International, Jerusalem, 1976), 2 Vols.

HVEM Kyoto, 1977: *High Voltage Electron Microscopy 1977. Proceedings of the Fifth International Conference on High Voltage Electron Microscopy*, Kyoto, 29 August to 1 September, 1977, T. Imura and H. Hashimoto, eds. (Japanese Society of Electron Microscopy, Tokyo, 1977); published as a supplement to *Journal of Electron Microscopy* **26** (1977).

Toronto, 1978: *Electron Microscopy 1978. Papers Presented at the Ninth International Congress on Electron Microscopy*, Toronto, J. M. Sturgess, ed. (Microscopical Society of Canada, Toronto, 1978), 3 Vols.

The Hague, 1980: *Electron Microscopy 1980. Proceedings of the Seventh European Congress on Electron Microscopy*, The Hague, P. Brederoo and G. Boom (Vol. I), P. Brederoo and W. de Priester (Vol. II), P. Brederoo and V. E. Cosslett (Vol. Ill), and P. Brederoo and J. van Landuyt (Vol. IV), eds. Vols. I and II contain the proceedings of the Seventh European Congress on Electron Microscopy, Vol. Ill those of the Ninth International Conference on X-Ray Optics and Microanalysis, and Vol. IV those of the Sixth International Conference on High Voltage Electron Microscopy (Seventh European Congress on Electron Microscopy Foundation, Leiden, 1980).

Hamburg, 1982: *Electron Microscopy 1982. Papers Presented at the Tenth International Congress on Electron Microscopy*, Hamburg (Deutsche Gesellschaft fur Elektronenmikroskopie, Frankfurt, 1982), 3 Vols.

HVEM Berkeley, 1983: *Proceedings of the Seventh International Conference on High Voltage Electron Microscopy*, Berkeley, 16–19 August, 1983, R. M. Fisher, R. Gronsky, and K. H. Westmacott, eds. Published as a Lawrence Berkeley Laboratory Report, LBL-16031, UC-25, CONF-830819.

Preface by Ernst Ruska

When electron microscopy came into being more than fifty years ago, light microscopy, already in existence for some 300 years, had long since attained a high degree of perfection. During those centuries, it had benefited numerous branches of science decisively while others had the light microscope to thank for their very existence. The scientists working in these fields were often conscious of the fundamental role of the microscope in their work. For some fifty years, the limit imposed on the resolution attainable with the light microscope by the wavelength of the light waves was likewise known. Considering that the waves associated with electrons would be some five orders of magnitude smaller than light waves, the development of an electron microscope seemed to offer the possibility at least of forming images of considerably better resolution than those the light microscope could provide. In a publication of 1932 it was already established that, even considering the technological difficulties to be expected, resolutions better than those of light microscope images could be anticipated.

With such a prospect, it must have seemed certain that the champions of microscope research in the various fields would welcome the development of the electron microscope so enthusiastically that the pioneers of the time in various countries would in turn receive generous financial assistance in a reasonably short time. But spontaneous support of this kind from microscope users scarcely materialized, and the financial means of the appropriate industries were not put at the disposal of the pioneers, for want of proof that there was adequate demand for such novel microscopes. The reasons for this lay certainly, to a considerable extent, in the serious reservations which obscured the real possibilities of electron microscopy for many years. Moreover, these reservations were perfectly comprehensible. An objection that was particularly difficult to refute was that the specimen preparation would be heated much too strongly by the electron energy absorbed within it. Nevertheless, it was found quite soon (1933/1934) that image contrast could arise not only from spatial variations in the absorption of electron energy but also from the difference in electron scattering from one point to another, scarcely any energy then being deposited in the specimen and converted into heat.

This knowledge was particularly important for the development of the electron microscope, which was just beginning, because it became reasonable to hope that specimens could be examined at high magnification but at a low enough temperature for there to be no danger of chemical modification. It also became clear, however, that if very little thermal energy was to be transferred to the specimen, and hence the image contrast created essentially by scattering, the thickness of the specimen would have to be almost a hundred times less than that attainable with the microtomes then used for light microscopy. The hope that it would one day be possible to cut such thin sections seemed as utopian as that of achieving electron microscope resolutions far better than that of the light microscope. Indeed, this problem severely limited the choice of objects for study in the electron microscope for some twenty years. Only with the successful development of the ultramicrotome, and also of other specimen-preparation techniques which permitted at least some kinds of objects to be obtained in sufficiently thin sections, did the electron microscope rapidly reveal itself to be useful in almost every discipline in which the light microscope had hitherto been an indispensible tool. Indeed, the widespread use of the latter had likewise been hampered, for some 200 years, by the lack of achromatic objective lenses. These did not come into use until 1825.

When writing this preface, I had not yet seen the material collected together here, contributed by those international pioneers of electron microscopy who are still living. I suspect, however, that in at least some of these recollections, as in my own conclusions, will figure the absence of encouragement on the part of experienced light microscopists and also, of course, the resulting doubts that tormented the pioneers themselves.

The technical development of the electron microscope consists almost exclusively of a struggle against the *bad* properties of electrons. Almost every property of these resembles, when we consider its influence on the development of the instrument, a coin, which notoriously possesses two sides. The property that is fundamental for microscopy, the shortness of the wavelength, is counterbalanced by the associated high energy. The latter also has advantages—for example, even highly magnified images remain visible on the fluorescent screen—but at the same time the decisive disadvantages of heating the specimen and generating X rays. The fact that electrons are deflected in magnetic fields is indispensible for the construction of good enough magnetic lenses but also created an obstacle to the achievement of the very highest resolution that has proved difficult to surmount. It was essential to hold the accelerating voltage and the lens currents extremely constant and to provide extremely good screening of the interior of the microscope against parasitic magnetic fields. Apart from the technical problems arising from the physical properties of electrons, other difficulties had to be overcome in the development of the electron microscope—the design of the specimen holder mechanism, for example. This and other mechanical difficulties resulted from the high magnification that must be attainable when the highest resolution is to be achieved.

The ever-increasing degree of perfection in the solutions of such problems as these, together with the laborious progress of specimen-preparation methods, has made the electron microscope a universal tool, taking it far beyond the imaginings of the pioneers of "the beginnings of electron microscopy."

Ernst Ruska

Afterword by Peter Hawkes

1937

1941

Figure 1 Ernst Ruska in 1937 and 1941. Courtesy of the Siemens Archives.

Ernst Ruska was born on Christmas Day 1906 in Heidelberg. I can add nothing to his book mentioned in the Foreword to the first edition and to the full biography by Ruska's secretary, Lotte Lambert and Tom Mulvey (Lambert & Mulvey, 1996). Frau Lambert's earlier article should also not be missed (Lambert, 1986). An account of the role of Ernst Ruska's brother Helmut by Hans Gelderblom and Detlev Krüger (2014) is also very relevant.

Ruska was awarded a half-share of the Nobel Prize in 1986, which produced the following reaction by Judith Reiffel, secretary of Elmar Zeitler, Ruska's successor at the Institute of Electron Microscopy in the Frtitz-Haber-Institute in Berlin and founder-editor of *Ultramicroscopy*:

> *"Just after your call the telephone rang again, and didn't stop ringing until five minutes ago. Because Ruska got the Nobel prize, sharing it with Binnig and Rohrer. Isn't that a gas? Ruska is on holiday, and the BZ (*Berliner Zeitung*) (yellow journal) sent a helicopter and found him taking a walk and schlepped him back to the hotel, where they interviewed him. Jesus, it's a good thing we've got a festschrift in press".*

Ernst Ruska's Nobel Lecture contains a wealth of details about his early years (Nobel Prize website and Ekspong, 1993); see Fig. 1.

Ernst Ruska died on 27 May 1988.

References

Ekspong, G. (Ed.). (1993). *Nobel Lectures in Physics 1981–1990*. Singapore: World Scientific.

Gelderblom, H. R., & Krüger, D. H. (2014). Helmut Ruska (1908–1973): His role in the evolution of electron microscopy in the life sciences, and especially virology. *Advances in Imaging and Electron Physics*, *182*, 1–94.

Lambert, L. (1986). Some anecdotes around and with Colonel Ernst August Ruska. *Ultramicroscopy*, *20*, 337–340.

Lambert, L., & Mulvey, T. (1996). Ernst Ruska (1906–1988). Designer Extraordinaire of the electron microscope: A Memoir. *Advances in Imaging and Electron Physics*, *95*, 2–62.

CHAPTER ONE

Electron-optical research at the AEG Forschungs-Institut 1928–1940☆

Dieter Typke
Formerly Munich, Germany

Contents

1. Introduction

The research on electrostatic electron-optical devices that was pursued in the *Zentrale Forschungs-Institut der AEG* (Central Research Institute of the AEG) in Berlin in the 1930s has not been covered in the previous (1985) edition of this book. Here, I try to give a rather short review of the intense work that was carried out in this laboratory. In the first part, short biographies of some of the people involved will be given. In the sec-

☆ Corresponding author: Peter Hawkes. e-mail address: hawkes@cemes.fr.

Advances in Imaging and Electron Physics, Volume 220
ISSN 1076-5670
https://doi.org/10.1016/bs.aiep.2021.08.001

ond part, I will deal with the work and achievements in electron-optical research during the decade before 1940.

AEG was founded in 1883 as the *Deutsche Edison-Gesellschaft für angewandte Elektricität*, and 1888 reordered and renamed as the *Allgemeine Elektricitäts-Gesellschaft*. The *Zentrale Forschungs-Institut der AEG* was founded in 1927 and headed by Carl Ramsauer.

Ramsauer was an excellent physicist with a broad knowledge in science and engineering, and he was an inspiring leader. As the research institute of a large electrical engineering company, one main task was applied research, particularly for improving technical solutions. However, basic research was also pursued on various subjects. Research results were reported in thc *Jahrbücher des Forschungs-Instituts der AEG* (yearbooks of the research institute of the AEG), and in numerous publications, lectures etc. Until 1942, nine volumes of the yearbook appeared, as well as roughly 1000 publications in scientific or technical journals. According to the table of contents of one yearbook (No. 3, 1933), fields of research covered acoustics, short-time-interval research, electrical engineering, tube-based techniques, electron beams, electron physics, atomic physics, physical chemistry, physics of materials, and electro-optics. The subject of this article, research on electron beams, was done in the physics department, headed by Ernst Brüche. An overview of the achievements, including many pictures, was given in the report *Zehn Jahre Elektronenmikroskopie* (Ten Years Electron Microscopy, ed. C. Ramsauer) in 1941.

2. Short biographies of some of the people at the institute

2.1 Carl Ramsauer

Carl Ramsauer (see Fig. 1), born in 1879, studied Mathematics and Physics in Munich, Tübingen, Berlin and Kiel, where he got his PhD in 1903. After four years at the *Kaiserliche Torpedo-Laboratorium* (Imperial Torpedo Laboratory) in Kiel, he joined the Radiological Institute in Heidelberg under Nobel Prize winner Philipp Lenard. He completed his *Habilitation* in 1909 and was appointed *extraordinarius* professor in 1915. In 1920, he discovered that slow electrons penetrate a gas more easily than fast ones, something that could not be explained in classical terms. This was the first experimental hint concerning the wave nature of electrons and – termed “The Ramsauer effect” – played an important role in the development of quantum mechanics.

Figure 1 A and B: Carl Ramsauer (A: Credit: Bundesarchiv; B: *Physikalische Blätter*); C: Max von Laue und Carl Ramsauer, 1942 (courtesy of *Archiv der Max-Planck-Gesellschaft, Berlin-Dahlem*).

In 1921, Ramsauer was appointed *ordinarius* (full) professor in Danzig, and in 1927 he became head of the *AEG Forschungs-Institut.* He appointed his student Ernst Brüche to direct the Physics Laboratory and strongly supported research and development of various electron-optical devices. In 1940 he became president of the *Deutsche Physikalische Gesellschaft* (*DPG*, German Physical Society). In this position, he helped to fend off political pressure by the *NSDAP* (Nazi party) on the DPG. After WWII, in 1945, he was appointed *ordinarius* professor at the *Technische Hochschule*[1] (Technical University) Berlin. Carl Ramsauer died in Berlin in 1955 (Brüche, 1955, 1975; Rukop, 1954).

2.2 Ernst Brüche

Ernst Brüche (see Fig. 2), born in 1900, grew up in Zoppot near Danzig. Before finishing school, in June 1918, he joined the military until the end of WWI. After he had received his *Abitur* (examination before leaving for university) in 1919, he started studying electrical engineering at the *Technische Hochschule* in Danzig. He changed to studying physics and mathematics in 1921 when Carl Ramsauer had been appointed professor of experimental physics. In 1924 he passed his exam to become a *Diplomingenieur* (graduate engineer), and in 1926 he earned the doctoral degree of *Dr.-Ing.* ("doctor of engineering") in Ramsauer's laboratory with a thesis *Über das Flächen- und Fadenmanometer* (On the area and thread manometer), representing methods of measuring the gas pressure in the presence of corrosive gases. He also pursued slow-electron scattering experiments (Brüche, 1930) and found the Ramsauer effect to be still valid for chain-like carbohydrates. A lifetime friendship developed between him and Ramsauer. After his *Habilitation* he became a private lecturer at Ramsauer's institute, and he went with him to the AEG Research Laboratory in Berlin upon its foundation in 1927. In 1929 he married a fellow student from Danzig, Dorothee, née Lilienthal. The couple stayed together, and Dorothee supported her husband throughout their lives. They had three daughters.

Early on, Brüche recognized the importance of Hans Busch's work, and developed a way of thinking about electron optics in analogy to light optics. Together with his collaborators he explored the basics as well as various possible applications of electron optics. In 1931, he produced the first large-scale electron-optical images of emitting cathode surfaces. Starting in

[1] The German *Technische Hochschulen* had university character; most of them were renamed *Technische Universitäten* by the end of the 20th century.

A

B

Figure 2 Ernst Brüche (A: 1943/44; *Physikalische Blätter*; B: 1970, on his 70th birthday; *Süddt. Lab. Mosbach*).

1938, together with H. Mahl and H. Boersch, he developed the electrostatic electron microscope. In 1941, he received, together with seven other pioneers of electron optics (Max Knoll, Ernst Ruska, Bodo von Borries, Hans Boersch, Hans Mahl, Manfred von Ardenne and Ernst Rackow), the Leibniz Silver Medal of the *Preußische Akademie der Wissenschaften* (Prussian Academy of Sciences). In 1943, he became an honorary professor at Berlin University.

In the beginning of 1945, the laboratory moved to Mosbach in Baden, where Brüche founded, with the help of the AEG and, later, the Carl Zeiss Company, the *Süddeutsche Laboratorien zur Entwicklung und Herstellung von Elektronenmikroskopen* (Southern German Laboratories for the Development and Manufacture of Electron Microscopes). In 1952, Brüche additionally founded the *Physikalische Laboratorium Mosbach* (Physics Laboratory Mosbach), which dealt with applications of electron microscopy. As early as 1944, on the advice of Ramsauer, then president of the *Deutsche Physikalische Gesellschaft DPG* (German Physical Society), Brüche had established the *Physikalische Blätter* (Physics Letters; Brüche, 1972). After WWII, he was able to reestablish them in 1946. For many years the *Physikalische Blätter* served as the bulletin of the *DPG*. In 1961, he became honorary professor

Figure 3 Alfred Recknagel, bronze relief at the University of Dresden (Wikipedia).

at Karlsruhe University. Ernst Brüche died in 1985 in Mosbach, about four months after his wife Dorothee (Rechenberg, 2000).

2.3 Alfred Recknagel

Alfred Recknagel (see Fig. 3), born in 1910, studied physics in Jena und Leipzig, where he completed his PhD thesis on *Berechnung der Elektronenterme der Stickstoffmolekel* (Calculation of the Electronic Characteristics of the Nitrogen Molecule) in Heisenberg's and Hund's Institute for Theoretical Physics. From 1934 until 1945, he did theoretical work at the AEG Research Laboratory on electron mirrors and electron-emission microscopes. In 1941, he wrote, together with Ernst Brüche, the monograph *Elektronengeräte* (Electron Instruments). Two years later he completed his *Habilitation* with a thesis on *Das Auflösungsvermögen des Elektronenmikroskops für Selbststrahler* (The resolution limit of the electron microscope for self-luminous emitters). After WWII, he worked as a physicist at the Carl Zeiss company in Jena, where he was involved in the development of an electrostatic electron microscope. He also lectured on electron physics at the University of Jena.

In 1948 he was appointed *ordinarius* professor and director of the *Institute für Experimental-Physik* at the *Technische Hochschule* Dresden. There he stayed until his retirement in 1975. Alfred Recknagel died in Dresden in 1994.

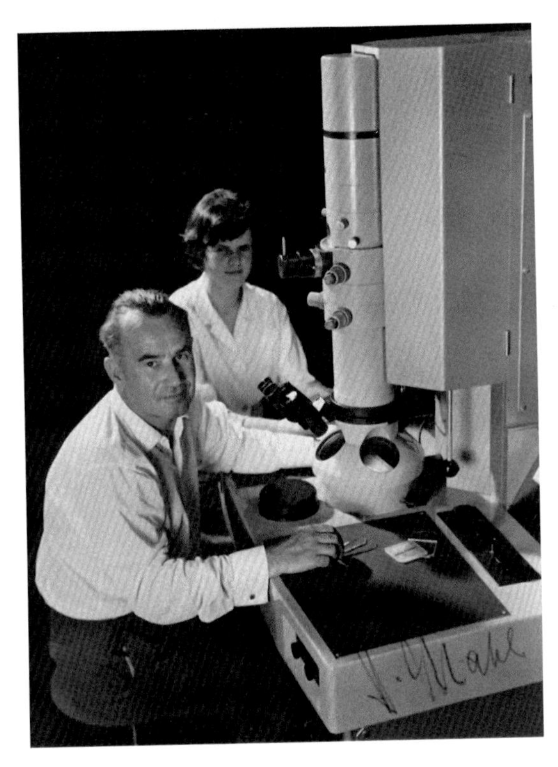

Figure 4 Hans Mahl at a Zeiss EM-9 magnetic TEM (courtesy of *ZEISS Archiv*).

2.4 Hans Mahl

Hans Mahl (see Fig. 4), born in 1909, studied physics at the *Technische Hochschule* in Munich, Danzig and Berlin. In 1934, he became a scientific collaborator at the AEG Research Laboratory and worked in Brüche's department on electron microscopy and electron-emission research. In 1937, he was promoted to *Dr.-Ing.* ('doctor of engineering'), with a thesis on *Elektronenoptische Kathodenabbildung in einer Gasentladung* (electron-optical imaging of cathodes in a gas discharge). He remained in the research laboratory until 1945, when Russian troops occupied the lab.

Together with Hans Boersch, Hans Mahl succeeded, in 1939, in building the first electrostatic *Durchstrahlungs-Übermikroskop* transmission super microscope (*Übermikroskop* meaning an instrument that provides a higher resolution than the light microscope). Mahl was also one of the first scientists to apply the electron microscope to problems in metallurgy, colloid chemistry, botany, bacteriology, and medicine. Due to the situation at the

Figure 5 Hans Boersch (Wikipedia).

end of WWII, the results were published years later (in 1951), together with Erich Gölz, in the monograph *Elektronenmikroskopie* (the manuscript had already been finished in 1944).

After WWII, Mahl worked from 1947 until 1953 at the *Süddeutsche Laboratorien* (*SDL*) in Mosbach. At that time, the *SDL* manufactured the first commercial electrostatic electron microscopes, which were sold as EM8-2 under the brand name AEG-Zeiss. After AEG had withdrawn from the consolidated companies, for many years Mahl was head of the department of electron microscopy and electron optics of the Carl Zeiss company in Oberkochen. He kept this position until his retirement in 1973. Hans Mahl died in 1988.

2.5 Hans Boersch

Hans Boersch (see Fig. 5) (see also Chapter 3), born in 1909, studied physics at the *Technische Hochschule* in Charlottenburg (Berlin) and at the University in Vienna. He got his PhD in 1935 with a thesis on *Bestimmung der Struktur einiger einfacher Moleküle mit Elektronen-Interferenz* (Determination of the structure of some simple molecules by means of electron interference). After that, he became a scientific collaborator at the AEG Research

Laboratory in Berlin, where he was responsible for the construction of the electrostatic *Schatten-Übermikroskop* (shadow super microscope). In 1941, he went to Vienna as an assistant at the *Erstes Chemisches Institut* (First Chemical Institute) of the University. He established a laboratory for *Strukturforschung und Übermikroskopie* and completed his *Habilitation* with a thesis on electron diffraction.

After intermediate stays at the University in Innsbruck and the *Institut des Recherches Scientifiques* in Tettnang, he moved to the *Physikalisch-Technische Bundesanstalt* (Federal Physico-Technical Institute) in Braunschweig. This was the newly created National Metrology Institute, analogous to the National Bureau of Standards in the USA. He also became an honorary professor at the *Technische Hochschule* in Braunscheig. It was there that he discovered the anomalous energy distribution of intense electron beams, later termed "The Boersch Effect".

In 1954, Boersch became a full professor and the head of the *Erstes Physikalisches Institut* at the *Technische Universität* in Berlin. At the Optical Institute of the *TU Berlin*, he established one of the first groups in Germany which dealt with lasers. In 1962, his PhD students Horst Weber und Gerd Herziger built their own solid-state and He–Ne lasers. Boersch stayed at *TU-Berlin* until his retirement. He died in 1986 in Berlin (Niedrig, 1987).

For a biography of **Otto Scherzer**, see the article "Otto Scherzer and his Contributions to Electron Microscopy" (volume 221). Scherzer joined the *Forschungs-Institut der AEG* for 3 years, from 1932 until 1935. (See Fig. 6.)

For several people who worked in the Physics Department of the *Forschungs-Institut*, no biographies are available; some of them who stayed there for at least several years are: Günther Dobke, Walter Henneberg, Helmut Johannson, H. Katz, W. Knecht, Walter Schaffernicht, H. R. Schulz.

3. Research on electron optics, 1928–1940

Soon after the foundation of the *Forschungs-Institut der AEG*, Ernst Brüche and his group started systematic investigations with the aim "to establish the geometrical optics for electrons" and "to build an electron microscope with very high magnification". The plan was that a resolution

Figure 6 Helmut Johannson and Otto Scherzer 1932.

well beyond that of the light microscope could be achieved. Under the category of microscopy, various other applications of electron optics were investigated. An impression of the intensity and breadth of the research can be obtained from the list of publications reproduced in Section 4 for the years 1929 until 1940 (from Ramsauer, 1941). For additional information, see also Brüche (1957, 1966).

An outside opinion on the work done in Brüche's laboratory was given by L.M. Myers in his book on Electron Optics (Myers, 1939): "Without doubt the greatest proportion of experimental and theoretical work during the last few years has been carried out at the A.E.G. Forschungsinstitut in Berlin, under the supervision of Dr. Brüche, consequently it is understandable that this introduction to the subject should follow very closely the writings of this investigator and his numerous collaborators. We have also followed the work of Zworykin and his collaborators at the R.C.A. Laboratories in America."

3.1 Gas-concentrated electron beams

Among the first experiments were studies of gas-concentrated electron beams in glass vessels (Fig. 7). Brüche had found a way to produce low-

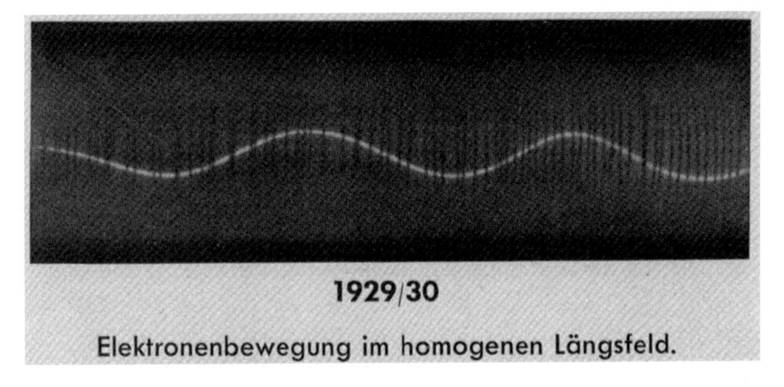

Figure 7 A gas-concentrated electron beam in a longitudinal homogeneous magnetic field (the wires of the electric coil are visible in front of the beam segments).

Figure 8 A gas-concentrated electron beam being strongly deflected near the magnetic pole of a model earth, termed *Terella* (little earth; Brüche & Ende, 1930).

energy (50 to 300 V accelerating voltage) beams of a constant (about 1 mm) diameter and a length up to about 1 m. They were called *Fadenstrahlen* (thread beams). They are held together by positive ions generated along the trajectory as well as by the surrounding electron cloud, and thereby visu-

Figure 9 Development of electrostatic einzel lenses.

alized by the light emitted by excited atoms. In this way, the influence of external magnetic or internal electrostatic fields on the electron trajectories could be studied and visualized. Examples were the screw-like movement in a homogeneous magnetic field produced by the electric current in a long coil or the strongly bent movement close to a magnetic pole (Fig. 8). With the latter experiments, the AEG laboratory was also part of the *Deutsche Nordlicht-Forschung* (aurora borealis research). Brüche gave impressive public lectures in which he showed some of these experiments. He and his wife were also invited to travel to Tromsö in Norway where aurora research was going on.

3.2 Basic studies of electron optics

A great effort went into basic studies of electron optics (Brüche, 1932; Brüche and Johannson, 1932; Brüche, 1933), such as the calculation of field distributions, the design of electrostatic lenses of both the immersion lens (Johannson, 1933a, 1933b) and the einzel lens (Johannson and Scherzer, 1932) type, as well as early calculations of lens aberrations (Scherzer, 1933). (See Figs. 9 and 10.)

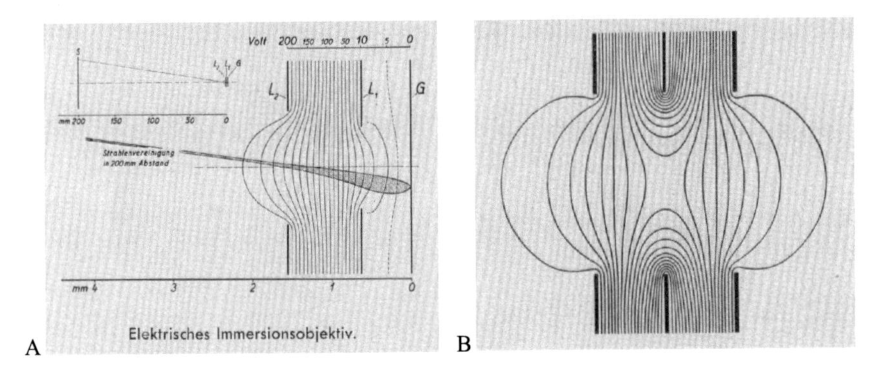

Figure 10 Electric potential field distributions of (A) an electrostatic immersion lens and (B) an einzel lens.

Figure 11 The first electrostatic emission electron microscope (glass vessel removed).

A result of these efforts was the book *Geometrische Elektronenoptik* (Geometric Electron Optics; Brüche and Scherzer, 1934). In its first part, the book deals with general foundations of electron optics; the second part deals with applications, including the Braun Tube, the Electron Microscope, and the Spectrograph.

3.3 Emission electron microscopes

The first electron microscope (see Fig. 11), which was realized in 1931, was of the emission type. With this instrument, Brüche and Johannson showed that they could obtain a reasonable image of the glowing cathode (Fig. 12)

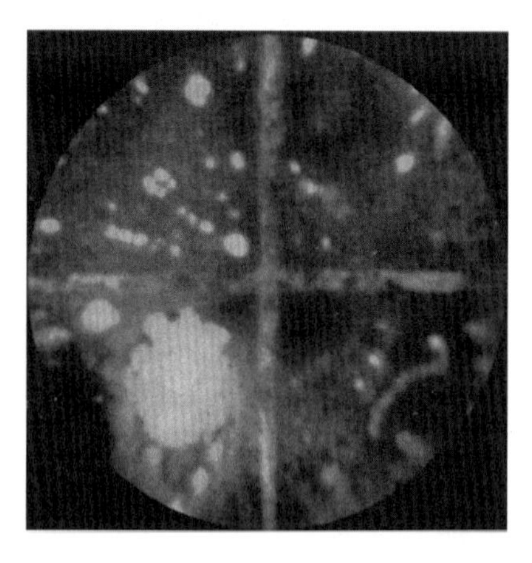

Figure 12 The first electron image of a glowing cathode that was recorded in the emission microscope (diameter of the field of view ca. 0.3 mm, Brüche and Johannson, 1932 [8]).

by means of an electrostatic immersion lens (something that had not been clear from the very beginning). The emission microscopes opened a whole new field of possible investigations of surfaces, such as crystalline structures of various metals, recrystallization, diffusion processes and contamination of surfaces. Electron emission could be produced by heating, light-emitting layers, electrons, or ions, thus opening the way to various types of investigations and applications.

In Figs. 13 and 14, some application examples are shown.

3.4 Image converters

A special application of emission instruments was the image converter. When projecting an image with light onto the cathode made up by a structureless light-emitting layer, the electron-optical system produces an image on the final fluorescent screen. This was the basis of the image converter, which was developed by W. Schaffernicht (1935). The main application was to convert infrared images into visible ones. After Schaffernicht had reported on this development at the *Physikertagung* in 1936, it was classified as a top-secret issue, and it became important during the WWII. Only after WWII, did he report about the development in a physics journal (Schaffernicht, 1948).

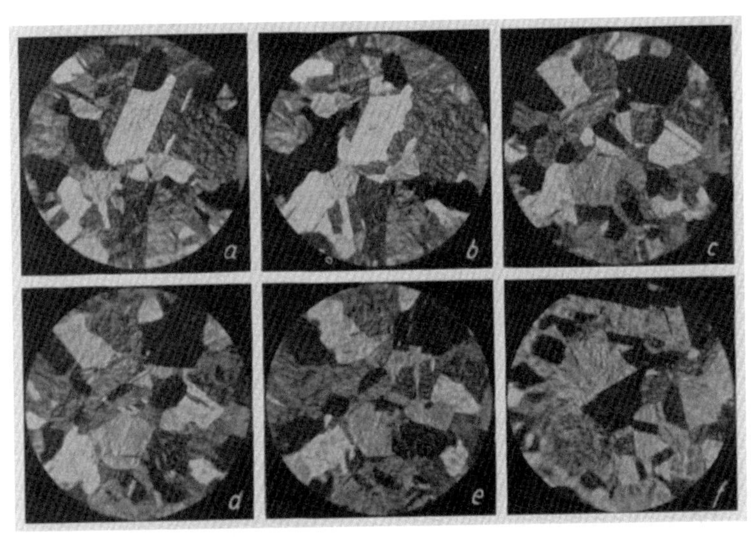

Figure 13 Cathode images showing the recrystallization of iron at 900 °C; the transition from the α-state to the γ-state. The process starts from the left side, which is at a slightly higher temperature (diameter of the field of view ca. 0.6 mm; Brüche and Knecht, 1935).

Figure 14 A picture (of Otto von Guericke, 17th century), projected onto a structureless light-emitting layer and converted into an electron image (Brüche, 1936b).

3.5 Braun Tubes

The Braun Tube had been invented as early as 1897. Due to its various possible applications, it was under steady development. Within a 10-

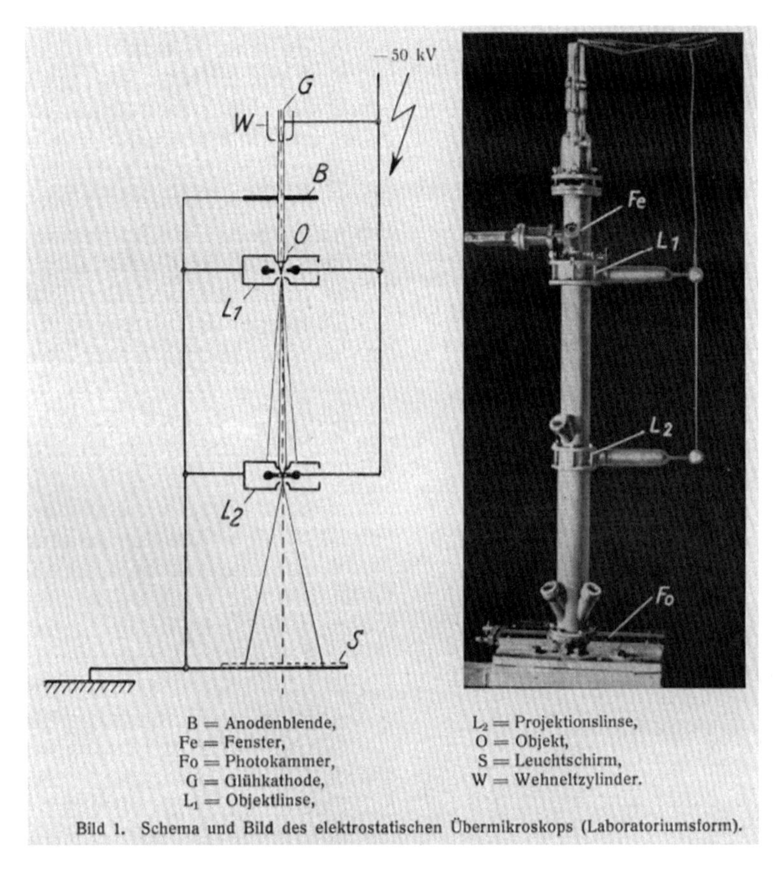

Figure 15 *Erstes elektrostatisches Abbildungs-Übermikroskop* (first electrostatic imaging super microscope, Mahl, 1939).

year period, 7 papers on Braun Tubes were published by Brüche's group. As one special application, an electron-beam oscilloscope was developed (Dantscher, 1933, 1935). It permitted the visualization of oscillations up to 30,000 Hz.

3.6 Transmission electron microscopes

Due to the broad spectrum of scientific interests, it took several years before Brüche's group seriously started the development of the transmission electron microscope. In yearbook No. 7, Brüche described some of the intermediate steps. From the beginning, Brüche had favored electrostatic systems, one reason being that they are more similar to light-optical ones. One clear advantage for using them in the electron microscope was, that

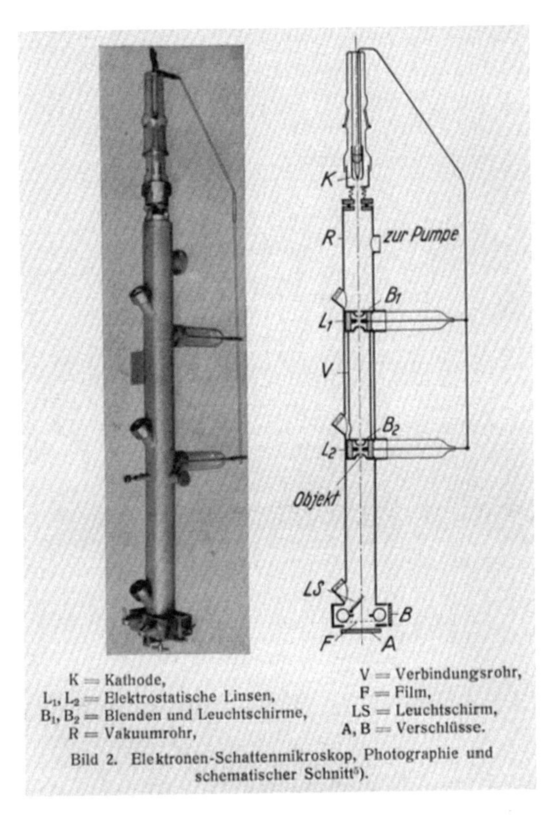

Bild 2. Elektronen-Schattenmikroskop, Photographie und schematischer Schnitt[5]).

Figure 16 *Elektronen-Schattenmikroskop* (electron shadow microscope, Boersch, 1939).

the lenses can be kept at the cathode potential or at a fixed ratio to it. Therefore, it is not necessary to keep the accelerating voltage extremely stable.

The last step was to build and test electrostatic einzel lenses of sufficiently short focal length that are suitable for rather high voltages.[2] Two different kinds of instrument were developed and successfully tested, a transmission EM by H. Mahl (1939) and a shadow projection EM by H. Boersch (1939). (See Figs. 15–17.) Both were equipped with two electrostatic lenses and could be operated at accelerating voltages of up to 60 kV. Both instruments were clearly able to reach a resolution exceeding that of the light

[2] Interestingly, B. von Borries and E. Ruska thought, even in 1938, that resolution beyond that of the light microscope would not be possible with an electrostatic microscope, but only with a magnetic one, for which they introduced the term *Übermikroskop* (super microscope; von Borries & Ruska, 1938).

Elektrostatisches AEG-Übermikroskop (Modell Ende 1940).

Das Gerät, das sich durch grundsätzliche und praktische Einfachheit auszeichnet, mit allen erforderlichen Hilfseinrichtungen: Pumpen, Netzanschlußgerät und Schaltpulten.

Figure 17 Complete microscope setup, 1940.

microscope; they were also termed *Übermikroskope* (super microscopes). In the latter instrument, the two lenses produced a strongly demagnified image of the cathode in front of the specimen, which was inserted below the second lens and projected onto the final screen.[3]

Investigations of various specimens from bacteriology, metallurgy and colloid chemistry demonstrated their capabilities. One of the TEM instruments was installed in the *Robert-Koch-Institut* (an institute for medical research) and intensively used for bacteriology and virus research. (See Figs. 18 and 19.)

[3] In his Nobel lecture in 1986, Ernst Ruska did not even mention Brüche and the work that had been done in the *Forschungs-Institut der AEG* at about the same time as his own development of the magnetic electron microscope.

Figure 18 Surface replica picture of hydronalium (Al-Mg alloy) (Mahl, 1940).

In the 1940s, WWII made scientific work more and more difficult. Some people stayed in the lab until the end of the war. However, because of the bomb threat, a part of the lab, including Brüche's group, was moved to Schönberg near Görlitz in Silesia in 1943.

After the war, Carl Ramsauer stayed in Berlin. As already mentioned, he became professor at the *Technische Universität Berlin*. Even before the end of the war, in the beginning of 1945, E. Brüche, with a few of the people of his group, moved via Dresden and Helmbrecht in *Oberfranken* to Mosbach, Baden. During the move from Silesia to the west, ten people of his group became victims of the bombing raid of Dresden in February 1945. Alfred Recknagel stayed in East Germany, first in Jena, until he was appointed professor in Dresden.

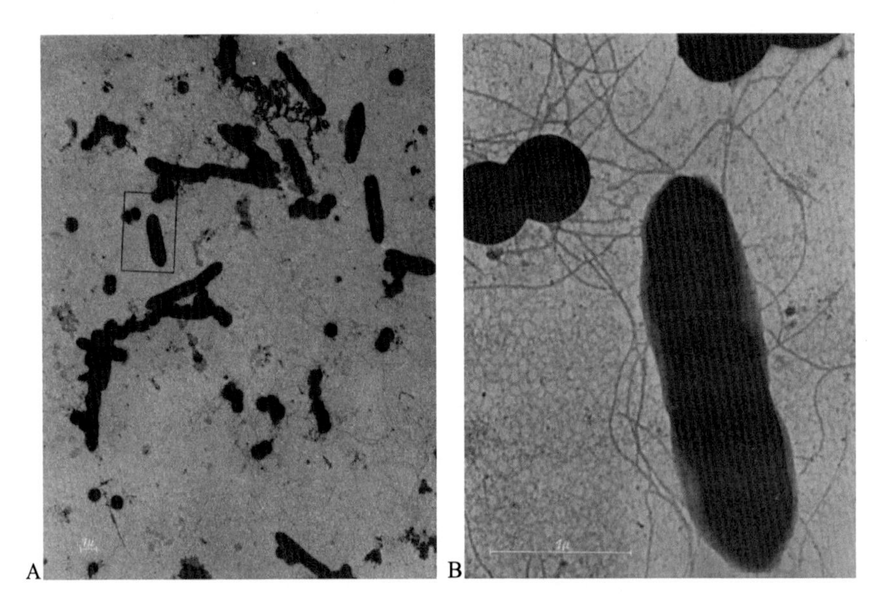

Figure 19 Two types of bacteria (*Staphylococcus* and *Tetanomorphus*); A: overview image (1000x original mag.); B. image at high magnification (9000x original mag.), surrounded by flagella (Jacob and Mahl, 1940).

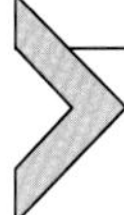

4. References from the book *Zehn Jahre Elektronenmikroskopie*

Unsere Zeitschriften-Veröffentlichungen über geometrische Elektronenoptik.

(Titel gekürzt.)

				Bd.	S.
		1930			
1	Brüche	Strahlen langsamer Elektronen	Petersen, Forschg. und Technik	—	24
		1932			
2	Brüche	Elektronenmikroskop	Naturwiss.	20	49
3	Brüche u. Johannson	Elektronenoptik und Elektronenmikroskop	Naturwiss.	20	353
4	Dobke	Eine neue Braunsche Röhre	Z. techn. Phys.	13	432
5	Brüche	Geometrie d. Beschleunigungsfeldes	Z. Phys.	78	26
6	Brüche u. Johannson	Kinematogr. Elektronenmikroskopie	Ann. Phys.	15	145
7	Schulz	Abbildung durch geschichtete Medien	Z. Phys.	78	17
8	Brüche u. Johannson	Einige neue Kathodenuntersuchungen	Phys. Z.	33	898
9	Johannson u. Scherzer	Über die elektr. Elektronensammellinse	Z. Phys.	80	183
		1933			
10	Scherzer	Theorie elektronenopt. Linsenfehler	Z. Phys.	80	193
11	Brüche	Grundlagen der Elektronenoptik	Z. techn. Phys.	14	49
12	Brüche	Optik der Braunschen Röhre	Arch. Elektrot.	27	266
13	Brüche	Geometrische Elektronenoptik	Jb. AEG-Forschg.	3	111
14	Brüche	Braunsche Röhre als Synchronoskop	Arch. Elektrot.	27	609
15	Brüche u. Johannson	Barium-Aufdampfkathode	Z. Phys.	84	56
16	Johannson	Immersionsobjektiv d. Elektronenoptik	Ann. Phys.	18	385
17	Johannson u. Knecht	Kombinierte Benutzung von Elektronenlinsen	Z. Phys.	86	367
18	Brüche	Abbildung mit lichtelektr. Elektronen	Z. Phys.	86	448
19	Richter	Emissionssubstanz auf Kathoden	Z. Phys.	86	697
20	Brüche u. Scherzer	Braunsche Röhre als elektronenoptisches Problem	Z. techn. Phys.	14	464
21	Brüche u. Johannson	Kristallographische Untersuchungen	Z. techn. Phys.	14	487
22	Henneberg	Massenspektrographie I	Ann. Phys.	19	335
		1934			
23	Henneberg	Massenspektrographie II	Ann. Phys.	20	1
24	Knecht	Komb. Licht- u. Elektronenmikroskop	Ann. Phys.	20	161
25	Brüche	Braunsche Röhre als Problem der Elektronenoptik	Arch. Elektrot.	28	384
26	Brüche u. Scherzer	Geometrische Elektronenoptik	Buch (Springer)	—	—
27	Henneberg	Achromatische elektr. Elektronenlinsen	Z. Phys.	90	742
28	Johannson	Immersionssystem als System der Braunschen Röhre	Z. Phys.	90	748
29	Brüche	Zu zwei Veröffentl. ü. Elektronenoptik	Z. Phys.	92	215
30	Johannson	Immersionsobjektiv II	Ann. Phys.	21	274
31	Henneberg	Massenspektrographie III	Ann. Phys.	21	390
32	Brüche u. Knecht	Eisenumwandlung	Z. techn. Phys.	15	461
33	Heß	Immersionslinse I	Z. Phys.	92	274
34	Pohl	Lichtelektrische Abbildung	Z. techn. Phys.	15	579
35	Fünfer	Voltmeter a. elektronenoptischer Grundlage	Z. techn. Phys.	15	582
36	Brüche u. Knecht	Auflösung des Immersionsobjektivs	Z. Phys.	92	462
37	Brüche	Schichtenuntersuchung	Kolloid-Z.	69	389

120

		1935		Bd.	S.
38	Brüche	Grundlagen d. angew. Elektronenoptik	Arch. Elektrot.	29	79
39	Henneberg	Low-Voltage Electron Microscope	J. E. E.	76	111
40	Mahl	Abbildung von Mineralien	Mineral. M.	46	289
41	Henneberg	Potential von Schlitz- u. Lochblende	Z. Phys.	94	22
42	Schaffernicht	Umwandlung von Licht- in Elektronenbilder	Z. Phys.	93	762
43	Brüche u. Knecht	Eisenumwandlung	Z. techn. Phys.	16	95
44	Henneberg	Das Elektronenmikroskop	ETZ	56	853
45	Brüche	Zur Braunschen Hochvakuumröhre	Arch. Elektrot.	29	642
46	Schenk	Emissionsverteilung a. kristalliner Kathode	Ann. Phys.	23	240
47	Henneberg	Auflösung des Elektronenmikroskops	Z. Instr. K.	55	300
48	Henneberg u. Recknagel	Chromatischer Fehler b. Bildwandler	Z. techn. Phys.	16	230
49	Mahl u. Pohl	Lichtelektrische Abbildungen	Z. techn. Phys.	16	219
50	Stabenow	Magnetische Linse ohne Bilddrehung	Z. Phys.	96	634
51	Glaser u. Henneberg	Potential von Schlitz- u. Lochblende	Z. techn. Phys.	16	222
52	Brüche u. Schaffernicht	Elektronenopt. Fragen auf dem Fernsehgeb.	ENT	12	381
53	Brüche	Elektronenoptisches Strukturbild	Z. Phys.	98	77
54	Henneberg u. Recknagel	Elektronenlinse, Elektronenspiegel und Steuerung	Z. techn. Phys.	16	621
55	Brüche u. Mahl	Thorierte Wolfram und Molybdän I	Z. techn. Phys.	16	623
56	Recknagel	Emissionskonstanten von Ein- und Vielkristallen	Z. Phys.	98	355
57	Mahl	Abbildung von emittierenden Drähten	Z. Phys.	98	321
		1936			
58	Brüche (Schaffernicht)	Fortschritte der Elektronenoptik	Jb. AEG-Forschg.	4	25
59	Schenk	Glühemissionsbild von Nickel	Z. Phys.	98	753
60	Brüche u. Mahl	Abbildung von thoriertem Wolfram II	Z. techn. Phys.	17	81
61	Brüche u. Recknagel	Modelle elektr. und magnet. Felder	Z. techn. Phys.	17	126
62	Behne	Immersionsobjektiv für schnelle Elektronen	Ann. Phys.	26	372
63	Behne	Folienabbildung	Ann. Phys.	26	385
64	Mahl u. Schenk	Einfluß der Gleitspurebenen	Z. techn. Phys.	101	117
65	Boersch	Primäres u. sekundäres Bild im Elektronenmikroskop I	Ann. Phys.	26	631
66	Brüche u. Recknagel	Dimensionsbeziehung I	Z. techn. Phys.	17	241
67	Brüche u. Mahl	Thoriertes Wolfram und Molybdän III	Z. techn. Phys.	17	262
68	Boersch	Primäres u. sekundäres Bild im Elektronenmikroskop II	Ann. Phys.	27	75
69	Behne	Zur Kenntnis der Immersionslinse III	Z. Phys.	101	521
70	Hottenroth	Über Elektronenspiegel	Z. Phys.	103	460
71	Brüche u. Henneberg	Geometrische Elektronenoptik	Erg. exakt. Nat.	15	365
72	Recknagel	Zur Theorie des Elektronenspiegels	Z. techn. Phys.	17	643
73	Mahl	Sauerstoffeinfluß auf die Glühemission	Z. techn. Phys.	17	653
74	Brüche	Exper. Elektronenoptik und Anwendung	Z. techn. Phys.	17	588
75	Brüche	Ausstellung Elektronenoptik	Z. techn. Phys.	17	622
76	Schaffernicht	Der elektronenoptische Bildwandler	Z. techn. Phys.	17	596
		1937			
77	Recknagel	Zur Theorie des Elektronenspiegels	Z. Phys.	104	381
78	Recknagel	Elektronenspiegel und Elektronenlinse	Beiträge zur Elektronenoptik	—	42
79	Mahl	Das elektronenoptische Strukturbild	Beiträge zur Elektronenoptik	—	73
80	Brüche	Geometrische Elektronenoptik	Schröter: Ferns.	—	87
81	Brüche u. Recknagel	Dimensionsbeziehung II	Z. techn. Phys.	18	139
82	Mahl	Feldemission geschichteter Kathoden	Naturwiss.	25	459
83	Hottenroth	Untersuchungen über Elektronenspiegel	Ann. Phys.	30	689

121

				Bd.	S.
84	Boersch	Abbildung von Dampfstrahlen	Z. Phys.	107	493
85	Recknagel	Zur Intensitätssteuerung	Hochfrequenzt. u. Elektroak.	51	66
86	Katz	Elektronendurchgang durch Metallfolien	Z. techn. Phys.	18	555
87	Mahl	Feldemission a. geschichteten Kathoden I	Z. techn. Phys.	18	559
88	Mrowka	Kristallgitterstruktur und Glühemission	Z. techn. Phys.	18	572
89	Boersch	Bänder bei Elektronenbeugung	Z. techn. Phys.	18	574
		1938			
90	Recknagel	Intensitätssteuerung v. Elektronenströmen	Hochfrequenzt. u. Elektroak.	51	66
91	Mahl	Elektronenopt. Kathodenabbild. in einer Gasentladung	Ann. Phys.	31	425
92	Brüche	Elektronenbewegung	Jb. AEG-Forschg.	5	27
93	Schaffernicht u. Steudel	Elektronengeräte	Jb. AEG-Forschg.	5	66
94	Mahl	Ionen- und Elektronenemission	Z. Phys.	108	771
95	Katz	Durchgang v. Elektronen durch Folien I	Ann. Phys.	33	160
96	Katz	Durchgang v. Elektronen durch Folien II	Ann. Phys.	33	169
97	Mahl	Feldemission a. geschichteten Kathoden II	Z. techn. Phys.	19	313
98	Boersch	Zur Bilderzeugung im Mikroskop	Z. techn. Phys.	19	337
		1939			
99	Mahl	Aufnahmen mit dem elektr. Übermikroskop	Naturwiss.	27	417
100	Boersch	Das Schattenmikroskop, ein neues Übermikroskop	Naturwiss.	27	418
101	Recknagel	Elektronenlinse mit Laufzeiterscheinungen	Jb. AEG-Forschg.	6	78
102	Neßlinger	Über Achromasie von Elektronenlinsen	Jb. AEG-Forschg.	6	83
103	Mahl	Elektrostat. Elektronenmikroskop hoher Auflösung	Z. techn. Phys.	20	316
104	Brüche u. Haagen	Übermikroskop der Bakteriologie	Naturwiss.	27	809
105	Boersch	Das Elektronen-Schattenmikroskop	Z. techn. Phys.	20	346
		1940			
106	Brüche	Verwendung elektr. und magnet. Felder	TFT	29	1
107	Mahl	Metallkundliche Untersuchungen	Z. techn. Phys.	21	17
108	Mahl	Stereoskopische Aufnahmen	Naturwiss.	28	264
109	Mahl	Das elektrostat. Elektronen-Übermikroskop und Anwendungen in der Kolloidchemie	Kolloid-Z.	91	105
110	Brüche	10 Jahre Entwicklung	Jb. AEG.-Forschg.	7	2
111	Brüche	Zweipolsystem als Ziel rein elektr. Abbildungsgeräte	Jb. AEG.-Forschg.	7	9
112	Recknagel	Fehler von Elektronenlinsen	Jb. AEG.-Forschg.	7	15
113	Kinder u. Pendzich	Neue magnetische Linse kleiner Brennweite	Jb. AEG.-Forschg.	7	23
114	Boersch	Problem der Bildentstehung	Jb. AEG.-Forschg.	7	27
115	Boersch	Elektronen-Schattenmikroskop	Jb. AEG.-Forschg.	7	34
116	Mahl	Das elektrostatische Übermikroskop	Jb. AEG.-Forschg.	7	43
117	Gölz	Spannungsfestigkeit der Elektrodenmetalle für die Linse des Übermikroskops	Jb. AEG.-Forschg.	7	57
118	Brüche u. Gölz	Einschleusung von Objekt und Platte	Jb. AEG.-Forschg.	7	60
119	Mahl	Übermikroskop in Kolloidchemie und Metallurgie	Jb. AEG.-Forschg.	7	67
120	Jakob u. Mahl	Übermikroskop in der Bakteriologie	Jb. AEG.-Forschg.	7	77
121	Mahl	Plastisches Abdruckverfahren	Metallwirtschaft	19	488
122	Döring u. Mayer	Geschwindigkeitsgesteuerte Laufzeitröhren	ETZ	61 61	685 713
123	Henneberg	Übermikroskop mit elektrostat. Linsen	ETZ	61	773
124	Jakob u. Mahl	Kapseldarstellung bei Anaerobiern	Arch. Zellforschg.	24	87
126	Kinder	Übermikroskopie mit höheren Spannungen	Z. techn. Phys.	21	222
127	Boersch	Fresnelsche Elektronenbeugung	Naturwiss.	28	709
128	Boersch	Fresnelsche Beugungserscheinungen im Übermikroskop	Naturwiss.	28	711
129	Mahl	Orientierungsbestimmung v. Aluminium-Einzelkristallen a. übermikroskopischem Wege	Metallwirtschaft	19	1082
130	Recknagel	Sphärische Aberration bei elektronenopt. Abbildung	Z. Phys.	117	67
131	Mahl	Übermikroskop. Elektronenbilder von Metalloberflächen	Z. angew. Photographie	2	58
132	Brüche	10 Jahre Elektronenmikroskopie bei d. AEG	AEG-Mittlg.	—	302

Acknowledgments

I would like to thank Peter Hawkes for inviting me to write this article. It was first meant to be part of the Scherzer article in volume 221 but after some consideration I decided to separate the subjects. Sources for this article were (1) Wikipedia biographies of people that had been involved, (2) several volumes of the *Jahrbücher des Forschungs-Instituts der AEG*, (3) the book *Zehn Jahre Elektronenmikroskopie*, edited by C. Ramsauer, and (4) various journal articles (mainly from *Physikalische Blätter*). I also would like to thank Mike Marko for copy-editing this article.

References

Brüche, E. (1930). Strahlen langsamer Elektronen. In PetersenW. (Ed.), *Forschung und Technik* (pp. 23–46). Berlin: Springer.
Brüche, E. (1955). Abschied von Carl Ramsauer. *Physikalische Blätter*, *12*, 49–54.
Brüche, E. (1957). 25 Jahre Elektronenmikroskop. *Naturwissenschaften*, *44*, 601–610.
Brüche, E. (1966). Forschungsreise durch das Gebiet der Elektronenoptik. *Optik*, *24*, 290–295.
Brüche, E. (1972). 25 Jahre Physik-Verlag in Mosbach. *Physikalische Blätter*, *28*. Anlage zu Heft 12.
Brüche, E. (1975). Erinnerungen an Carl Ramsauer. *Physikalische Blätter*, *28*, 405–408.
Brüche, E., & Ende, W. (1930). Demonstrationsversuche zu Störmers Polarlichttheorie. *Physikalische Zeitschrift*, *31*, 1015–1016.
Dantscher, J. (1933). *Der Elektronenstrahloszillograph*. AEG-Mittlg. (p. 9).
Dantscher, J. (1935). Über die neuere Entwicklung des Elektronenstrahloszillographen. *Archiv für Elektrotechnik*, *29*, 833–841.
Myers, L. M. (1939). *Electron optics*. London: Chapman and Hall.
Niedrig, H. (1987). Nachruf auf den Pionier der Elektronenmikroskopie: Hans Boersch 1909–1986. *Optik*, *75*, 172–174.
Ramsauer, C. (Ed.). (1941). *Zehn Jahre Elektronenmikroskopie*. Berlin: Springer.
Rechenberg, H. (2000). Vom "Übermikroskop" zu den Physikalischen Blättern. *Physikalische Blätter*, *56*, 75–77.
Rukop, H. (1954). Carl Ramsauer 75 Jahre. *Physikalische Blätter*, *10*, 76.
Schaffernicht, W. (1948). Der Bildwandler. *Physikalische Blätter*, *4*, 4–10.
von Borries, B., & Ruska, E. (1938). *Wissenschaftliche Veröffentlichungen aus den Siemens-Werken*, *17*(1), 99.

On the history of scanning electron microscopy, of the electron microprobe, and of early contributions to transmission electron microscopy☆

Manfred Von Ardenne[a], **Peter Hawkes (Afterword)**[b,*], **and Tom Mulvey (Appreciation)**[c]

[a]Formerly Forschungsinstitut Manfred von Ardenne, Dresden-Weisser Hirsch, German Democratic Republic
[b]CEMES-CNRS, Toulouse, France
[c]Formerly University of Aston, Birmingham, United Kingdom
*Corresponding author. e-mail address: hawkes@cemes.fr

Contents

☆ Part of this article is a translation of an earlier historical article by the author, published in *Optik* 50, 177 (1978). Permission to publish this English version has been granted by Dr. E. Menzel and the Wissenschaftliche Verlagsgesellschaft, Stuttgart, and is gratefully acknowledged.

Advances in Imaging and Electron Physics, Volume 220
ISSN 1076-5670
https://doi.org/10.1016/bs.aiep.2021.08.002

1. Scanning electron microscopy, electron microprobe

The electron microprobe, which forms the basic element of the two versions of scanning electron microscopes (SEM for surface imaging or STEM for transmission imaging) and electron beam microanalyzers, came into being in 1937 (von Ardenne, 1938a, 1938b, 1940a) by a simple reversal of the ray path of the electron microscope. From this fact it follows that the history of scanning electron microscopy and electron beam microanalysis, at least in its early beginnings, stands in close relationship to the present somewhat abbreviated account of the history of the electron microscope. After the de Broglie equation had shown us that highly accelerated electrons have an associated wavelength about four to five orders of magnitude smaller than the wavelength of visible light, and after H. Busch had founded geometrical electron optics in 1926 with his proof of the lens properties of a magnetic coil, the idea of placing electron lenses one behind the other to imitate the imaging process in the light microscope was "in the air" (around 1930). Thus it came about that this problem was attacked almost simultaneously from several sides. In this initial phase the names B. von Borries, E. Brüche, F. G. Houtermans, H. Johannson, M. Knoll, L. Marton, E. Ruska, and O. Scherzer should be recorded. The history of the development of the high-resolution transmission electron microscope, from the first improvisations in the early 1930s, is the history of a long chain of contributions to the optimization of magnified electron-optical imaging on the basis of experiments and theoretical indications. The optimization of the electron microscope to a resolving power more than 100 times better than that of the light microscope (today clearly better by a factor of 1000) was mainly carried out by three German research groups:

1. The group inspired by the subsequent Nobel Prize winner D. Gabor in the High Voltage Institute of the Technische Hochschule Berlin (Prof. A. Matthias) in Babelsberg; this included B. von Borries, E. Driest, M. Knoll, F. Krause, H. O. Mueller, and E. Ruska, whose know-how flowed into the Siemens electron microscope development through von Borries and Ruska. For the history of electron microscopy from the point of view of this group, see, for example, von Borries and Ruska (1944) and Gabor (1957).
2. The working group led by E. Brüche at the Research Institute of the AEG in Berlin with H. Boersch, E. Kinder, H. Mahl, and A. Pendzich. For the history of electron microscopy from this point of view, see, for example, Brüche (1941, 1943).

3. The small group led by M. von Ardenne with H. Reibedanz and E. Lorenz in Berlin-Lichterfelde, which was dependent on the Siemens circle and whose results were fed into the Siemens development on a contractual basis. For the history of electron microscopy from the point of view of this group, see, for example, von Ardenne (1940b, 1944).

All these efforts were officially honored in 1941, in recognition of their important contributions to the development and optimization of high-resolution electron microscopes, by the bestowal of the Silver Leibnitz Medals of the Prussian Academy of Sciences on von Ardenne, Boersch, von Borries, Knoll, and Ruska (von Auwers & Vahlen, 1942).

The history of scanning electron microscopy, the principal events of which are summarized in Table 1 (Crewe & Wall, 1971; Everhart & Thornley, 1960; Hillier & Ramberg, 1947; Knoll, 1935; Krisch et al., 1976; Oatley et al., 1965; Ruska, 1934; Ruska & von Borries, 1932; von Ardenne, 1937, 1940c, 1972; von Ardenne & C. Lorenz, 1933; von Ardenne et al., 1942; Zworykin et al., 1942, 1945), is easier to review and describe (in the initial phase there was only one working group), although, as is clear from the table, several contributions to the optimization of the transmission electron microscope were also decisive for the development of the scanning electron microscope.

For the practical realization and optimization of the two versions of the scanning electron microscope (and also of the transmission electron microscope) only a few constructional components were needed in addition to the physical and constructional principles. These specific points have been given in entries 1, 2, 3, 7, 8, 10, 11, 13, and 14 of Table 1, with the corresponding literature references. Particular credit is to be attached to the magnetic polepiece lens of short focal length (Ruska, 1934; Ruska & von Borries, 1932) of Ruska and von Borries (entry 1). In this connection (increase of resolving power) the further development shown in Fig. 1 should be pointed out, namely, the singlefield condenser–objective lens (von Ardenne, 1944) (entry 11). In our construction the entire polepiece could be replaced from the side. In addition to the system illustrated there also existed further polepiece inserts with larger polepiece separations and also with larger polepiece bores, whose measurements came close to those of the geometries in use today. By means of this lens design the resolving power of the electron microscope, was improved in 1943 to 1.2 nm. The next step forward of primary importance could be considered to be the improvement of the objective lens in 1946 by the introduction of a stig-

Table 1 Facts about the history of the scanning electron microscope (SEM) and the electron microprobe.

Year	Names	Facts	Resolution (δ)	References (comments)
1932	E. Ruska, B. von Borries	1. Invention of the short-focal-length magnetic polepiece lens (reduction of image aberrations)	—	Ruska (1934); Ruska and von Borries (1932) (predecessor: iron-clad coil of D. Gabor)
1933	M. von Ardenne	2. First imaging of surfaces by means of secondary electrons of a raster-shaped (x, y) deflected electron beam; beam energy 1200 eV	0.3 mm	von Ardenne (1937); von Ardenne and C. Lorenz (1933) (structurized semiconductor surface)
1934	M. Knoll	3. Experimental arrangements and circuit as in 1; beam energy 3000 eV	0.1–1 mm	Knoll (1935) (siliconized sheet iron)
1937	M. von Ardenne	4. Invention of the scanning electron microscope; concept for surface imaging and transmission imaging	—	von Ardenne (1972) (February 16, 1937)
1937	M. von Ardenne	5. First scanning electron microscope for the imaging of object surfaces with large depth of focus	≈ 100 nm	von Ardenne (1938a, 1938b, 1940a)
1938	M. von Ardenne	6. First scanning electron microscope for transmission imaging of specimens, even of considerable thickness	≈ 40 nm	von Ardenne (1938a, 1938b, 1940a) (bright field, dark field)

continued on next page

Table 1 *(continued)*

Year	Names	Facts	Resolution (δ)	References (comments)
1938	M. von Ardenne	7. Reduction of sensitivity to vibration of the electron microscope by the principle of pressing the specimen holder onto the lens polepiece		von Ardenne (1938b, 1940a) (about twice the former resolution)
1940	M. von Ardenne	8. Introduction of the stereo method in electron microscopy	—	von Ardenne (1940c)
1938–1942	M. von Ardenne	9. Layouts, experiments, and proposals for electron microprobe analysis with the scanning electron microscope	—	von Ardenne (1938a, 1938b); von Ardenne et al. (1942)
1942	V. K. Zworykin, J. Hillier, R. L. Snyder	10. Introduction of the secondary-electron detector with 50 nm postacceleration (9 keV), fluorescent screen, and photomultiplier	50 nm	Zworykin et al. (1942) (improvement of signal-to-noise ratio)
1942	M. von Ardenne, E. Ruska	11. First single-field condenser–objective lens with specimen plane between the polepiece (and side-entry specimen changer)	1.2 nm	von Ardenne (1944) (see Fig. 1)

continued on next page

Table 1 *(continued)*

Year	Names	Facts	Resolution (δ)	References (comments)
1944	—	12. Obliteration of the von Ardenne scanning electron microscope installation of by means of an air raid on March 25, 1944	—	This and other consequences of the war terminated work of von Ardenne on SEM
1946	J. Hillier, E. G. Ramberg	13. Introduction of the stigmator for the correction of axial astigmatism	0.6 nm	Hillier and Ramberg (1947); Zworykin et al. (1945)
1960	T. E. Everhart, R. F. M. Thornley	14. Improvement of the secondary-electron detector of 10 by the introduction of organic scintillators with light guides		Everhart and Thornley (1960); also see Hill and Gopinath (1977) (improvement of signal-to-noise ratio)
1965	C. W. Oatley	15. The "Stereoscan" apparatus of the Cambridge Instruments Company appears; the first commercial SEM (surface type)	20 nm	Oatley et al. (1965)
1970	A. V. Crewe, J. Wall	16. Introduction of the field emission electron gun into the scanning electron microscope (transmission type)	0.5 nm	Crewe and Wall (1971)
From 1970		17. The Elmiskop ST 100 F-Siemens microscope with field emission electron gun available commercially (transmission type, and also surface imaging)	$\geq$ 0.2 nm	Krisch et al. (1976)

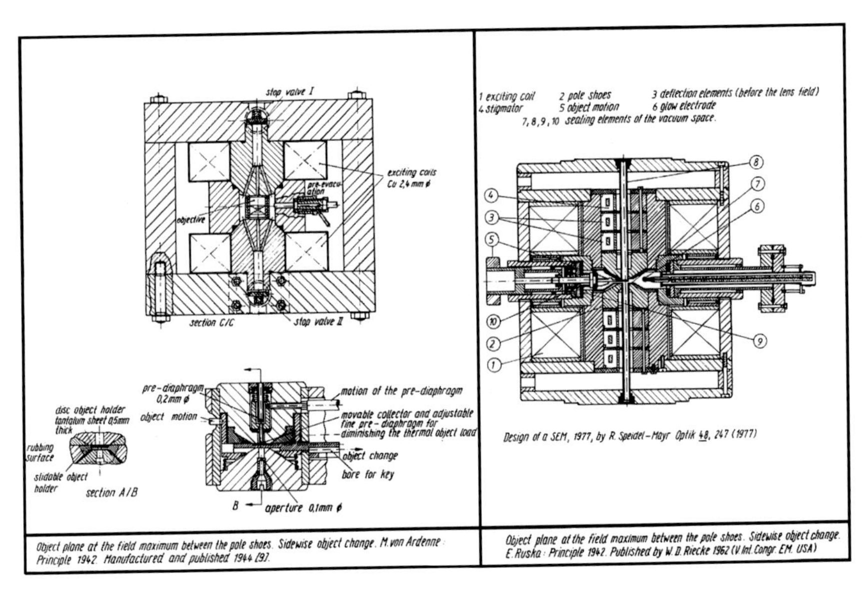

Figure 1 Versions of the single-field condenser–objective lens from the years 1944 ($\delta =$ 1.2 nm) and 1977.

mator for the correction of axial astigmatism (Hillier & Ramberg, 1947; Zworykin et al., 1945) (entry 13).

The other constructional elements and principles could be taken over from related techniques and specialities such as high- and low-voltage electron beam oscillograph techniques, the television technique, the technique of electron-optical image converters, and vacuum techniques. Mention should be made here of the three-electrode electron gun with beam crossover (von Ardenne, 1962); iron-clad magnetic coils with lens properties (Gabor, 1927); electronic scanning and image reproduction with line rasters and parallel-connected deflector systems (von Ardenne, 1931); the energy intensification, by several orders of magnitude, of low-energy electrons emitted by a cathode, by means of postacceleration and the conversion of their energy in specially selected phosphors (von Ardenne, 1936); photomultipliers for wide-band amplification of weak light signals (fluorescence) (Zworykin et al., 1936); cameras with air locks for photography in vacuum (Hochhäusler, 1929); and the technology needed for the construction of metal vacuum apparatus (Espe & Knoll, 1936).

In several reviews of the development of scanning electron microscopy (Reimer & Pfefferkorn, 1973), the simple imaging of surfaces by means of secondary electrons with the aid of a raster-shaped deflected electron beam

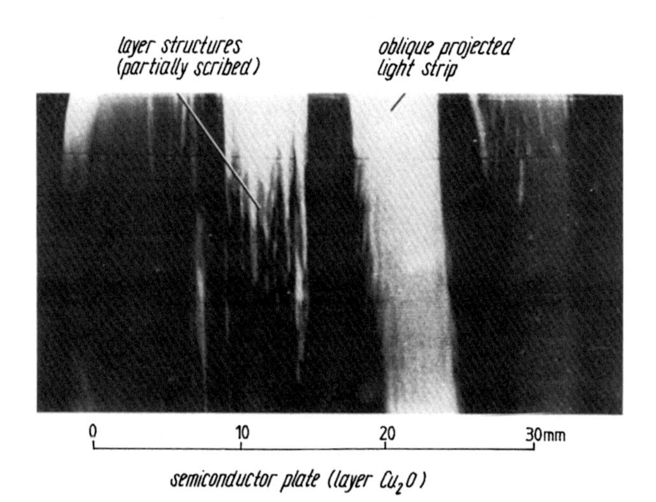

Figure 2 First imaging of surfaces by secondary electrons resulting from a raster-scanned electron beam. Image reproduction by means of a synchronous sweeping beam modulated by the signal of the secondary electrons. [From von Ardenne (von Ardenne & C. Lorenz, 1933).]

over the object was given a high priority as the *principle* of the scanning electron microscope (Reimer & Pfefferkorn, 1973) or possibly as its forerunner (Schneider et al., 1974). This kind of surface imaging was realized in 1933 by M. von Ardenne and in 1934 by M. Knoll. One of our original images made in this way in 1933 is shown in Fig. 2.

At that time images were formed of surface structures such as craters and scratches on a light-sensitive semiconductor plate (forerunner of the Vidicon camera tube development). In a second photograph of the same specimen, the broad white band visible to the right of our image, caused by the oblique projection of a light strip, was absent.

The first speculative sketch, preserved over the vicissitudes of time, of the principle and concept of both versions of the scanning electron microscope is shown in Fig. 3. An indication of the ray path and of three basic methods of capturing from the object-modulated electron rays in those early days (von Ardenne, 1938a) is shown in Fig. 4. A photograph of the first scanning electron microscope for the imaging of surfaces is shown in Fig. 5. An electron microprobe of about 50–100 nm in diameter, scanned in a rectangular line raster over the object, was formed by a two-stage reduction of the electron source crossover with the aid of magnetic polepiece lenses. The deflection took place immediately in front of the second reducing lens. The image signal was produced by means of a specially

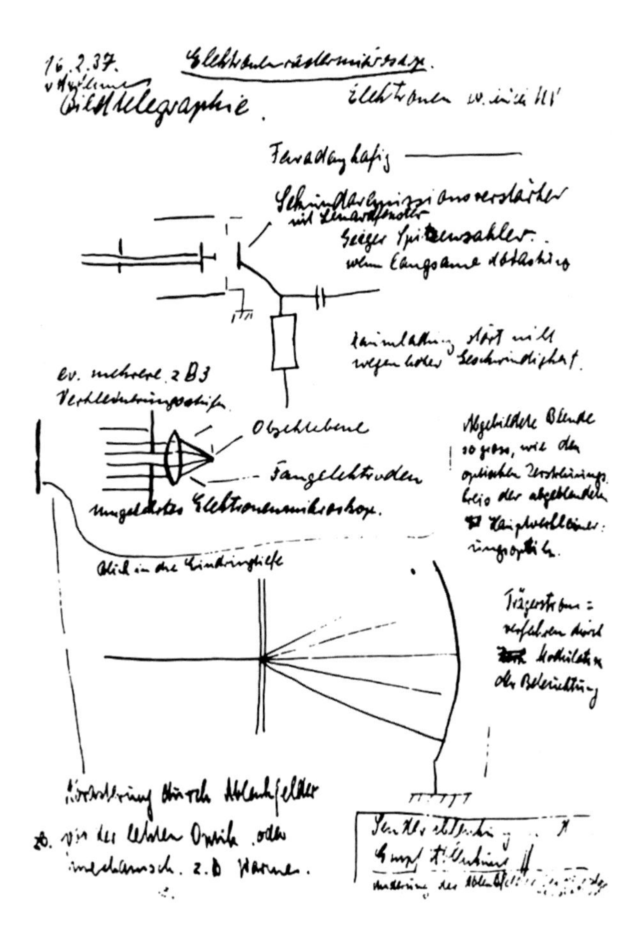

Figure 3 First surviving record of the fundamental ideas of the scanning electron microscope. Top: version for transmission imaging; middle: version for surface imaging.

developed low-capacitance detector system for the secondary electrons released from the specimen. This signal was further amplified in a wide-band television amplifier. The image was reproduced on a television tube with a long-persistence screen visible on the left with a television-type scanning system (for low resolution) and/or with very strongly reduced line and frame frequency (for high resolution). By an appropriate aperturing of the second reduction lens, its spherical aberration could be matched to the small probe diameter required, so that at the specimen a very small aperture existed; this distinguished electron microscopy from optical microscopy.

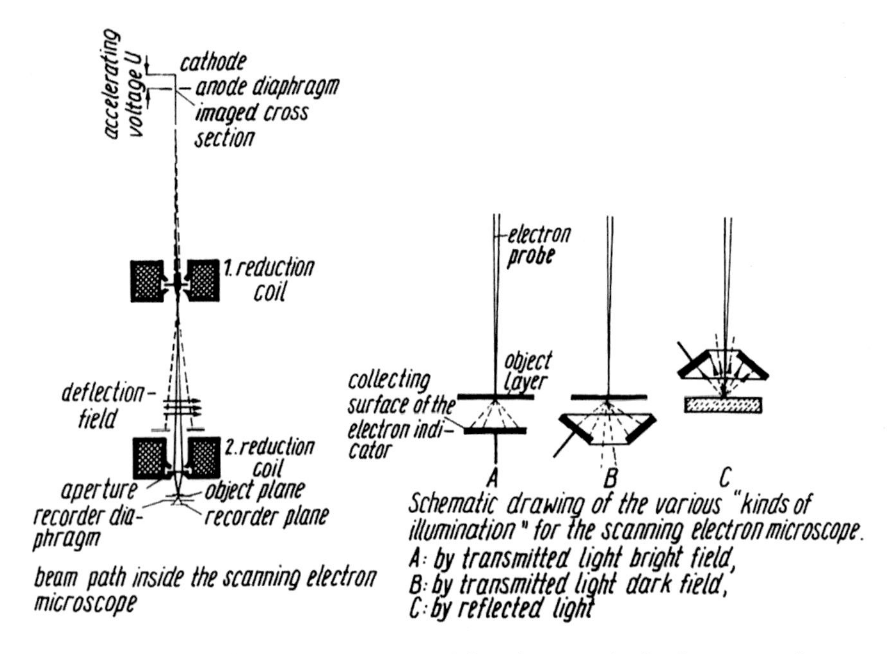

Figure 4 Scanning electron microscope and three basic methods of capturing electron rays modulated by the object. Electron microprobe. [From von Ardenne (1937).]

The enormous and unusual depth of focus of this method of imaging was very striking even during our first investigations, in which we had deliberately chosen diatoms because of their large axial extent. Unfortunately, these investigations of the version for surface imaging could not be pursued for very long since it was a matter of fulfilling our contractual obligations, with the aid of the same laboratory equipment, to develop the version for scanning transmission imaging.

Thus came into being, at the beginning of 1938, the first scanning electron microscope, shown in Fig. 6, for the imaging of specimens in transmission. At the time, this version seemed to be at the forefront of the subject, since it offered the possibility, even with the irradiation of relatively thick specimen foils (e.g., microtome sections), of keeping the chromatic aberration of the image very small. In the first investigation with the scanning transmission electron microscope, the recording of the electrons leaving the specimen was carried out photographically. The integration time per image element could be made suitably long, in the interests of high resolution, by reducing the speed of the photographic recording drum. The necessary mechanical stability was achieved by the principle

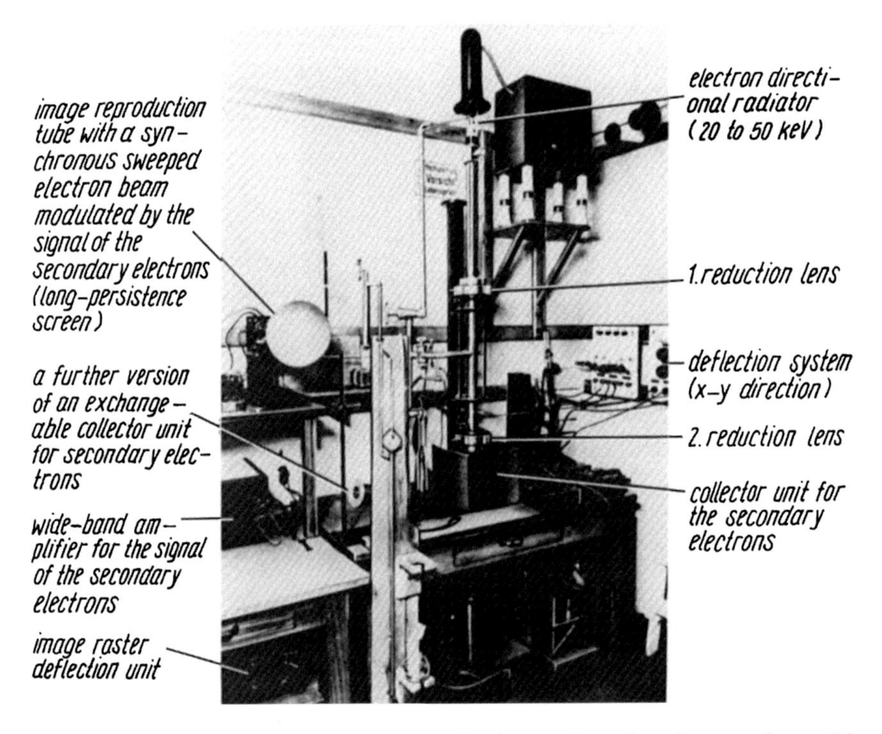

Figure 5 First scanning electron microscope for the imaging of specimen surfaces with a high depth of focus and a high resolution ($\delta \approx 100$ nm). Imaging of diatoms. [From von Ardenne (1937).]

of firmly pressing the specimen holder onto the polepiece surface of the second reduction lens. The photographic recording system is described in detail in von Ardenne (1938b, 1940a). In later investigations electron detectors for bright-field or dark-field images according to the principles already described in von Ardenne (1938b, 1940a), with secondary-electron multipliers of various types, were fitted or designed. Unfortunately, these investigations were never brought to full fruition, since the Lichterfelde scanning electron microscope apparatus was destroyed in an air raid in the spring of 1944. With this event ended our contribution to the development of scanning electron microscopy. Only several years later did other research groups take up the baton once more. Then, for the first time, through the availability of the factors shown in entries 14–17 of Table 1—especially by the British electron opticians in Cambridge—the worldwide breakthrough of scanning electron microscopy was realized.

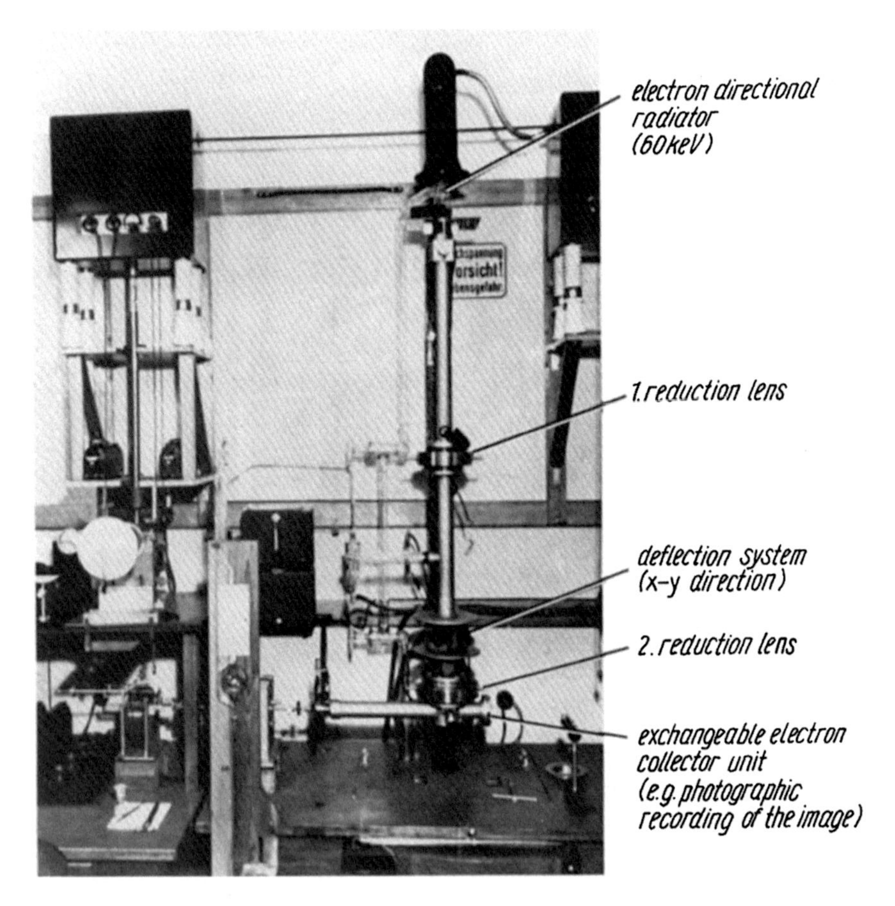

Figure 6 First scanning electron microscope for the imaging of specimens in transmission with a small chromatic aberration ($\delta = 40$ nm). [From von Ardenne (1938a, 1938b).]

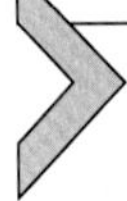

2. Early contributions to transmission electron microscopy

2.1 The first stereo electron microscope

Our researches on the imaging of surfaces with the help of a scanning electron microscope showed us impressively the importance of specimen imaging with a very small objective aperture (the imaging of specimens spatially extended with a high depth of focus). Therefore, early in 1940 it occurred to the author to make the first experiments with stereoscopic imaging and a high depth of focus of three-dimensional micro-objects by means of an electron microscope equipped for these investigations with a specimen swivel unit (von Ardenne, 1940c). At that time a strong motivation was the observation that for the light microscope, the stereomethod is inevitably imperfect for imaging object structures at the resolution limit since they require objectives with an especially big aperture, that is, with an especially low depth of focus. Already the first stereo-images, here shown in Fig. 7, as taken from the appendix of von Ardenne (1940a), confirm that stereo-imaging by an electron microscope with a high depth of focus is in principle superior to stereo-imaging by a light microscope with a higher resolution.

2.2 Exchangeable objective systems for stereo-imaging, specimen heating, specimen cooling, specimen reactions, and specimen analysis

When the contract with Siemens concerning the patents and the development of the scanning electron microscope was concluded, we were not permitted to continue research on the development of a conventional transmission electron microscope. The author found this prohibition (effected by Ruska) to be contrary to scientific ethics and, in 1939, secretly developed the universal electron microscope for bright-field, dark-field, and stereograph modes (von Ardenne, 1940d), described in detail in von Ardenne (1940a, 1940d). The author is much obliged to the director of the Kaiser Wilhelm Institute for Physical Chemistry in Berlin-Dahlem at that time for the promotion of this development in its secret initial phase. In 1939 it was clearly discernible that our small team in Lichterfelde could win this race against the development teams of the Siemens and AEG companies only by very flexible and adaptable instrument design. This recognition led to the construction of objective and projective systems that can easily be exchanged sidewise from the microscope tube. The objective system comprised both polepieces for the short-focus magnetic lens, the object system, and arrangements for special methods. Above and below the exchangeable objective system the blocking elements are placed in order to provide a vacuum lock in this part of the instrument. Thanks to this simple constructional principle, many objective systems, all with different objective data and aims, could be manufactured and tested in rapid succession without time-consuming reconstructions of the microscope body proper. This principle offered us a great advantage as against the instruments of Siemens and AEG with their compact design. Thus, soon after 1939 the objective systems for stereo-imaging (von Ardenne, 1940c), for

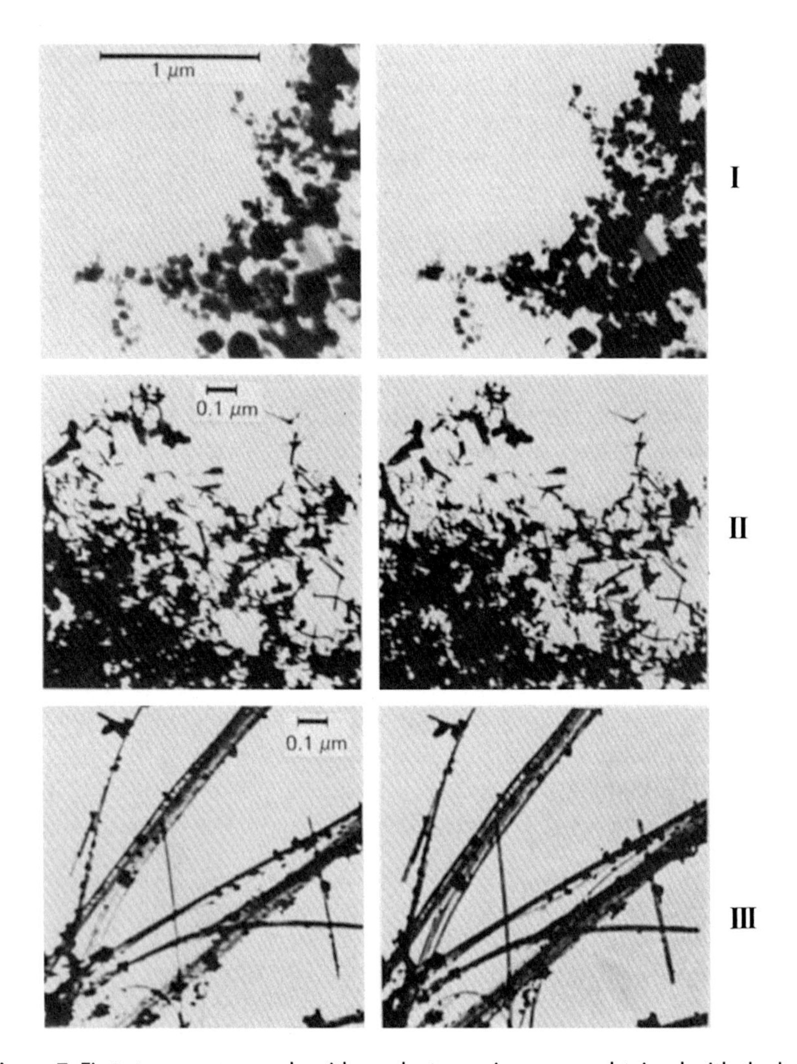

Figure 7 First stereograms made with an electron microscope, obtained with the help of the universal electron microscope of M. von Ardenne in Berlin-Lichterfelde in the winter 1939/1940: (I) The first stereogram obtained with an electron microscope. Magnesium oxide smoke with cubic crystals partially reacted through. Swivel angle, 4°. (Imaging by M. von Ardenne.) (II) Zinc oxide crystals from the vapor of an arc discharge between zinc electrodes. Swivel angle, 4°. (Imaging by M. von Ardenne.) (III) Unused palladinized asbestos catalyst. This image revealed the surprising fact that asbestos consists of very fine ribbons (only 3 μm thick in some places) split many times. Swivel angle, 6°. (Imaging by M. von Ardenne.) (IV) Strongly exposed, undeveloped silver bromide grains. The left partial picture was made later. Thus the grain variations became immediately visible due to the electron action over a longer time, similar to the stereocomparator. Swivel angle, 4°. (Imaging by M. von Ardenne.)

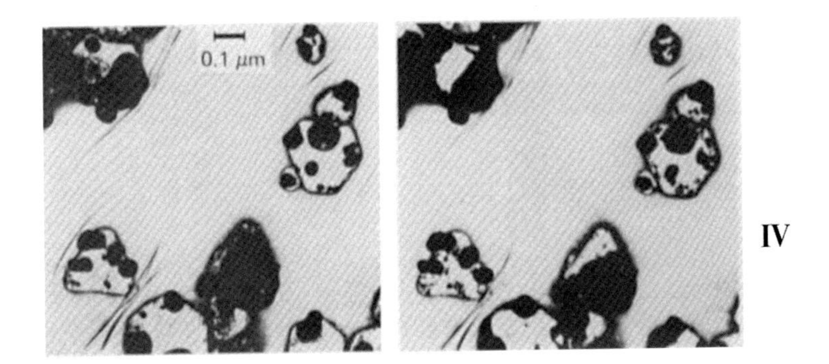

Figure 7 *(continued)*

specimen heating up to more than 2000°C (von Ardenne, 1941a), for specimen cooling by hydrogen with a reduced pressure, for chemical reactions with gaseous reacting agents (von Ardenne, 1942), and for specimen examination to study relatively radioresistant bio-objects (von Ardenne, 1941b) were built. Furthermore, from this constructional principle, inserts for a pressure subdivision inside the microscope tube (von Ardenne, 1943a) and for microbeam electron diffraction (von Ardenne et al., 1942) arose. By the same constructional principle the single-field condenser–objective lens mentioned above, with the object plane at the field maximum between the polepieces, was also realized in 1944 (von Ardenne, 1944).

2.3 The beginning of extra-high-voltage electron microscopy

To increase the resolving power for the imaging of objects with a greater mass thickness, at the end of 1940 the author started experiments for raising the electron energy up to 300 and 1000 keV, respectively. Thus, in Lichterfelde the instrument shown in Fig. 8 was created (von Ardenne, 1941c). The high-tension dielectric strength of the electron-accelerating system of this instrument had been increased up to almost 300 kV shortly before its destruction by a bombing raid in 1943.

The design of the instrument illustrated, concerning its dimensions and the reserves in the objective system, was chosen in such a manner that it could be mounted onto the lower end of the discharge tube of a one-million-volt electron-beam generator that was already operating in 1943. [This fact is pointed out in von Ardenne (1943b), p. 237.] In Lichterfelde the path of extra-high-voltage electron microscopy was prepared and trodden, but the events and the end of World War II prevented its continuation.

Figure 8 View of a universal electron microscope for electron energies from 200 up to 300 keV. (From M. von Ardenne, 1941–1943.)

2.4 Electron microcinematography with the aid of a vacuum film camera

In order to be able to image moving objects in electron microscopic researches with the full resolution, the vacuum film camera shown in Fig. 9 was manufactured in 1942 (von Ardenne, 1943a). At that time the need for cinematographic recording of electron microscope pictures arose especially from the use of the electron microscope for specimen heating. The moving processes and object changes provoked by a temperature increase were captured on a number of electron-microscopic films with a high resolution. Some of these films were projected in the Harnack House of the Kaiser-Wilhelm-Gesellschaft in 1943 (processes involving melting uranium particles in their oxide skin, ceramic and metal–ceramic sintering, and chemical microreactions).

Figure 9 View of the vacuum film camera before moving it into the photochamber of the universal electron microscope. By covering the semilunar rotary diaphragm, discernible at the image, with luminescent material, the filmed process could be observed continuously.

2.5 The beginning of the electronic preparation technique, first specimen holder foils, first thin sections cut by the wedge-cut microtome, first specimen staining with the help of the osmium method

In the early days of electron microscopy it was usual to fasten the object particles to be imaged simply at the edge of the specimen holder bore or to evaporate them on. An essential, practical step forward was the introduction, in 1937, of thin specimen holder foils into the preparation technique of electron microscopy (von Ardenne, 1940a, p. 269ff.). The first preparations of this type, based on a method which was initiated by Langmuir–Blodgett and improved by W. Trenktrog, were demonstrated by the author to the Siemens team within the contractual relationship.

In order to open up the possibility of rapid imaging with high resolution for biological objects with a greater mass thickness and to serve for the investigation of the living cell and its organelles, the wedge-cut microtome was developed in 1938 in Lichterfelde. This was the first ultramicrotome (von Ardenne, 1940a, p. 259ff.).

The most important elements of the wedge-cut microtome are indicated in Fig. 10. Fig. 11 reminds us of the appearance of a light-optic image of a wedge section from the year 1938. The aperture angle of the

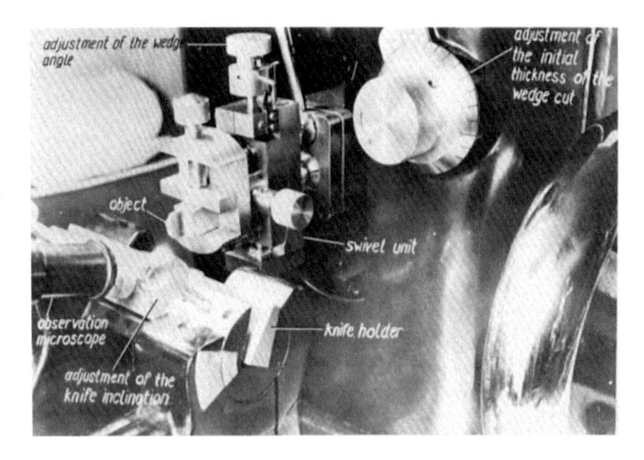

Figure 10 Photo of the most important elements of the wedge-cut microtome. [From von Ardenne (1938a, 1938b).]

wedge section was set with the aid of the swivel unit, which is assembled on a standard Minot microtome. By the wedge-cut method the first ultrathin sections (section thickness about 2×10^{-4} mm) could be obtained, which initiated the cytologic research that later became very successful.

In no time the results were such that electron-microscopic investigations clearly needed staining of the specimens by heavy metals. Thus, we succeeded (von Ardenne et al., 1941) in introducing the staining of biological objects especially by a treatment with OsO_4, a method that later quickly prevailed in the field of the electron-microscopic preparation techniques.

2.6 Discoveries with the aid of the electron microscope, the filiform nature of myosin, the fiber structure of the developed silver bromide grain

After the development of the electron microscope with high resolution in its various versions, after the creation of essential elements of the electron-microscopic preparation technique, and after the construction of special microtomes for preparing ultrathin sections, electron microscopy entered a new phase. During World War II, almost each image yielded views into a microcosmic world never directly seen until then. The wonderful era of the great electron-microscopic discoveries in various fields of natural science and, especially, the investigation of the microstructure of the living cell and its organelles was dawning. Today, about 40 years later, the author remembers with distressing feelings the fact that the events of the war (de-

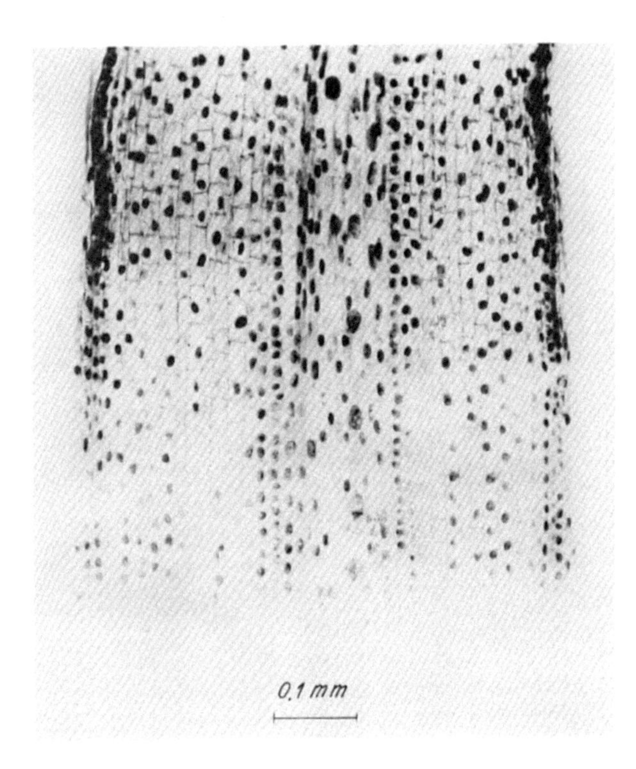

Figure 11 Wedge cut of the middle piece of an onion root tip at an initial thickness of 2×10^3 mm and a cut aperture of *l* : 50 (von Ardenne, 1940a).

struction of three pioneer plants of the electron microscopy of those days and of the one-million-volt dc plant by air raids) and postwar fate forced him to leave the field shortly before the harvest. However, it was a comfort that the work in the assigned new field (the development of procedures for industrial isotope separation) served to establish rapidly the nuclear equilibrium and thus to help maintain peace. In spite of the situation described, in the early stage of electron microscopy the author was able to take part in two discoveries. These were the discovery and first imaging of the *filiform structure of the myosin molecule* (von Ardenne & Weber, 1941) and the discovery of the *fiber structure of the developed silver bromide grain* with the aid of the original image seen in Fig. 12 (von Ardenne, 1940e).

It lay well beyond the limits of human fantasy to surmise, in the early days of the development of this electronic method, to what heights transmission electron microscopy, scanning electron microscopy, and the elec-

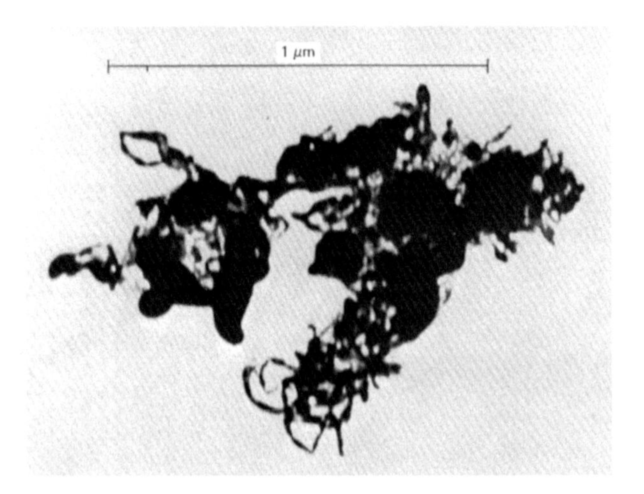

Figure 12 Strongly exposed, developed, and unfixed silver bromide grains with thin (silver?) filaments 10–15 μm thick. (Imaging by M. von Ardenne.)

tron probe microanalysis technique developed from it would rise in the science of our day.

3. Conclusion

A chain of contributors has been presented in chronological order, who from the point of view of today should probably be regarded as those most important in the overall development of transmission electron microscopy, scanning electron microscopy, and the electron microprobe.

4. Afterword by Peter Hawkes

Manfred von Ardenne was born on 20 January 1907 in Hamburg and died on 26 May 1997 in Dresden. As a young man, he built a simple cat's whisker wireless set and even published a book on *Fernsehempfehlung*, which was translated into English and read by the schoolboy Tom Mulvey. He used his inheritance to create a private laboratory, the Forschungslaboratorium für Elektronenphysik in Berlin–Lichterfelde, where his electron microscopes were built. During the wartime years, he attempted to raise the accelerating voltage to 300 kV, and potentially to 1 MV, but this instrument was destroyed in a bombing raid. At the end of the war, he spent 10 years in Russia, in Sukhumi on the Black sea coast, returning to Dresden (then in

(a)

(b)

Figure 13 (a) Manfred von Ardenne in the 1930s. (b) Manfred von Ardenne and Ellis Cosslett in 1973. Photographs kindly provided by von Ardenne's son Alexander.

East Germany) in 1954, where he was appointed professor at the Technische Hochschule. At the same time, he created a (private) research institute, the Forschungsinstitut Manfred von Ardenne. This foundered after German reunification in 1991 but resurfaced as Von Ardenne Anlagentechnik GmbH. In his later years, he entered the field of cancer research.

Looking forward to Chapter 6 by Gaston Dupouy, we draw attention to von Ardenne's articles on the observation of living material in the electron microscope (von Ardenne, 1939b, 1941b; von Ardenne & Friedrich-Freksa, 1941, not all of which are cited above). Dupouy was aware of these and the librarian of the Laboratoire d'Optique Electronique in Toulouse translated at least one of them into French.

Manfred von Ardenne wrote several autobiographical volumes, of which the most recent is *Erinnerungen Fortgeschrieben* (Droste, Düsseldorf 1997). Figs. 13a and 13b show him in the 1930s and in 1973, arm in arm with Ellis Cosslett.

Finally, for lovers of the novels of Theodor Fontane, we note that von Ardenne's grandmother, Baroness Elisabeth von Ardenne (1853–1952), is arguably the model for Effi Briest.

5. Appreciation by Tom Mulvey[1]

Manfred von Ardenne was born in Hamburg on Sunday 20 January 1907. He died on 26 May 1997 at the age of 90 in Dresden, after a short period of weakening physique, but retaining all his critical faculties and self-confidence to the end. His career, in advanced engineering, technology and scientific research, was remarkably varied and unpredictable.

Manfred was the eldest of five children. His father, Egmont Baron von Ardenne, was an Army Officer, a First Lieutenant (Oberstleutnant). Manfred esteemed him highly and was always grateful that his father gave him complete freedom and encouragement to choose a career for himself. He was very tolerant of his son's activities, intervening only when things became dangerous, as when Manfred's later experiments at home with explosives began to alarm the neighbors. His mother Adela, née Muntzenbecher, was a devoted mother to all the five children and untiring in her efforts to look after them. Manfred was particularly fond of his mother and regarded her as a close friend. Both father and mother dedicated themselves to the children's upbringing, aiming to present to them a practical example of what a family should be like. There were sanctions that could be applied but corporal punishment, for example, then commonly used by German parents to establish their authority, was out. Manfred soon emerged as a strikingly self-confident person, full of initiative, characteristics that persisted right up to his death.

From earliest youth, he was fascinated by any form of technology, watching dredgers on the canal or being impressed by the internal speaking tube system in his grandfather's house, especially by the whistle on the end of the tube that alerted the listener that a caller was on the line! He was likewise interested in watching the moon and stars; his memory went back to the age of three, when in 1910 his father pointed out to him Halley's Comet, with its luminous trail. A few years later there was an eclipse of the sun. His father gave him a small telescope and dark glasses to observe the whole process. This made a big impression, stimulating his later interest in all forms of optics. When Manfred was five years old, his father bought a camera with a ground glass screen for focusing. Manfred couldn't understand why the image on the screen was upside down and he began to think about it. His parents also provided a well-stocked lumber room

[1] Reproduced from *Journal of Microscopy* 188, 1997, 94–95 by kind permission of the Royal Microscopical Society.

where many things could be made, including home-made rafts. His father, returning home one evening, was surprised to see his young son, unable to swim, drifting slowly down the canal. Fortunately he was able to shout instructions on how to bring the craft to the bank.

As a young man in the 1930s, von Ardenne became well known in the U.K. and elsewhere as an author of books for amateur radio and television set constructors. The English language versions became widely available in public libraries. This activity was only one of a number of ways that von Ardenne used to raise funds for his experiments. He also acted as a professional photographer at weddings etc. He was always extremely successful as a fund raiser! At his Grammar School he was never interested in studying for examinations, but he constructed a telescope and a sextant for the use of the other pupils. Some of the Masters reported that his knowledge of physics and chemistry was way ahead of his classmates but that he didn't bother to study any other subjects. His father then transferred him to a *Realschule* where physics was given a higher profile. Manfred took out his first patent in 1923, in wireless telegraphy, at the age of 16; he was still at school! He also published his first book, *Das Funk-Ruf-Buch* (Call-sign Book) in the same year, another fund-raising venture! This book worried the Establishment. Having mastered the Morse Code and built himself a powerful radio receiver at his home in Berlin, to say nothing of his secret illegal transmitter, he was able to list the call-signs and wavelengths of transmitters receivable in Berlin in the wavelength range 200–20 000 m. The official Call-sign Register excluded sensitive wavelengths, such as those used by the police and the secret military stations. Manfred's book listed everything, making it very attractive. He was soon informed officially that he must recall the book. He ignored the request as he needed the money! Remarkably, no official action was taken.

After some studies in physics and chemistry at Berlin University, a new and important phase began as a brilliant researcher in electron optics and electron microscopy. He set up his own Manfred von Ardenne Research Laboratory, where he did contract research in cathode ray tubes and the new transmission electron microscope being developed commercially at Siemens und Halske in Berlin, with whom he obtained a research contract. He soon invented and built a scanning transmission electron microscope (STEM) and also invented the point projection X-ray microscope, but was unable to get funds to build it. It was left to Cosslett and Nixon at Cambridge to bring this to fruition. Likewise, a practical form of scanning electron microscope (SEM) had to wait many years to be brought to fruition by C. W. Oatley and his team at Cambridge University. The STEM itself had to wait for Albert Crewe in the U.S.A. to make it a complete success by incorporating a field emission gun. In 1941 von Ardenne held the world record for high-resolution TEM with his 'Universal TEM' and was one of a group of six German pioneers of electron microscopy to receive individual Leibnitz Silver Medal awards for their contribution to electron microscopy.

In 1943, a copy of Ardenne's practically orientated book *Elektronenmikroskopie* was secretly transferred in a German wartime submarine to Japan, where it was quickly translated into Japanese. The book had an instant and electrifying effect on the electron diffraction and electron optics community, who immediately began an intensive programme to build electron microscopes. Von Ardenne can truly be called the founder of the huge Japanese electron microscope industry of today!

On 25 March 1944, the Manfred von Ardenne Laboratory in Berlin was bombed by the RAF; all the electron microscopes were destroyed and the building was badly damaged. Rebuilding started immediately and it was ready again for service by the end of the year! After the storming of this part of Berlin by the Soviet Army, von Ardenne, his team and all their equipment were transported to Sukumi on the shores of the Black Sea, to continue their work there. It was not to be. When the atomic bomb was dropped on Hiroshima in August 1945, the Sukumi Institute had to stop work on electron microscopy and concentrate on producing ^{235}U. Von Ardenne took up isotope production, for which he later received the Stalin Prize. He avoided working on the bomb itself as this would mean that he would not be allowed to return to Germany.

He returned in the 1950s to Dresden in East Germany, where he developed his remarkable 'Krebs-Mehrfach-Therapie' (Cancer multistage Therapy), where a patient wearing an oxygen mask generates 200 W of electric power while riding a standbicycle under medical supervision. He believed that such treatment would enable him to live to the age of 90. As usual, he was right.

References

Brüche, E. (1941). Vom Mikroskop zum Übermikroskop. *Schweizer Archiv für Angewandte Wissenschaft und Technik*, 7, 46.

Brüche, E. (1943). Zum Entstehen des Elektronenmikroskopes. *Physikalische Zeitschrift*, *44*, 176.

Crewe, A.V., & Wall, J. (1971), Grenoble, 1970 Vol. 1.

Espe, W., & Knoll, M. (1936). *Werkstoffkunde der Hochvakuumtechnik*. Berlin and New York: Springer-Verlag.

Everhart, T. E., & Thornley, R. F. M. (1960). Wide band detector for micro-microampere low-energy electron currents. *Journal of Scientific Instrumests*, *37*, 246.
Gabor, D. (1927). Oszillographieren von Wanderwellen mit dem Kathodenoszillographen. *Forschungshefte der Studiengesellschaft für Höchstspannungsanlagen*, *1*.
Gabor, D. (1957). Die Entwicklungsgeschichte des Elektronenmikroskopes. *Elektrotechnische Zeitschrift. Ausgabe A*, *78*, 522.
Hill, M. S., & Gopinath, A. (1977). Channel plate multiplier as an emissive mode detector in the SEM. *Review of Scientific Instruments*, *48*, 806.
Hillier, J., & Ramberg, E. G. (1947). The magnetic electron microscope objective: Contour phenomena and the attainment of high resolving power. *Journal of Applied Physics*, *18*, 48.
Hochhäusler, P. (1929). Ein- und Ausführung von Platten und Filmen an Kathodenstrahloszillographen ohne Störung des Hochvakuums. *Elektrotechnische Zeitschrift*, *50*, 820.
Knoll, M. (1935). Aufladepotential und Sekundäremission elektronenbestrahlter Körper. *Zeitschrift für technische Physik*, *16*, 467.
Krisch, B., Muller, K. H., Schliepe, R., Thon, F., & Willasch, D. (1976). *Elmiskop ST 100 F—ein Durchstrahlungs-Rasterelektronenmikroskop höchster Leistung*. Siemens Druckschrift.
Mulvey, T. (1997). Baron Manfred von Ardenne (1907–1997). *Journal of Microscopy (Oxford)*, *188*, 94–95 (reproduced above).
Oatley, C. W., Nixon, W. C., & Pease, R. F. (1965). Scanning electron microscopy. *Advances in Electronics and Electron Physics*, *21*, 181.
Reimer, L., & Pfefferkorn, G. (1973). *Raster-elektronenmikroskopie*. Berlin and New York: Springer-Verlag.
Ruska, E. (1934). Über ein magnetisches Objektiv für das Elektronenmikroskop. *Zeitschrift für Physik*, *89*, 90.
Ruska, E., & von Borries, B. (1932). German Patent B154,916.
Schneider, V., Schwarz, W., Dunger, B., & Bahr, J. (1974). *Rasterelektronenmikroskopie. Pressedienst Wissenschaft Freie Universität Berlin*, *2*, 306.
von Ardenne, M. (1931). Über neue Fernsehsender und Femsehempfänger mit Kathodenstrahlröhren. *Fernsehen*, *2*, 65.
von Ardenne, M. (1936). Über die Umwandlung von Lichtbildern aus einem Spektralgebiet in ein anderes durch elektronenoptische Abbildungen von Photokathoden. *Elektrische Nachrichten-Technik*, *13*, 230.
von Ardenne, M. (1937). Über Versuche mit lichtempfmdlichen Halbleiterschichten in Elektronenstrahlröhren. *Hochfrequenztechnik und Elektroakustik*, *50*, 145.
von Ardenne, M. (1938a). Das Elektronen-Rastermikroskop. Theoretische Grundlagen. *Zeitschrift für Physik*, *109*, 553.
von Ardenne, M. (1938b). Das Elektronen-Rastermikroskop. Praktische Ausführung. *Zeitschrift für technische Physik*, *19*, 407.
von Ardenne, M. (1939a). Die Keilschnittmethode, ein Weg zur Herstellung von Mikrotomschnitten mit weniger als 10^{-3} mm Stärke für elektronenmikroskopische Zwecke. *Zeitschrift für Wissenschaftliche Mikroskopie*, *56*, 8.
von Ardenne, M. (1939b). Über die Möglichkeit der Untersuchung lebender Substanz mit Elektronenmikroskopie. *Zeitschrift für technische Physik*, *20*, 239.
von Ardenne, M. (1940a). *Elektronen-Übermikroskopie*. Berlin and New York: Springer-Verlag (Translated editions were published during the war in the Soviet Union, the USA, and Japan).
von Ardenne, M. (1940b). Elektronen-Übermikroskopie—ein Nachbargebiet der Fernsehtechnik. *Telegraph-, Fernsprech-, Funk- und Fernseh-Technik*, *29*, 367.
von Ardenne, M. (1940c). Stereo-Übermikroskopie mit dem Universal-Elektronenmikroskop. *Naturwissenschaften*, *28*, 248.

von Ardenne, M. (1940d). Über ein Universal-Elektronenmikroskop für Hellfeld-, Dunkelfeld- und Stereobildbetrieb. *Zeitschrift für Physik, 115*, 339.
von Ardenne, M. (1940e). Analyse des Feinbaues stark und sehr stark belichteter Bromsilberkömer mit dem Universal-Elektronenmikroskop. *Zeitschrift für Angewandte Photographie in Wissenschaft und Technik, 2*, 14.
von Ardenne, M. (1941a). Erhitzungs-Übermikroskopie mit dem Universal-Elektronenmikroskop. *Kolloid-Zeitschrift, 97*, 257.
von Ardenne, M. (1941b). Elektronen-Übermikroskopie lebender Substanz. *Naturwissenschaften, 29*, 521, 523.
von Ardenne, M. (1941c). Über ein 200 kV-Universal-Elektronenmikroskop mit Objektabschattungsvorrichtung. *Zeitschrift für Physik, 117*, 657.
von Ardenne, M. (1942). Reaktionskammer-Übermikroskopie mit dem Universal-Elektronenmikroskop. *Zeitschrift für Physikalische Chemie. Abteilung B, 52*, 61.
von Ardenne, M. (1943a). Elektronenmikrokinematographie mit dem Universal-Elektronenmikroskop. *Zeitschrift für Physik, 120*, 397.
von Ardenne, M. (1943b). Über eine Atomumwandlungsanlage für Spannungen bis zu 1 Million Volt. *Zeitschrift für Physik, 121*, 236.
von Ardenne, M. (1944). Über ein neues Universal-Elektronenmikroskop mit Hochleistungsmagnet-Objektiv und herabgesetzter thermischer Objektbelastung. *Kolloid-Zeitschrift, 108*, 195.
von Ardenne, M. (1962). *Tabellen zur angewandten Physik, vols. 1 and 2*. Dtsch. Verlag Wiss. Berlin (p. 142).
von Ardenne, M. (1972). *Ein glückliches Leben für Technik und Forschung*. Munich: Kindler Verlag (p. 120).
von Ardenne, M., & C. Lorenz, A. G. (1933). German Patent L84,500.
von Ardenne, M., & Friedrich-Freksa, H. (1941). Die Auskeimung der Sporen von *Bacillus vulgatus* nach vorheriger Abbildung im 200-kV-Universal-Elektronenmikoskop. *Naturwissenschaften, 29*, 523–528.
von Ardenne, M., & Weber, H. H. (1941). Elektronenmikroskopische Untersuchung des Muskeleiweißkörpers "Myosin". *Kolloid-Zeitschrift, 97*, 322.
von Ardenne, M., Friedrich-Freksa, H., & Schramm, G. (1941). Elektronenmikroskopische Untersuchung der Präcipitinreaktion von Tabakmosaikvirus mit Kaninchenantiserum. *Archiv für die Gesamte Virusforschung, 2*, 80.
von Ardenne, M., Schiebold, E., & Günther, F. (1942). Feinstrahl-Elektronenbeugung im Universal-Elektronenmikroskop. *Zeitschrift für Physik, 119*, 352.
von Auwers, O., & Vahlen, T. (1942). Zur Überschreitung der Grenze des lichtmikroskopischen Auflösungsvermögens. *Forschung Fortschritte, 18*, 291.
von Borries, B., & Ruska, E. (1944). Neue Beiträge zur Entwicklungsgeschichte der Elektronenmikroskopie und der Übermikroskopie. *Physikalische Zeitschrift, 45*, 314.
Zworykin, V. K., Morton, G. A., & Malter, L. (1936). The secondary emission multiplier—A new electronic device. *Proceedings of the IRE, 24*, 351.
Zworykin, V. K., Hillier, J., & Snyder, R. L. (1942). A scanning electron microscope. *ASTM Bulletin, 117*, 15.
Zworykin, V. K., Morton, G. A., Ramberg, E. G., Hillier, J., & Yance, A. W. (1945). *Electron optics and the electron microscope*. New York: Wiley.

CHAPTER THREE

A pioneer of electron microscopy: Hans Boersch (1909–1986)☆

Heinz Niedrig
Institut für Optik und Atomare Physik, Technische Universität Berlin, Berlin, Germany
e-mail address: heinz.niedrig@t-online.de

Contents

Hans Boersch was born in Berlin on June 1, 1909. After attending school and graduating from high school (Abitur) in Berlin-Lankwitz, he began studying physics in 1930 at the Technische Hochschule Berlin-Charlottenburg (now Technische Universität Berlin) with G. Hertz, M. Vollmer, R. Becker and at the Friedrich-Wilhelms-Universität (now Humboldt-Universität Berlin) with M. von Laue, W. Nernst, E. Schrödinger, and others. After his intermediate diploma (Vordiplom), in 1933 he accepted an offer from H. Mark to work under his supervision at the University of Vienna on electron diffraction in gases. This was Boersch's first contact with electron optics, which was quite new at that time. Two years later this work was completed with the defense of his dissertation (Boersch, 1935). (See Fig. 1.)

Returning to Berlin, Boersch accepted a position as "training engineer" at the AEG Research Institute in Berlin-Reinickendorf (director C. Ramsauer) in 1935. There, under the direction of E. Brüche, "Geometric Electron Optics" was developed, parallel to the work of M. Knoll, B. von Borries, and E. Ruska at the Technische Hochschule Berlin. According to Boersch, there was only occasional contact between the two groups. Here Boersch was significantly involved in the further development of the electrostatic electron microscope in particular. He investigated electron-optical image formation for the first time from a wave-optical

☆ Translated by Tolga Wagner and Fredrik Otto from *Elektronenmikroskopie* No. 29 (2009) 15–21 with kind permission of the Deutsche Gesellschaft für Elektronenmikroskopie.

Advances in Imaging and Electron Physics, Volume 220
ISSN 1076-5670
https://doi.org/10.1016/bs.aiep.2021.08.003

Figure 1 Prof. Dr. Hans Boersch as director of the First Institute of Physics at the Technical University of Berlin (early 1960s).

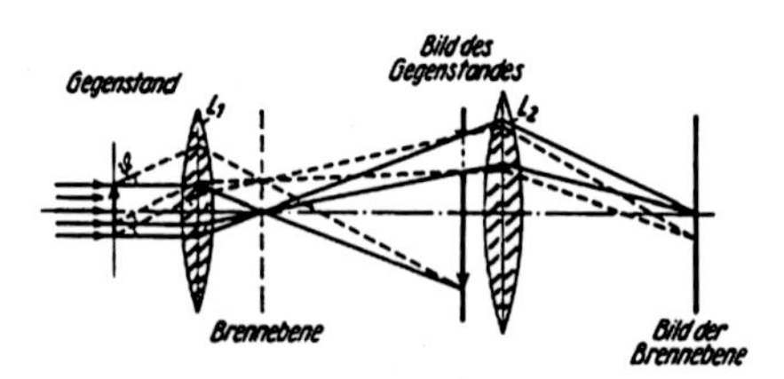

Figure 2 Selected-area diffraction with "Boersch's beam path" for two-stage imaging: The diffraction mode for the area of the object selected by the aperture in the intermediate image is drawn. By changing the lens current in the projector lens L2, it is possible to switch in a simple way between the imaging and diffraction ray paths (Boersch, 1936, 1940a, 1940b).

point of view; for this purpose, he transferred Abbe's microscope theory to the electron microscope and developed the so-called "Boersch beam path", later named after him, for selected-area diffraction at small, controllable object areas (Boersch, 1936, 1940a; Fig. 2), which is realized today in every commercial transmission electron microscope.

In 1938, Boersch made the first attempts to divide the imaging process into two separate steps (Boersch, 1938) in the sense of "two-wavelength microscopy", a predecessor of holography (Boersch, 1967). In 1939, Boersch developed the electron shadow microscope (Boersch, 1939a, 1939b,

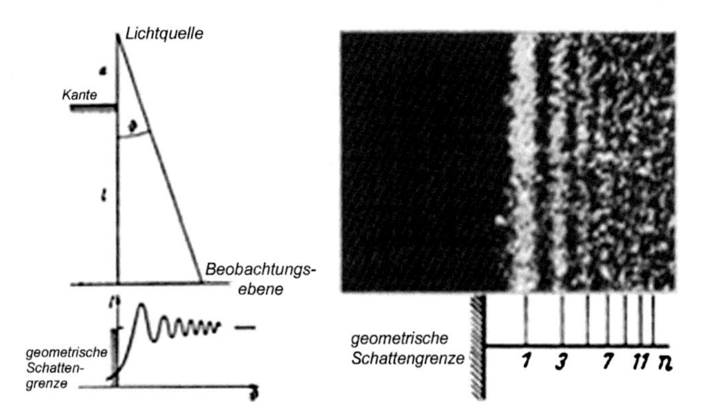

Figure 3 Fresnel electron diffraction at the edge of an amorphous alumina foil (34 keV). *Left*: Schematic of the experimental setup and theoretical intensity distribution of the Fresnel diffraction pattern. *Right*: Photographically recorded Fresnel electron diffraction figure. n = 1, 3, 5,...: number of Fresnel diffraction maxima (Boersch, 1940a).

1940b) using an electron-optical two-stage reduced electron source to produce a high-resolution shadow image of microscopic objects. With this instrument, he perceived Fresnel diffraction of electrons at an edge (Boersch, 1940c, 1940d; Fig. 3), providing definitive evidence for the wave nature of electrons independent of diffraction from periodic microstructures (crystal lattices). Fresnel diffraction phenomena also arise in the transmission electron microscope by defocusing (Boersch, 1943); the number of Fresnel diffraction fringes observable in this process now serves as a measure of the coherence of the illuminating electron beam.

In 1941 Boersch left the AEG Research Institute in Berlin and went to the University of Vienna as an Assistant. After his habilitation in 1942 (Fresnel electron diffraction), he worked there as a lecturer until 1946. During this time, he succeeded in using an electrostatic ion ultramicroscope to pass below the resolution limit of light for the first time, now with ions (Boersch, 1942). After the end of the war and a short detour through the University of Innsbruck, he was able to work from the end of 1946 in Tettnang in Württemberg at the Kaiser Wilhelm Institute for Metallurgy (originally in Stuttgart), which had been taken over by the French occupation forces, and set up a laboratory for electron optics. There he continued working on ion microscopy (Boersch, 1947a, 1947b) and developed high-resolution electron energy filters (retarding-field mesh filters, Boersch, 1948, 1949) and later (in Braunschweig) filter lenses (Boersch, 1951a, 1951b, 1951c, 1953a, 1953b) for contrast enhancement by separa-

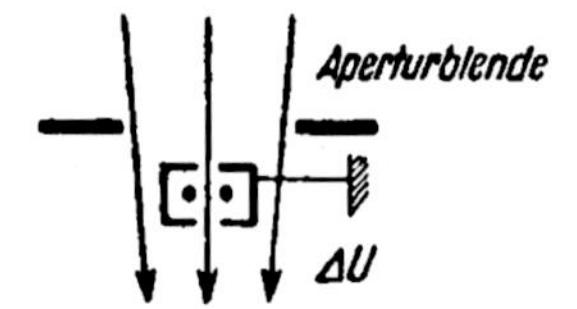

Figure 4 Proposal of H. Boersch for phase-shifting of the zero-beam with respect to the diffracted-beams for the transformation of the electron-microscopic phase-contrast into amplitude-contrast according to Zernike with the help of the phase-shifting effect of an electrostatic miniature-lens (phase-plate) (Boersch, 1947a, 1947b).

tion of inelastically scattered electrons during electron-optical imaging and diffraction.

In Tettnang, Boersch also investigated in theoretical works the possibility of imaging single atoms in the electron microscope by elastic and inelastic scattering (Boersch, 1946a, 1946b, 1946c, 1947a, 1947b). In these publications, especially in that of 1947, he proposed for the first time the realization of phase contrast imaging in electron microscopy (as in light optics according to Zernike by transformation of phase contrast into amplitude contrast by means of phase plates) and electrostatic microlenses for phase shifting of unscattered electrons in the zero beam with respect to those in the diffraction orders (Fig. 4), which, however, were technically not realizable at that time. In the meantime, the present possibilities of electron beam and ion beam lithography have made it possible to produce and use such Boersch phase plates in the form of electrostatic micro-einzel lenses as shown in Fig. 4 (Schultheiß et al., 2006).

At the special request of M. von Laue, Boersch accepted a position at the end of 1948 as senior government councillor at the newly founded National Metrology Institute of Germany (Physikalisch-Technische Bundesanstalt PTB) in Braunschweig and was also appointed honorary professor of the Technical University there. In addition to his work on filter lenses (Boersch, 1951a, 1951b, 1951c, 1953a, 1953b), Boersch discovered—with a retarding-field method for measuring the energy distribution of electron beams—-the anomalous broadening of the energy distribution in high intensity electron beams, now known as the Boersch effect (Boersch, 1953a, 1953b, 1954, Fig. 5); this is an important resolution-limiting phenomenon in electron microscopy and electron beam lithography. In total, Boersch published nearly 50 papers on his work until 1954.

In 1954, Boersch accepted an appointment as full professor and successor to C. Ramsauer at the Technical University of Berlin, where he

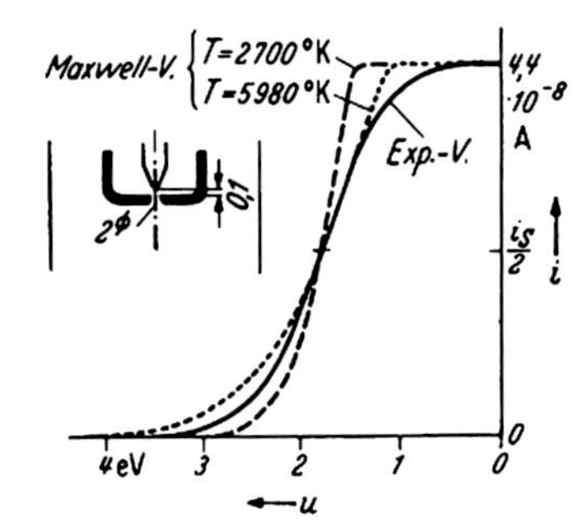

Figure 5 Experimental energy distribution of the electron beam of a hairpin cathode compared with a Maxwell distribution for cathode temperatures of 2700 K and 5980 K (curve fitted to experimental distribution): the experimental curve shows an anomalous energy broadening (Boersch effect) (Boersch, 1954).

served as director of the First Institute of Physics until his retirement in 1974. When he moved into the physics building of the former Technische Hochschule Berlin-Charlottenburg (today: Ernst-Ruska-Bau), which had been erected under Gustav Hertz between 1930 and 1932, another extremely productive phase of his life began. The modular principle of a vertical electron-optical bench (Boersch, 1951b, 1951c), which he had brought with him from Braunschweig, was of great importance for the rapid development of the institute. Its standardized components (cathodes, lenses, specimen, and aperture holders, screen and camera components, etc.) were manufactured in large quantities by the institute workshop, so that soon every graduate and doctoral student had his own electron-optical apparatus (in Institute jargon: "Boersch tower") adapted to the problem to be investigated and was not dependent on the use of commercial electron microscopes. In a short period of time, this resulted in important research results, e.g. the electron lithographic generation of unsupported nickel and carbon lattices with lattice constants down to 0.1 μm together with H. Hamisch and K. H. Löffler (Boersch et al., 1959), or the light- and electron-optical visualization of ferromagnetic regions in thin films and their investigation (data storage! Boersch & Lambeck, 1960; Boersch & Raith, 1959; Boersch et al., 1960) together with M. Lambeck, H. Raith, and D. Wohlleben. Under Boersch, the I. Physikalisches Institut developed an extraordinarily broad research activity under the general theme "Interaction of electrons, ions, and light with matter and electromagnetic fields" and quickly attracted international attention, e.g., by contributions on electron polarization (together with W. Raith, K. Trad-

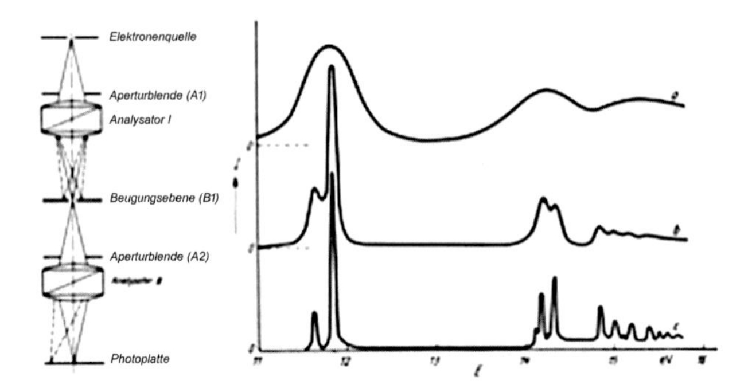

Figure 6 Wien filter arrangement for electron beam monochromatization and energy loss analysis. *Right*: enhancing the resolving power of an energy-loss spectrum of 25 keV electrons in Ar gas by increasing the monochromatization of the primary beam. Energy width a: 0.7 eV, b: 0.1 eV, c: 0.04 eV (Boersch et al., 1964a, 1964b).

owsky, R. Schliepe, J. Lemmerich, and others), on Lorentz microscopy (together with W. Raith, H. Weber, O. Bostanjoglo, and others), on electron interferometry on magnetic structures, and many others. For example, the experimental proof of the phase shift of electron waves by the magnetic vector potential without local field influence was achieved together with H. Hamisch, K. Grohmann, and D. Wohlleben (Boersch et al., 1961a, 1961b). Some of this work was spurred by competition, especially with Möllenstedt's Institute in Tübingen, which often led to exciting discussions at the national and international congresses on electron microscopy.

Highlights of the experimental art at Boersch's institute were, e.g., the development of the Wien filter (crossed electric and magnetic field in the center of an electrostatic lens) into a very high resolution electron energy loss spectrometer with a relative energy resolution of 10^{-7} by monochromatization of the primary electron beam in the group of J. Geiger (Boersch et al., 1964a, 1964b, Fig. 6); the pioneering work on electron microscopy at extremely low temperatures (2 to 5 K) with liquid helium object stages (Boersch et al., 1964a, 1964b); and the use of superconducting magnetic lenses (1965, Boersch et al., 1966a, 1966b) developed at the Institute together with O. Bostanjoglo, K. Grohmann, H. Niedrig, and others. This resulted in the discovery of anomalous electron transparency of crystalline layers at low temperatures with O. Bostanjoglo and H. Niedrig (1964, Boersch et al., 1964a, Fig. 7) and the observation of single magnetic flux quanta in superconducting hollow cylinders, together with B. Lischke (Boersch & Lischke, 1970, Fig. 8).

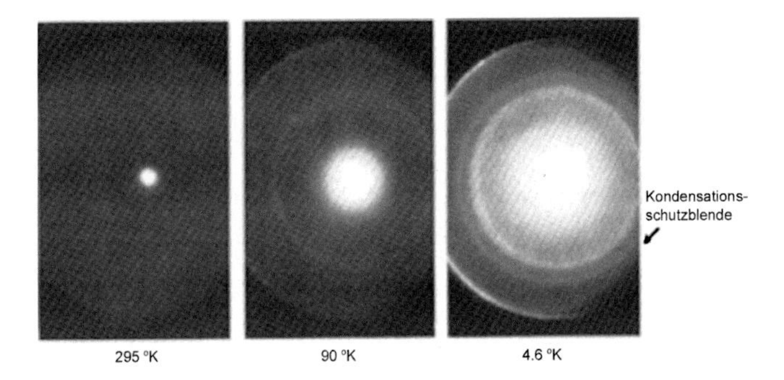

Figure 7 Anomalous electron transparency at low temperatures: transmission diffraction patterns of a 100 nm thick Pb film with 30 keV electrons (Boersch et al., 1964a).

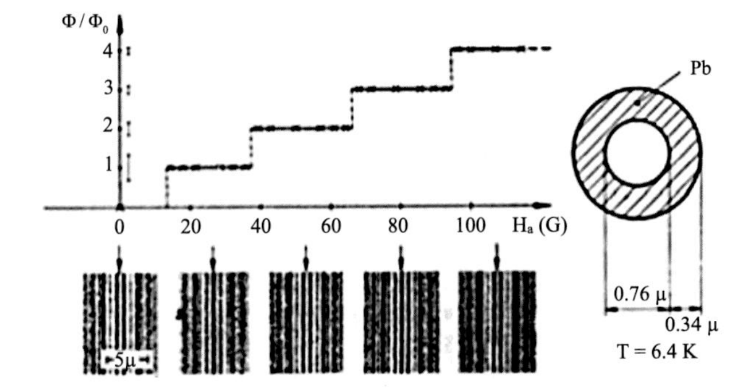

Figure 8 Electron interferometric detection of magnetic flux quantization in superconducting hollow cylinders: Frozen magnetic flux as a function of external magnetic field H_a in a Pb hollow cylinder. As the field increases, the interference intensities change discontinuously, indicating the quantum nature of the frozen flux in integer multiples of $hc/2e$ (Boersch & Lischke, 1970).

Since, according to Boersch, there is only a formal and an instrumental difference between electron optics and light optics, but no difference in principle, he had also always carried out light-optical investigations on problems of electron optics, e.g., the light-optical imaging of magnetic fine structures with the Faraday effect (rotation of the plane of polarization of light by magnetic fields in thin layers and their utilization for contrast formation), together with M. Lambeck (Boersch & Lambeck, 1960, Fig. 9), and in parallel the electron-optical investigation of Weiss domains in thin iron films (Lorentz microscopy: exploitation of electron deflection in the magnetic field of Weiss domains for contrast formation between differently oriented domains), together with H. Raith, W. Raith, H. Weber, D. Wohlleben (Boersch & Raith, 1959; Boersch et al., 1960, 1961a, 1961b,

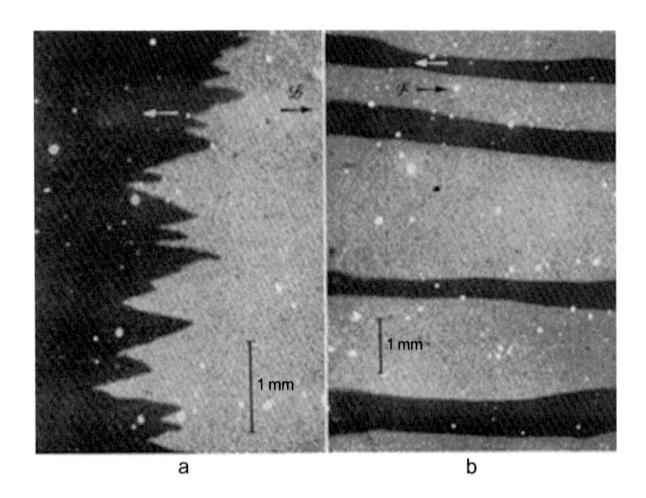

Figure 9 Light-optical imaging of magnetic domains (Weiss domains) with the Faraday effect [38]: iron layer after different magnetic pretreatment in homogeneous fields. a: DC fields; b: DC and AC fields (Boersch & Lambeck, 1960).

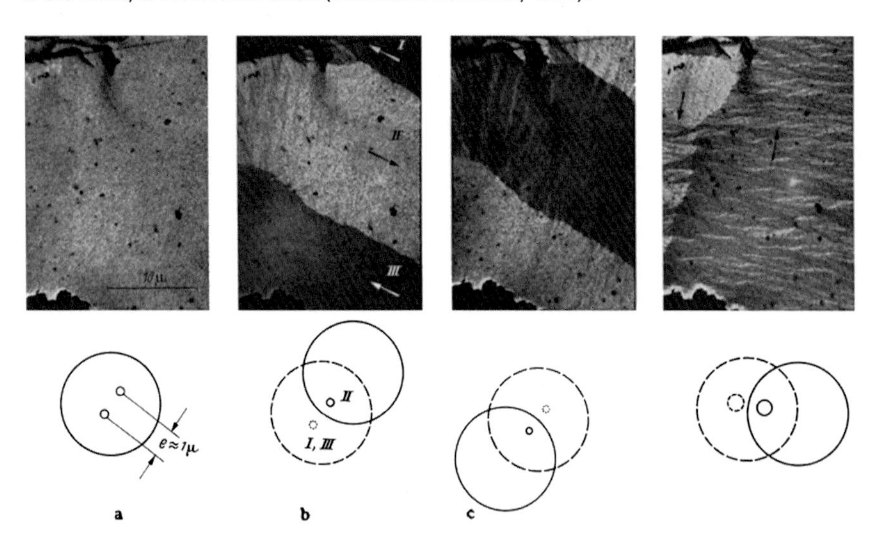

Figure 10 Electron-optical imaging of magnetic domains (Weiss domains) by Lorentz microscopy at different settings of the contrast aperture (Boersch et al., 1960).

Fig. 10), later at low temperatures in the group of O. Bostanjoglo and others (e.g., Bostanjoglo, 1970).

In parallel with the above-mentioned high-resolution electron energy-loss measurements on gases, Boersch also investigated the light emission resulting from electron impact excitation with a newly developed light spectrograph (Reich, 1963), where analogies between excitation from light and electron impact could be confirmed (Boersch & Reich, 1965). More than 260 publications on the topics of electron diffraction, electron scattering, electron microscopy (including observations at low temperatures), electron energy losses, light emission from gases and solids by electron impact, Lorentz microscopy, electron interferometry, electron mirror microscopy, ion scattering, light diffraction, Faraday and Kerr effect on ferromagnetics, surface micro-flashovers, etc., were produced until Boersch's retirement. Once, Boersch reported on his electron beam as if he was drawing up a balance sheet (1989 published posthumously, Boersch, 1989). The guiding formula behind Boersch's research program was the diffraction formula $\Theta = \lambda/d$, called "Boersch's world formula" by his co-workers.

When the laser became known in the early 1960s as a light source with previously unimaginable coherence, H. Boersch was one of the first in Germany to recognize that a new era had dawned for optics and immediately began—together with G. Herziger and H. Weber, and later also with H.-J. Eichler—to set up a working group for laser research, which soon developed into one of the most important in Germany. Here, too, scientific successes were quickly achieved, e.g. in the development of highly stable lasers, for extremely sensitive length measurements (length changes down to the lower limit of 1/1000 atomic diameter) for example, in the generation of very short light pulses, or in the development of high-power argon ion lasers. His sense of the technical and scientific potential of the laser was astonishing. Instrumentation and materials processing were already key areas at the I. Physikalisches Institut in the early 1960s, and today they are the most important fields of application of laser technology worldwide. As early as 1962, the possibilities of micromaterial processing were investigated, which led to the production of ultra-fine holes of 0.8 micrometers (Boersch et al., 1963). From this group alone, more than 80 papers on ruby and He–Ne lasers, ion lasers, dye lasers, laser amplifiers, coupled resonators, and light beats were published until Boersch's retirement in 1974. The German laser industry, a world leader, "Uhrensteinbohren" in Switzerland, and two Fraunhofer institutes focusing on laser materials processing (Dresden, Aachen) have been significantly influenced by Boersch students and their collaborators.

A splendid, wide-ranging experimental-physics lecture for the students of most of the natural science and technical faculties of the TU Berlin was one of the activities that Hans Boersch took particularly seriously; he spent a lot of time with his collaborators, setting up old and new experiments in a didactically meaningful way and ensuring that they were visible up to the last row of the auditorium of almost a thousand listeners ("The Boersch Circus"). He turned many of these into a series of educational short films about physics experiments.

In order to illustrate the personality of Hans Boersch, excerpts from an honorary speech by W. Raith on the occasion of his 70th birthday are quoted here (Raith, 1979): "Behind the great scientific productivity of the institute was Boersch's enormous personal commitment.... In addition to unannounced visits by him [to the laboratories], in the first years there were also official weekly rounds which took place like a chief physician's visit in a hospital; the assistants marched behind and learned how Boersch wanted the experiments to be carried out: start experimenting quickly! Better to learn from experimental failures than to lose precious time with too long preliminary considerations at the desk!... Boersch influenced all of us with his strong personality. Discussions in the institute were passionate. With Boersch's readiness to defend his opinion with great commitment against all objections at any time and on any question, we had difficulty in holding opposing views to him; on the other hand, we could practice self-assertion in the process.... We were also impressed by his way of doubting seemingly established findings and of refusing to be overawed even by famous colleagues in the field.... Of course, his irreverence rubbed off on us; Boersch was therefore never respectfully admired by his students, but always quite critically appreciated.... [Later] I could see that what we learned from Boersch then is still very useful today!"

The lasting impact of a great scientist is demonstrated not only by the fact that his work is still cited, but also by the quality and success of his students. More than 270 diploma students and nearly 60 doctoral students were trained at the First Physical Institute under Boersch. Today for the most part retired, most of them had reached leading positions in industry or became university professors.

In 1941, Hans Boersch received the Silver Leibniz Medal of the Prussian Academy of Sciences "in recognition of his scientific and technical achievements in the development of the electron microscope" (together with Manfred v. Ardenne, Bodo v. Borries, Ernst Brüche, Max Knoll, Hans Mahl, and Ernst Ruska).

From 1957 to 1959, Boersch was dean of the former Faculty of General Engineering at the TU Berlin. From 1963 to 1965 he was chairman of the German Society for Electron Microscopy. In 1965, Boersch received an invitation to join the University of Cologne, which he declined. In 1975 he became an honorary member of the German Society for Electron Microscopy and in 1985 an honorary member of the TU Berlin. He died on June 9, 1986. His great competitor in electron microscopy, Gottfried Möllenstedt, on the occasion of a memorial colloquium on April 30, 1987 (which is preserved as an audio document in the possession of the author), acknowledged the fundamental importance of Boersch's work and concluded his lecture with the words: "I would like to bow down before the greatness of this man."

Acknowledgments

I thank my colleagues Prof. Wilhelm Raith and Prof. Horst Weber for their critical review and their important additions.

References

Boersch, H. (1935). Bestimmung der Struktur einiger einfacher Moleküle mit Elektroneninterferenzen. Dissertation Universität Wien. *Monatshefte für Chemie. Abteilung IIb*, *65*, 311–337.

Boersch, H. (1936). Über das primäre und sekundäre Bild im Elektronenmikroskop. II. Strukturuntersuchung mittels Elektronenbeugung. *Annalen der Physik*, *5*(27), 75–80.

Boersch, H. (1938). Zur Bilderzeugung im Mikroskop. *Zeitschrift für technische Physik*, *19*, 337–338.

Boersch, H. (1939a). Das Schattenmikroskop, ein neues Elektronen-Übermikroskop. *Naturwissenschaften*, *27*, 418.

Boersch, H. (1939b). Das Elektronen-Schattenmikroskop I. Geometrisch-optische Versuche. *Zeitschrift für technische Physik*, *20*, 346–350.

Boersch, H. (1940a). Übermikroskope. 5. Das Elektronen-Schattenmikroskop. *Jahrbuch der AEG-Forschung*, 7, 34–42.

Boersch, H. (1940b). Übermikroskope. 4. Das Problem der Bildentstehung. *Jahrbuch der AEG-Forschung*, 7, 27–33.

Boersch, H. (1940c). FRESNELsche Elektronenbeugung. *Naturwissenschaften*, *28*, 709–711.

Boersch, H. (1940d). FRESNELsche Beugungserscheinungen im Übermikroskop. *Naturwissenschaften*, *28*, 711.

Boersch, H. (1942). Über hochauflösende Abbildung mittels Ionenstrahlen (Ionen-Übermikroskopie). *Naturwissenschaften*, *30*. Heft 46/47.

Boersch, H. (1943). Fresnelsche Beugung im Elektronenmikroskop. *Physikalische Zeitschrift*, *44*, 202–211.

Boersch, H. (1946a). Über die Möglichkeit der Abbildung von Atomen im Elektronenmikroskop. I. (Kontrastbildung durch elastische Streuung). *Monatshefte für Chemie*, *76*, 86–92.

Boersch, H. (1946b). Über die Möglichkeit der Abbildung von Atomen im Elektronenmikroskop. II. (Kontrastbildung durch unelastische Streuung). *Monatshefte für Chemie*, *76*, 163–167.

Boersch, H. (1946c). Über die Möglichkeit der Abbildung von Atomen im Elektronenmikroskop. III. (Kontraste von Kristallgittern und elektronenmikroskopisches Phasenkontrastverfahren). *Monatshefte für Chemie*, *78*, 163–171.

Boersch, H. (1947a). Ionenübermikroskopie. *Experientia*, *3*, 1–15.

Boersch, H. (1947b). Über die Kontraste von Atomen im Elektronenmikroskop. *Zeitschrift für Naturforschung*, *2a*, 615–633.

Boersch, H. (1948). Über die Beseitigung der inkohärenten Streuung in elektronenmikroskopischen Abbildungen und Elektronenbeugungsdiagrammen durch Elektronenfilter. *Naturwissenschaften, 35*, 26–28.

Boersch, H. (1949). Ein Elektronenfilter für Elektronenmikroskopie und Elektronenbeugung. *Optik, 5*, 436–450.

Boersch, H. (1951a). Über die Bildentstehung im Elektronenmikroskop. In *"Electron Physics". Proc. NBS Semicentennial Symposium on Electron Physics*. Nov. 5–7, 1951; Published 1954.

Boersch, H. (1951b). Ein Hochvakuum-Pumpstand für das Laboratorium. *Zeitschrift für Physik, 130*, 513–516.

Boersch, H. (1951c). Eine Vakuum-Bank. *Zeitschrift für Physik, 130*, 517–520.

Boersch, H. (1953a). Gegenfeldfilter für Elektronenbeugung und Elektronenmikroskopie. *Zeitschrift für Physik, 134*, 156–164.

Boersch, H. (1953b). Energieverteilung von Glühelektronen aus Elektronenstrahlerzeugern. *Naturwissenschaften, 40*, 267–268.

Boersch, H. (1954). Experimentelle Bestimmung der Energieverteilung in thermisch ausgelösten Elektronenstrahlen. *Zeitschrift für Physik, 139*, 115–146.

Boersch, H. (1967). Holographie und Elektronenoptik. *Physikalische Blätter, 23*, 393–404.

Boersch, H. (1989). Das I. Physikalische Institut der Technischen Universität Berlin. *Optik, 81*, 42–50. Personal survey by H. Boersch of his published works on the theme of electron diffraction, published posthumously.

Boersch, H., & Lambeck, M. (1960). Mikroskopische Beobachtung gerader und gekrümmter Magnetisierungsstrukturen mit dem Faraday-Effekt. *Zeitschrift für Physik, 159*, 248–252.

Boersch, H., & Lischke, B. (1970). Direkte Beobachtung einzelner magnetischer Flußquanten in supraleitenden Hohlzylindern. I. *Zeitschrift für Physik, 237*, 449–468.

Boersch, H., & Raith, H. (1959). Elektronenmikroskopische Abbildung Weißscher Bezirke in dünnen ferromagnetischen Schichten. *Naturwissenschaften, 20*, 574.

Boersch, H., & Reich, H.-J. (1965). Oszillatorenstärken aus der durch schnelle Elektronen angeregten Lichtemission. I. Messungen an Helium. *Optik, 22*, 289.

Boersch, H., Hamisch, H., & Löffler, K. H. (1959). Elektronenoptische Herstellung freitragender Mikrogitter. *Naturwissenschaften, 21*, 596.

Boersch, H., Raith, H., & Wohlleben, D. (1960). Elektronenoptische Untersuchungen Weißscher Bezirke in dünnen Eisenschichten. *Zeitschrift für Physik, 159*, 388–396.

Boersch, H., Hamisch, H. J., Grohmann, K., & Wohlleben, D. (1961a). Experimenteller Nachweis der Phasenschiebung von Elektronenwellen durch das magnetische Vektorpotential. *Zeitschrift für Physik, 165*, 79.

Boersch, H., Raith, W., & Weber, H. (1961b). Die magnetische Ablenkung von Elektronenstrahlen in dünnen Eisenschichten. *Zeitschrift für Physik, 161*, 1.

Boersch, H., Herziger, G., Anger, K., & Weber, H. (1963). Herstellung von Blendenbohrungen mit Laserstrahlung. *Physikalische Verhandlungen, 14*, 162.

Boersch, H., Bostanjoglo, O., & Niedrig, H. (1964a). Temperaturabhängigkeit der Transparenz dünner Schichten für schnelle Elektronen. *Zeitschrift für Physik, 180*, 407–414.

Boersch, H., Geiger, J., & Stickel, W. (1964b). Das Auflösungsvermögen des elektrostatischmagnetischen Energieanalysators für schnelle Elektronen. *Zeitschrift für Physik, 180*, 415–424.

Boersch, H., Bostanjoglo, O., & Grohmann, K. (1966a). Supraleitender Hohlzylinder als magnetische Linse. *Zeitschrift für Angewandte Physik, 20*, 193.

Boersch, H., Bostanjoglo, O., & Lischke, B. (1966b). Helium-Kühlbauteil für hochauflösende Elektronenmikroskopie und supraleitender Hohlzylinder als magnetische Linse. *Optik, 24*, 460.

Bostanjoglo, O. (1970). Low temperature reversal in anisotropic ferromagnetic films. *Physica Status Solidi, 34*, 247.

Raith, W. (1979). Zum 70. Geburtstag von Hans Boersch. *Physikalische Blätter*, *35*, 226–229.
Reich, H.-J. (1963). Ein 50-cm-Vakuumspektrograph für mehrere Konkavgitteraufstellungen und Detektionsarten. *Zeitschrift für Angewandte Physik*, *16*, 299.
Schultheiß, K., Pérez-Willard, F., Gerthsen, D., Barton, B., & Schröder, R. R. (2006). Fabrication of a Boersch phase plate for phase contrast imaging in a transmission electron microscope. *Review of Scientific Instruments*, 77, Article 33701, 4pp.

CHAPTER FOUR

Raymond Castaing (1921–1998)

C. Colliex[a,*], P.W. Hawkes[b], and P. Duncumb FRS (Obituary)[c]

[a]Laboratoire de Physique des Solides (LPS), CNRS UMR 8502, Bâtiment 510, Université Paris-Sud, Université Paris-Saclay, Orsay, France
[b]CEMES-CNRS, Toulouse, France
[c]Formerly Tube Investments Research Laboratories (TIRL), Hinxton Hall, Cambridge, United Kingdom
*Corresponding author. e-mail address: christian.collliex@u-psud.fr

Contents

1. Introduction

The scientific biography of Raymond (Raimond) Castaing by Christian Colliex, which first appeared in the *Comptes Rendus Physique* **20** (2019) 746–755 is reproduced in full. By way of introduction, it is preceded by a short obituary by Peter Duncumb and myself that appeared in the *Proceedings of the Royal Microscopical Society* (**33**, 1998, 289–290). We are very grateful to Elsevier and to the Royal Microscopical Society for permission to reprint these articles.

2. Obituary

On 10 April 1998, Raimond Castaing died. One of the most colourful and imaginative figures of the French electron physics community, he was well known in the UK, where he frequently attended Royal Microscopical Society and Institute of Physics meetings and could be relied on to ask the kind of questions that provoke a lively, sometimes even acrimonious, debate.

Castaing was born on 28 December 1921 in Monaco, and he was deeply marked by the happiness of his early childhood. When he was eight, his par-

Advances in Imaging and Electron Physics, Volume 220
ISSN 1076-5670
https://doi.org/10.1016/bs.aiep.2021.08.004

ents moved back to the family home in Condom in south-western France. For some years, he did not notice that mention of his origins was found mildly amusing by native English speakers but when he did notice and found out why, that mild British amusement was drowned by Gallic guffaws!

He passed the entrance examination to the Ecole Normale Supérieure in Paris after a spell at the Lycée Fermat in Toulouse, but subsequently became an *ouvrier sondeur* (prospector for an oil company) to avoid being sent for forced labour to Germany. (To appreciate why he liked to recall this, you need to know that the French used often to refer to the microanalyser as 'la sonde [probe] de Castaing'.)

In January 1947 he joined the ONERA, then at Le Bouchet outside Paris, where he built the first X-ray micro analyzer as his PhD project. A preliminary report was presented at the 1949 Electron Microscope Congress in Delft. In his paper he gave the expected resolution as about 1 μm, and afterwards a very distinguished participant came up to him and said "it's a very nice method; it's a pity it won't be any use". Castaing said afterwards "I wasn't very far from agreeing with him, but as we say in France 'If you have drawn the wine you have to drink it', and I had no option but to carry on. Many careers have resulted from his decision to do so.

His PhD thesis, published in 1952, was a classic, and has to be the most widely quoted reference in microprobe analysis. In it he lays down much of the basics of the theory which is still used today, giving the name $\phi(\rho z)$ to the distribution with depth of X-rays generated in the target and $f(\chi)$ to the fraction emerging in the direction of the detector. One effect he missed initially was the loss of intensity due to electron backscattering, but he was the first to recognize it a few years later and energetically explained the principles to me (using both hands to do so) as he was driving me [Peter Duncumb] through Paris in 1960. It was an unforgettable experience! It was left to others to work out the theory in detail, and by that time Castaing had taken up new challenges in ion beam microanalysis.

He was not only a theoretician. His instrument design of 1954 formed the basis of the first commercial microprobe made by Cameca, and its descendants are in production today. Likewise the applications he carried out, both at ONERA and in conjunction with the IRSID Laboratories, laid the foundations for metallurgical applications worldwide.Thus the impact of his work created a microprobe community spreading through metallurgy, mineralogy, biology and medicine, and including even archaeology, art and the environment. Benefits can be counted in every continent. The growth of microprobe analysis illustrates well how an understanding of the world around us on a micrometre scale can influence our daily lives.

In 1966, his work and the interest shown in it on both sides of the channel were recognized by the award of the Holweck Medal and Prize. These are awarded alternately to a French and a British physicist by the Société Française de Physique and the Institute of Physics. His speech on that occasion is to be found in the *Physics Bulletin* for 1967 (pp. 93–97) and gives an extremely vivid picture of the man and his work. It is at once scientifically serious, highly entertaining and the reverse of pompous or self important:

> In short, after a few months severe training, I had acquired the basic virtue of every experimenter who becomes involved in this fascinating field of electron optics, and I would have been able to hold my own with an experienced charwoman; but on the other hand I had taken a solemn vow never to do any washing up out of my laboratory time.

We are not told what Madame Castaing felt about this. And he concludes by likening his practical skills to those of Uncle Podger. . . .

He was honoured nationally and internationally in a host of ways. No doubt the highest honour was membership of the Académie des Sciences, where he joined many other well-known names of electron optics and electron physics, among them Louis de Broglie, Gaston Dupouy, Pierre Grivet and André Guinier. The booklet that records this ceremony brings him vividly to life, for the speeches overflow with affection and respect, while Castaing's own speech is indescribable: a mixture of childhood memories, recollections of friendships and of professionally satisfying moments. One paragraph is typical:

> It is generally agreed in that it is scents that have the greatest power of evoking the past. For me, the fragrance of geraniums takes me back to my first youth but so too, to tell the truth, does that of an onion tart known as la pissaladière. It used to be sold in wedges on street stalls; when I went to the market with my mother, I always managed to convince her to buy me a portion to eat in the street, even though she was terrified at the thought of all the dangerous germs that might have settled on it.

Castaing was active in the French Physical Society and was the first President of the Société Française de Microscopie Electronique. He and his pupils were regular contributors to the Society journal and he could always be relied on for a preface or similar ephemera. (He was good enough to write such a preface for *Electrons et Microscopes*, and in the proofs his name appeared as Raymond Castaing. He altered this to Raimond and I [PWH] phoned him to query this since his Christian name is spelt with a y in his earlier publications. "Well", he said, "I'm a member of the Académie now and they have a committee that decides how French should be spelt. I can hardly not follow their rules.") Castaing was elected an honorary fellow of the RMS in 1982.

It would be easy to fill pages with lists of the important posts that Castaing occupied, and with anecdotes and reminiscences about him. He was a major figure of French physics and especially of electron and ion physics. His sense of science and his sense of humour will be much missed. We offer our profound condolences to his family.

P.W. Hawkes and P. Duncumb

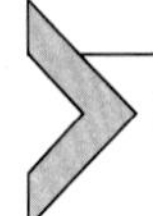

3. The "father" of microanalysis: Raymond Castaing, creator of a generation of scientific instruments, still in worldwide operation by Christian Colliex[1]

Seeing the invisible has been a permanent quest for the human being and has stimulated a large number of efforts to create tools to push further our field of vision into the domain of the small and ultrasmall. It is possible to start with the first optical microscope, built in the late seventeenth century by Antoni von Leeuwenhoek in Holland, which allowed him to discover unicellular organisms. A major step forward has been, around 1930, the use of electron beams instead of photon beams, which has opened a breakthrough in spatial resolution as a consequence of the strong wavelength reduction of the associated waves. The first electron microscope, built by Knoll and Ruska in Berlin, generated a lot of technical developments which brought the accessible spatial resolution down to 10 nm in 1939. Together with the preparation of thin specimens, a first generation of images was produced, revealing the sub-micron structure of particles, metals, and ceramics.

At the same time (1937), in two notes published in the *Comptes rendus*, Guinier (1937) and Jacquet (1937) described some experimental progress that will generate some non-negligible impact into the research domains explored in the following years. Jacquet introduced a method relying on

[1] Reproduced from C. Colliex, *Comptes Rendus Physique* **20** (2019) 746–755, published by Elsevier Masson SAS.

electrolytic reactions to produce a perfect polishing of the surfaces of thin metallic layers. As for Guinier, in the preparation of his thesis at the "École Normale Supérieure", University of Paris to be defended in 1939, he built a monochromatic X-ray illumination system providing diffraction patterns of much increased quality. With this system in hand, he could solve crystalline structures on samples made of large crystals. And in 1938, he described (Guinier, 1938a) a new type of diffraction pattern recorded on a series of aluminum–copper alloys, exhibiting unusual streaks that he interpreted as due to the coalescence of Cu clusters of typical sizes 150 Å in diameter and 3 to 4 Å in thickness. A following publication of these results by Guinier in *Nature* (Guinier, 1938b) provoked a comment in the same issue of the *Nature* journal by Preston (1938) reporting similar results in support of their interpretation of a second phase in these age-hardened alloys. This was the birth of the Guinier–Preston (GP) zones to be further discovered in many alloys.

This subject reappeared after the war when Guinier asked one young student to enrich these observations in the diffraction world with direct imaging in an electron microscope, the performance and domain of use of these instruments having also progressed during this period. This student, Raymond Castaing, had been hired at that time (1947) as an "ingénieur des petites études" at ONERA ("Office National d'Études et de Recherches Aéronautiques") to prepare a doctoral thesis under the supervision of André Guinier. Previously, Castaing had been admitted at the "École Nationale Supérieure", Paris, in 1940, joined the French resistance during the war, and graduated in 1946 as an "agrégé" in physics, the highest grade for teaching physical sciences. During the year 1949, Castaing and Guinier published three notes in the *Comptes rendus* (Castaing, 1949; Castaing & Guinier, 1949a, 1949b), reporting the technique used in electron microscopy (the observation of oxide replicas of the specimen's surface) and its application to the study of platelets created in aluminum–copper and aluminum–magnesium–silicon alloys hardened by heat treatments. If this printing technique has provided hints of the presence of platelets in epitaxy with the [100] lattice planes of the matrix, it has failed to directly image the first gathering of Cu atoms as suggested by X-ray diffraction techniques to be of very small size below the resolution of the technique. Five years later, after having improved the polishing technique (electrolysis followed by ion irradiation) (Castaing & Laborie, 1953), clean thin foils of alloys were obtained and examined in a transmission electron microscope. Castaing and Lenoir could then image with weak contrast Guinier–Preston zones in an Al–Cu 4% foil in its first stage of heating (Castaing & Lenoir, 1954).

Beforehand, during the early years of his thesis preparation, another idea had emerged between Guinier and Castaing: would it be possible to push the technique beyond imaging these small objects, the zones of copper precipitation in the aluminum matrix, i.e. analyze them. The basic idea was to direct onto a particular point on the specimen surface a finely focused electron beam, the electron probe, and to collect and analyze the wavelength of the emitted X-rays under the electron impact in order to determine the local chemical composition. Guinier had established himself at that time as an expert in X-ray diffraction. At ONERA, Castaing could use a CSF 3M electron microscope with electrostatic lenses resulting from the work of Grivet and then made commercially available. He first devoted most of his attention to the production of 30-kV electron beams focused into small probe sizes (about 1 μm in diameter) and carrying a maximum current (typically 10^{-8} A) in order to generate X-ray emission fluxes sufficiently intense to be measured. This search required efforts to understand, model, measure, and reduce the effects of the aberrations of the electron lenses, in particular those associated with astigmatism (Castaing, 1950a, 1950b, 1950c).

The first demonstration of the successful use of these fine electron probes for the elemental analysis of solid specimens was made public during international conferences of electron microscopy in Delft (1949) and Paris (1950) and published, co-authored by Castaing and Guinier in the associated proceedings (Castaing & Guinier, 1949c, 1950). This latter contribution describes the design of the instrument realized by Castaing with his CSF microscope (see Fig. 1). A new objective lens made of two reducing electrostatic lenses, which, together with the astigmatism corrector, delivers a current varying from 10^{-8} A to 10^{-7} A in a spot of 0.5 to 2 μm. The X-rays leave the column through a thin Al window and are analyzed with a home-built curved quartz spectrometer and a Geiger counter as the detector. Typically, on a Cu sample, an intensity of the $K\alpha_1$ line greater than 100 counts per second is recorded for a 1-μm probe of 30-kV electrons. Fig. 2 shows a view of the whole microscope equipped with this wavelength-dispersive spectrometer.

Surprisingly, during the next couple of years, we cannot find in the open literature, including the *Comptes rendus*, any text published by Castaing relative to this new instrument as an analyzing tool and its performance or field of use. By the way, there is a short note in which Castaing mentions that it enables one to record diffraction patterns of the emitted X-

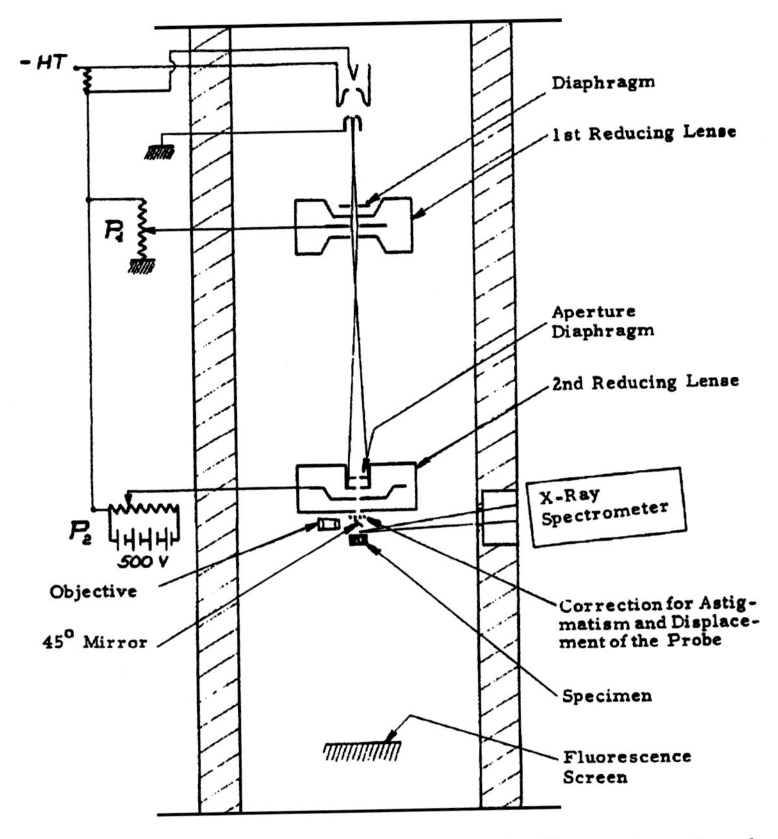

Figure 1 Scheme of the first electron microprobe realized by transformation of a CSF electrostatic microscope (from R. Castaing's Ph.D. manuscript, 1951).

rays, the Kossel lines, to identify crystalline structures (Castaing, 1951). As a matter of fact, he concentrates all his efforts into writing down his doctorate manuscript. The oral presentation took place at Paris University in June 1951 and the written version, entitled "Application des sondes électroniques à une méthode d'analyse ponctuelle, chimique et cristallographique", was published by ONERA in 1952 (Castaing, 1952) (Fig. 3).

This is actually a very rich text. Beyond a detailed description of the instrument together with the required technical innovations to obtain its best performance in terms of resolution and signal, it contains the basic theory of the physical processes involved in the X-ray emission processes. How far can we relate the intensity of an emitted X-ray line to the concentration of the corresponding element in the analyzed volume? This is the basis of quantitative analysis. Castaing's thesis introduces and indicates how to calcu-

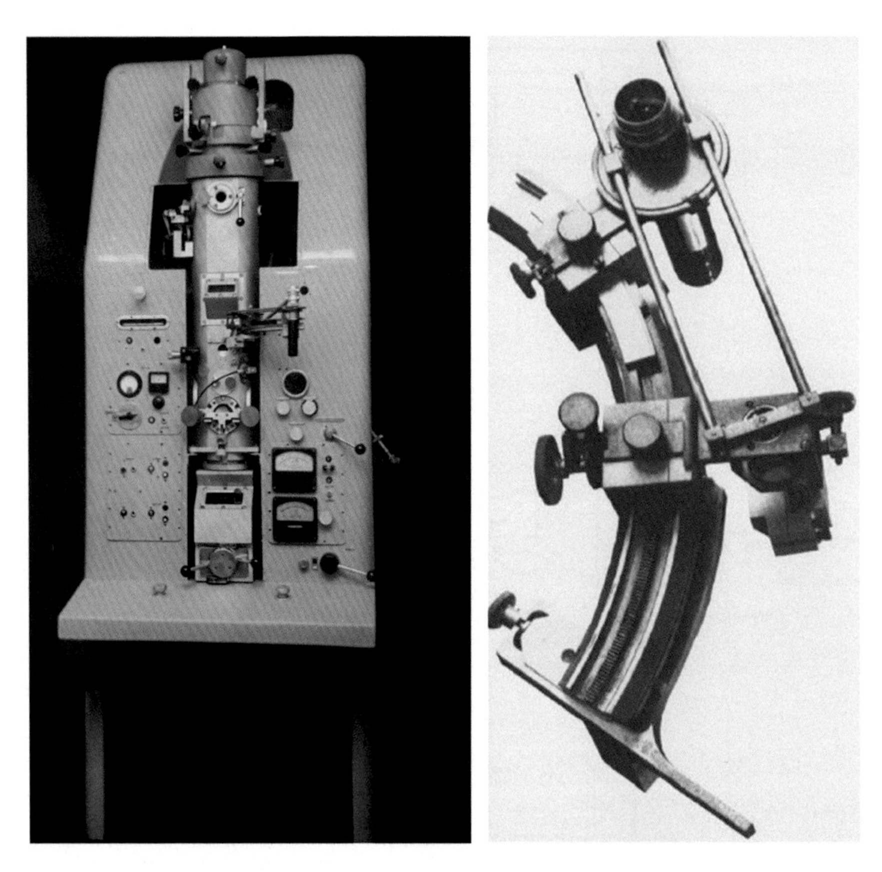

Figure 2 Global view of the instrument schematized in Fig. 1 and detail of the spectrometer (from R. Castaing's thesis, 1951).

late the main terms to be considered for a satisfactory quantitative analysis: (i) the distribution in depth of the characteristic X-ray emission $\phi(\rho z)$, and the adverse correction factors to be applied to the intensity of the X-ray line, notably (ii) the absorption of the emitted X-rays in the analyzed material itself (absorption), or (iii) their excitation by another characteristic or background radiation (fluorescence). This will be the basis of the largely used ZAF method developed fifteen years later by Philibert and Tixier (1968). Meanwhile, it is interesting to point out Castaing's special interest for the study of the distribution in depth of the X-ray emission from the surface of a material, which he investigated experimentally with Descamps (Castaing & Descamps, 1953).

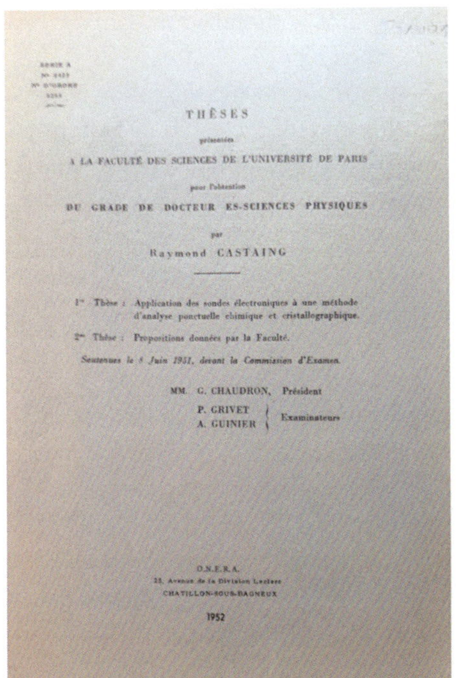

THÈSES

présentées

A LA FACULTÉ DES SCIENCES DE L'UNIVERSITÉ DE PARIS

pour l'obtention

DU GRADE DE DOCTEUR ES-SCIENCES PHYSIQUES

par

Raymond CASTAING

1re Thèse : Application des sondes électroniques à une méthode d'analyse ponctuelle chimique et cristallographique.

2me Thèse : Propositions données par la Faculté.

Soutenues le [illegible] Juin 1951, devant la Commission d'Examen.

MM. G. CHAUDRON, Président

P. GRIVET, A. GUINIER — Examinateurs

O.N.E.R.A.

25, Avenue de la Division Leclerc

CHATILLON-SOUS-BAGNEUX

1952

Figure 3 Cover of Raymond Castaing's Ph.D. manuscript, and the young laureate in 1951.

At that point, it is interesting to mention that the content of the thesis of Castaing, written in French as were all his publications until that date, was recognized sufficiently rich to deserve a translation in English, made by Pol Duwez and David Wittry at Caltech and published as a technical report under the support of the US Department of Army in 1955 (Castaing, 1955)!!! The interest of the US community for the work of Castaing may have grown after a meeting at the National Bureau of Standards dedicated to Electron Physics, which took place in November 1951. But the Proceedings came out only in 1954. They contain two texts from Castaing written in English: one is dedicated to the principle and corrections in microanalysis by means of an electron probe and can be read as a long summary of his thesis. The second describes first applications in the metallurgical domain including the use of the selective reflection of the emitted X-rays on well-oriented crystal planes for crystallographic identification and orientation of the probed volume of matter (Castaing, 1954a). In the discussion following the second paper, it is interesting to read questions by famous scientists at that time, Marton and Gabor, and they deal with radiation damage and contamination.

The role of Guinier becomes again important during the following period. He convinces IRSID (the French Institute of Research in Siderurgy) to support the design and building of a new prototype to further explore applications in metallurgy, using such arguments as "inclusions as small as a few microns can be analyzed". The first machine of this new generation was built at ONERA and delivered in 1955 to IRSID, where it was operated by Philibert. But a similar one was installed at ONERA and there operated by Castaing himself (Fig. 4).

This machine is noticeably different from the first prototype: a magnetic lens is used to focus the beam on the specimen; it incorporates an optical microscope for direct localization of the beam on the specimen surface and two X-ray spectrometers are placed under vacuum in front of the specimen, therefore improving the collection of X-rays. This is really the "first" micro-analyzer. It will be made commercially available as the MS58 instrument in 1958 by the French company CAMECA and delivered in various places in France (CEA, CNET, BRGM) as well as in the USA. Castaing gave a first description of the present state of the realization of this new instrument during the International Electron Microscopy Conference in London in 1954 (Castaing, 1954b), and published together with Descamps an extended report on the different parameters involved in the emission process that can govern the accuracy of a quantitative measurement (Castaing & Descamps, 1955). Finally, one can find a full review text on Electron Probe Microanalysis in English in 1960 (Castaing, 1960).

During this period, many users from quite different research fields demonstrated the power of this instrument in domains far from the originally and most explored one in metallurgy (Castaing et al., 1957): for instance, in geology (Castaing & Fredriksson, 1958) and in medicine (Galle, 1964), which then became great customers of microanalytical characterizations. In the following years, CAMECA developed successive generations of X-ray microanalyzers, the most famous one being the Camébax instruments. One can read today on the CAMECA website the presentation of their most recent X-ray microanalyzer: "Since pioneering Electron Probe MicroAnalysis in the 1950s, CAMECA has released several generations of microprobes, all with a proven valuable track record for analytical performance and reliability. The new SXFive-TACTIS builds on this legacy to deliver enhanced imaging and quantitative analysis in a user-friendly environment."

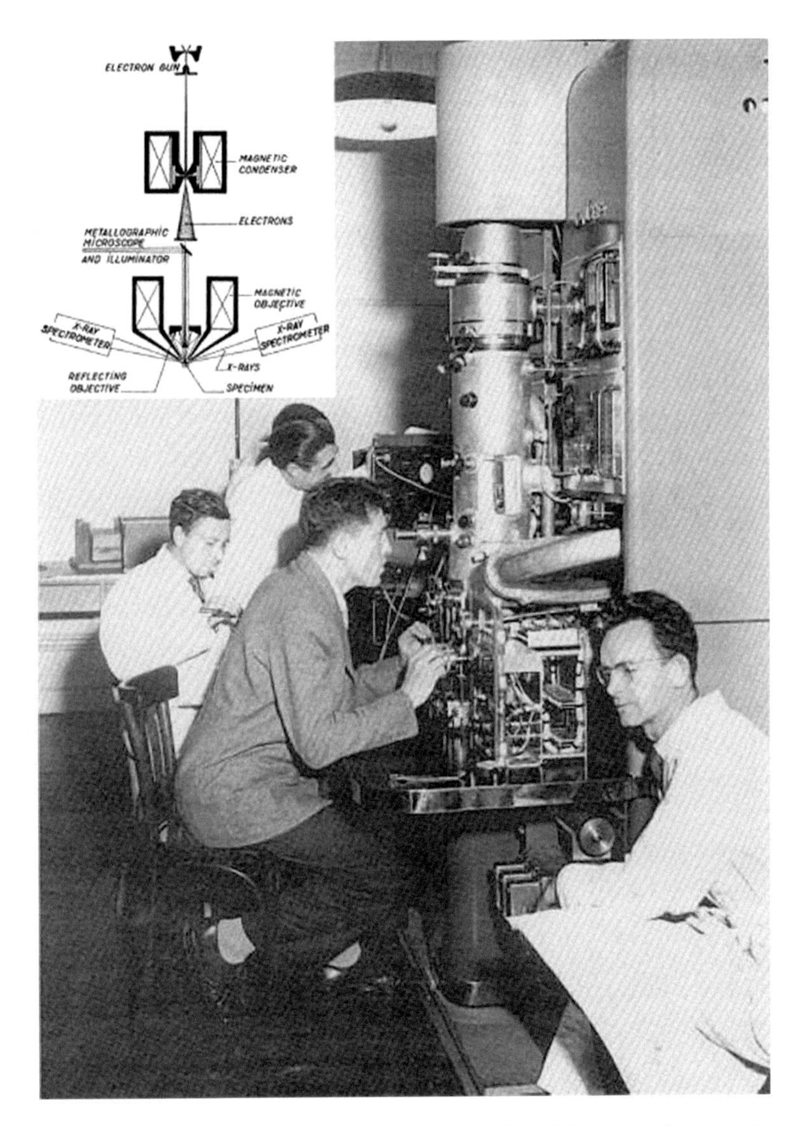

Figure 4 Raymond Castaing operating the microprobe of the second generation developed at ONERA in the mid-fifties.

Back to the fifties, the personal situation of Castaing changed rapidly as he entered the academic career as a young "maître de conférences" at the University of Toulouse in 1952, quickly followed by a professor position at the University of Paris (1956). He then moved in 1959 to the newly created "Faculté des Sciences d'Orsay" where, together with Friedel and Guinier, he founded the “Laboratoire de Physique des Solides” (LPS). This was the starting point of a new period of intensive research and innovation in the domain of... microanalysis.

Castaing was then following his quest of the design of an instrument delivering best compositional maps. He was not that fond of the idea of scanning a probe on the specimen, although this was the solution used in the X-ray microprobe. He was much more enthusiastic for particle optics systems delivering filtered images. In the LPS at Orsay, with two students, Slodzian, who had followed him from Toulouse, and Henry, he investigated and created two new research tools delivering chemical maps: (i) secondary-ion microscopy and spectroscopy (SIMS) and (ii) electron energy-loss imaging and spectroscopy (EELS). We can find the first reports on these two approaches in two notes in the *Comptes rendus* published in 1962 (Castaing & Henry, 1962; Castaing & Slodzian, 1962a).

The secondary-ion microscope uses analytical signals carried by the secondary ions produced under the sputtering of the sample surface by primary ions, as suggested in a first demonstration of the potential of such a method for mapping the spatial distribution of the elements constituting the specimen (Castaing et al., 1960). These secondary ions are analyzed by a mass spectrometer that generates images of the elemental composition, but also of the isotope distribution. This instrument, the principle of which is shown in Fig. 5a, has required the development of the first imaging mass spectrometer relying on the focusing properties of the fringing field of a magnetic sector and a novel ion-to-electron image converter to obtain adequate images. A more complete description of this prototype and its first images was published at the same time in the *Journal de Microscopie* (Castaing & Slodzian, 1962b), the publication of the French Society of Electron Microscopy (SFME), the first President and Founder of which was, by the way, Raymond Castaing. These images demonstrated the theoretical resolution limit, around 0.5 μm, imposed by aberrations arising from the angle and energy spread of the sputtered ions. Slodzian's machine was then realized commercially by CAMECA as the SMI 300 ion microscope, introduced on the public market in 1968. Several generations of instruments have followed, bringing CAMECA to the leadership on the world market for this type of analyzer recognized as the most sensitive elemental and isotopic surface analysis (de Chambost, 2011). With the NanoSIMS using a nanoprobe design realized at ONERA under supervision by Slodzian, it provides an extremely high sensitivity for all elements from hydrogen to uranium and above (down to the ppb level for many elements), together with a high

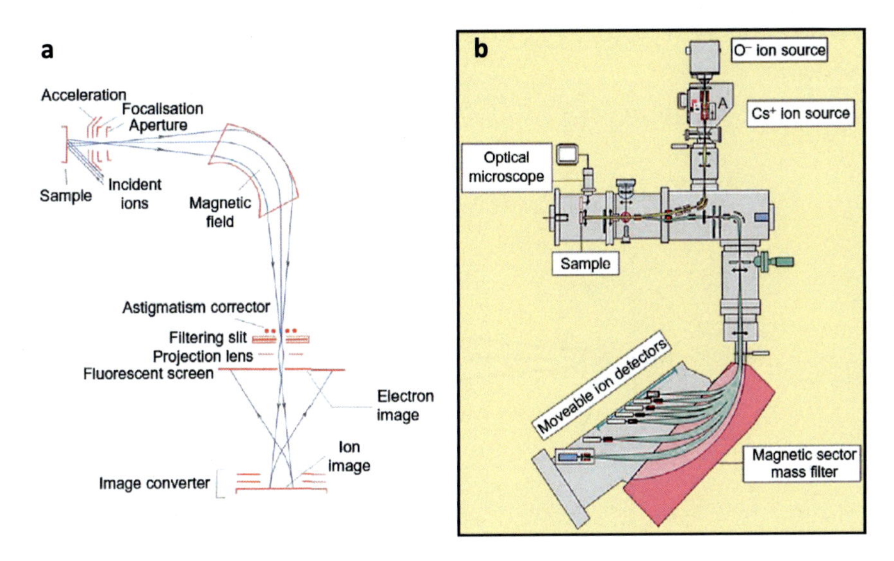

Figure 5 (a) Diagram of the first secondary ion-microanalyzer built in 1962 (from Castaing & Slodzian, 1962a); (b) prototype of the NanoSIMS developed by CAMECA in the mid-nineties, equipped with a mass spectrometer of the Mattauch and Herzog types.

lateral resolution imaging down to 40 nm. This is by definition a destructive technique as it relies on the sputtering of the surface layers, which can become a very powerful tool for in-depth analysis of trace elements with a depth resolution ranging from the sub-nanometer scale to tens of nanometers. Fig. 5 compares the principles of the instruments built by Slodzian for his thesis in 1963 and for NanoSIMS in 1997.

The second major project led by Castaing in the early sixties together with his student Henry was to insert an electron energy spectrometer within the column of a transmission electron microscope. One could thus measure the energy loss suffered by the primary electrons across the specimen and use those electrons having lost a given energy and therefore contributing to a specific excitation, to realize energy-filtered images. The electron-optical system designed and built for this purpose is a dispersive system made of the association of a magnetic sector with an electrostatic mirror, as shown in Fig. 6. Such a design exhibits, for a satisfactory excitation of the mirror, two couples of stigmatic points, one real, R_1 and R_3, and one virtual, V_1 and V_3. An energy loss spectrum is formed at the level of R_3 and a filtered image is formed at the level of V_3 when a slit is

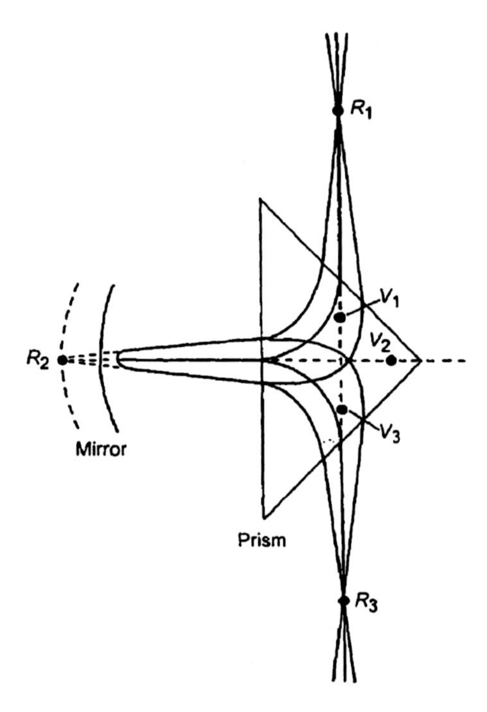

Figure 6 Scheme of principle of the energy filtering device, known as the Castaing and Henry filter, developed for electron energy-loss analysis and filtering on the column of a conventional transmission electron microscope (from Castaing & Henry, 1962).

introduced in R_3 to select an energy window corresponding to the desired energy loss, E with a given energy width δE.

In a following note (Castaing et al., 1965), Castaing et al. demonstrated that this technique could constitute a new method for qualitative microanalysis at high resolution, as illustrated in Fig. 7. This is demonstrated in the thesis work by Lucien Henry (1964) and Ali El Hili (1967). Fig. 7 shows how energy-loss spectra (a) can be used to produce chemical maps. The specimen is a partially oxidized thin foil of aluminum with two characteristic spectra showing plasmon peaks at 15 eV in Al and at 6.5 and 21 eV in Al_2O_3. The left-hand-side image in (b) corresponds to a zero-loss image; it reveals the presence of a continuous layer; the central one filtered with a 1.5 eV slit around 6.5 eV displays, with a strong white contrast, the presence of alumina on given areas, while the right-hand-side one, filtered with a width of 1.5 eV around the 15-eV line, confirms the presence of an aluminum foil between the oxide areas. In the following years, Castaing and

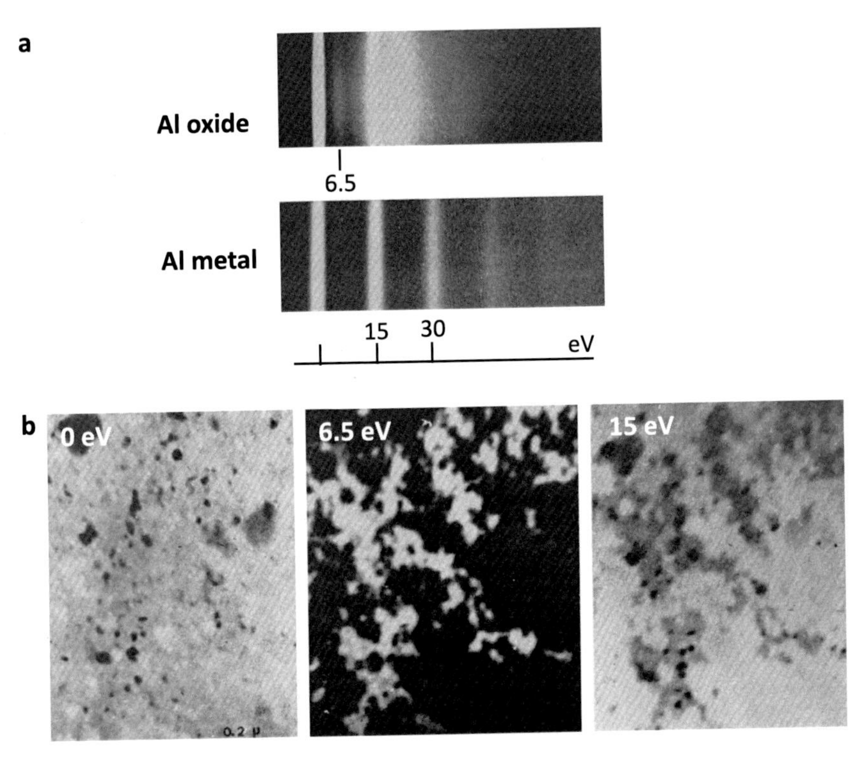

Figure 7 Examples of electron energy-loss spectra and associated energy filtered images recorded on a partially oxidized aluminum thin foil (from Lucien Henry thesis work, 1964).

his coworkers deepened the understanding of the origin of the diffraction contrast carried by inelastically scattered electrons (Castaing et al., 1966).

This was the starting point of a long story accompanying the growing role of electron energy loss spectroscopy which became a very popular acronym – EELS – in the mid-70s as an undisputable micro-and nano-analytical tool. Colliex and Jouffrey in Raymond Castaing's lab in Orsay installed an energy-loss filter and spectrometer of Castaing and Henry type on a modern TEM at that time, a Hitachi HU11B, and then demonstrated the role of the EELS signals associated with specific core-level excitations as true signatures of the presence of given elements under the impact of the primary electrons in the microscope. This research was also published in the *Comptes rendus* in 1970 (Colliex & Jouffrey, 1970a, 1970b). Many years later, using an alternative approach, i.e. recording a whole electron energy-loss spectrum for all the positions of a sub-nanometer electron probe, we succeeded in identifying individual single atoms of rare-earth elements (Suenaga et al., 2000). And today, all electron microscopes of the latest generations are equipped with an energy-loss analyzer and filter, the Gatan company being the major supplier of such devices in the world.

Being the creator of the three major microanalytical tools, the X-ray Microprobe, the Secondary Ion Mass Spectrometer, and the Electron Energy-Loss spectrometer, Raymond Castaing is undoubtedly the "father" of microanalysis. His influence has spread over many diversified fields of research from metallurgy to biomedicine. One can find in the recent literature reviews of the latest developments of these techniques (Aronova & Leapman, 2012; Gloter et al., 2017; Rinaldi & Llovet, 2015; Yang & Gilmore, 2015) and can come across examples of applications in domains as different as dentistry (Gonzalez-Jaranay et al., 2007), earth mantle (Kumamoto et al., 2017), atmospheric particles (Huang et al., 2017), or the detailed atomic and electronic structures at oxide interfaces, which are responsible of some of the most intriguing behaviors (Fitting-Kourkoutis et al., 2010) now observed in condensed-matter physics (colossal magnetoresistance, colossal ionic conductivity, ferroelectricity...). In recognition of these superb realizations and achievements, his role has fairly been recognized by many awards and in particular the gold medal of CNRS in 1975 and his election to the French Academy of Sciences in 1977, a fully deserved honor celebrated with the attribution of his personal academician sword (Fig. 8).

4. Endnote

On 2 June 1998, the sword worn by members of the Académie des Sciences was presented to Castaing and a booklet records the tributes by J.-P. Kahane, P. Contensou, G. Slodzian and A. Guinier as well as the very moving 'Remerciements' by Castaing, in which he admits that history was his weakest subject at school; he was glued to the screen during a showing of Henry V with Laurence Olivier, wondering who was going to win at Agincourt. Before these tributes and thanks, however, are pictures of the sword. The shields of Monaco and Condom are represented as well as a bunch of grapes (Armagnac) and a chronometer (recalling his children's swimming trophies). On the top of the pommel is a chrysanthemum, a reminder of his father's garden. Elsewhere there is a rainbow (spectral analysis), a derrick (a pun on the word 'sonde') and a sphere, for Castaing held

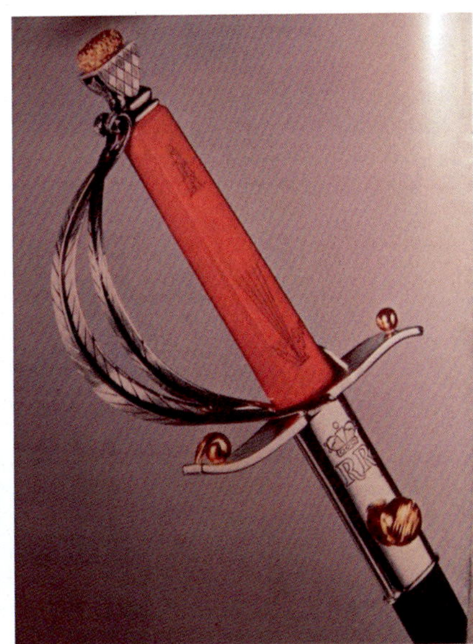

Figure 8 A well-known portrait of Raymond Castaing accompanied by a view of the personal sword he received as a testimony of his membership at the French Academy of Sciences.

the record for putting the shot at the Ecole Normale Supérieure in the rue d'Ulm.

Castaing's doctoral thesis was printed by the ONERA and not published in the *Annales de Physique*, as was usual at the time. An English translation was made by Pol Duwez and David. Wittry in 1955. At Marton's suggestion, a full account of 'Electron probe microanalysis' appeared a few years later in *Advances in Electronics and Electron Physics* **13**, 1960, 317–386.

References

Aronova, M. A., & Leapman, R. D. (2012). Development of electron energy-loss spectroscopy in the biological sciences. *MRS Bulletin*, *37*, 53–62.

Castaing, R. (1949). Recherches en microscopie électronique sur les précipitations dans les alliages d'aluminium. *Comptes Rendus Hebdomadaires des Séances de l'Académie des Sciences Paris*, *228*, 1341.

Castaing, R. (1950a). Une méthode de détection et de mesure de l'astigmatisme d'ellipticité. *Comptes Rendus Hebdomadaires des Séances de l'Académie des Sciences Paris*, *231*, 835.

Castaing, R. (1950b). Détection et mesure directe de l'astigmatisme d'ellipticité d'une lentille électronique. *Comptes Rendus Hebdomadaires des Séances de l'Académie des Sciences Paris*, *231*, 894.

Castaing, R. (1950c). Lentille corrigée de l'astigmatisme et son utilisation pour l'obtention de sondes de grande brillance. In *Comptes rendus du premier congrès international de microscopie électronique* (pp. 148–154).

Castaing, R. (1951). Méthode d'analyse cristallographique ponctuelle. *Comptes Rendus Hebdomadaires des Séances de l'Académie des Sciences Paris, 232*, 1948.
Castaing, R. (1952). Application des sondes électroniques à une méthode d'analyse ponctuelle chimique et cristallographique (PhD thesis). Paris: ONERA.
Castaing, R. (1954a). Microanalysis by means of an electron probe: Principle and corrections and applications of the electron probe microanalyzer. In *NBS Circular: Vol. 527. Proceedings of Electron Physics* (pp. 305–309).
Castaing, R. (1954b). État actuel du microanalyseur à sonde électronique. In *Proc. International Conference on Electron Microscopy* (pp. 300–304).
Castaing, R. (1955). Application of electron probes to local chemical and crystallographic analysis. Thesis translated by Duwez, P., & Wittry, D.B., the-mas.org/wp-content/castaing-thesis-clearscan.pdf.
Castaing, R. (1960). Electron probe microanalysis. *Advances in Electronics and Electron Physics, 13*, 317–386.
Castaing, R., & Descamps, J. (1953). Sur la répartition en profondeur de l'émission X d'une anticathode. *Comptes Rendus Hebdomadaires des Séances de l'Académie des Sciences Paris, 237*, 1220.
Castaing, R., & Descamps, J. (1955). Sur les bases physiques de l'analyse ponctuelle par spectrographie X. *Journal de Physique et Le Radium, 16*, 304–317.
Castaing, R., & Fredriksson, K. (1958). Analysis of cosmic spherules with an X ray microanalyser. *Geochimica Et Cosmochimica Acta, 14*, 114.
Castaing, R., & Guinier, A. (1949a). Sur les images au microscope électronique des alliages aluminium-cuivre durcis. *Comptes Rendus Hebdomadaires des Séances de l'Académie des Sciences Paris, 228*, 2033.
Castaing, R., & Guinier, A. (1949b). Etude au microscope électronique du vieillissement des alliages aluminium–magnesium–silicium. *Comptes Rendus Hebdomadaires des Séances de l'Académie des Sciences Paris, 229*, 1146.
Castaing, R., & Guinier, A. (1949c). Application des sondes électroniques à l'analyse métallographique. In *Proc. Conf. on Electron Microscopy, Delft* (pp. 60–63).
Castaing, R., & Guinier, A. (1950). Sur l'exploration et l'analyse élémentaire d'un échantillon par une sonde électronique. In *Comptes Rendus du Premier Congrès International de Microscopie Electronique Paris* (pp. 391–397R).
Castaing, R., & Henry, L. (1962). Filtrage magnétique des vitesses en microscopie électronique. *Comptes Rendus Hebdomadaires des Séances de l'Académie des Sciences Paris, 255*, 76.
Castaing, R., & Laborie, P. (1953). Examen direct des métaux par transmission au microscope électronique. *Comptes Rendus Hebdomadaires des Séances de l'Académie des Sciences Paris, 237*, 1330.
Castaing, R., & Lenoir, G. (1954). Étude micrographique à haute résolution des premiers stades du vieillissement d'un alliage aluminium-cuivre. *Comptes Rendus Hebdomadaires des Séances de l'Académie des Sciences Paris, 239*, 972.
Castaing, R., & Slodzian, G. (1962a). Premiers essais de microanalyse par émission ionique secondaire. *Comptes Rendus Hebdomadaires des Séances de l'Académie des Sciences Paris, 255*, 1893.
Castaing, R., & Slodzian, G. (1962b). Microanalyse par émission ionique secondaire. *Journal de Microscopie, 1*, 395–410.
Castaing, R., Philibert, J., & Crussard, C. (1957). Electron probe microanalyzer and its application to ferrous metallurgy. *Transactions of the American Institute of Mining, Metallurgical, and Petroleum Engineers Incorporated, 209*, 389–394.
Castaing, R., Jouffrey, B., & Slodzian, G. (1960). Sur les possibilités d'analyse locale d'un échantillon par utilisation de son émission ionique secondaire. *Comptes Rendus Hebdomadaires des Séances de l'Académie des Sciences Paris, 251*, 1010.

Castaing, R., El Hili, A., & Henry, L. (1965). Microanalyse qualitative par images électroniques filtrées. *Comptes Rendus Hebdomadaires des Séances de l'Académie des Sciences Paris, 261*, 3999.

Castaing, R., El Hili, A., & Henry, L. (1966). Quelques effets dynamiques dans la diffusion des électrons par les réseaux cristallins. *Comptes Rendus Hebdomadaires des Séances de l'Académie des Sciences Paris B, 262*, 1051.

Colliex, C., & Jouffrey, B. (1970a). Contribution à l'étude des pertes d'énergie dues à l'excitation de niveaux profonds. *Comptes Rendus Hebdomadaires des Séances de l'Académie des Sciences Paris B, 270*, 144.

Colliex, C., & Jouffrey, B. (1970b). Images filtrées obtenues avec des électrons ayant subi des pertes d'énergie dues à l'excitation de niveaux profonds. *Comptes Rendus Hebdomadaires des Séances de l'Académie des Sciences Paris B, 270*, 673.

de Chambost, E. (2011). A history of Cameca (1954–2009). *Advances in Imaging and Electron Physics: Vol. 167*. Elsevier (pp. 1–119).

El Hili, A. (1967). *Étude expérimentale de l'influence de la diffusion inélastique sur la formation des contrastes en microscopie électronique* (Thèse, Orsay).

Fitting-Kourkoutis, L., et al. (2010). Atomic-resolution spectroscopic imaging of oxide interfaces. *Philosophical Magazine, 90*, 4731–4749.

Galle, P. (1964). Mise au point d'une méthode de microanalyse des tissus biologiques au moyen de la microsonde de Castaing. *Revue Française d'Études Cliniques et Biologiques, 9*, 203.

Gloter, A., Badjeck, V., Bocher, L., Brun, N., March, K., Marinova, M., Tencé, M., Walls, M., Zobelli, A., Stéphan, O., & Colliex, C. (2017). Atomically resolved mapping of EELS fine structures. *Materials Science in Semiconductor Processing, 65*, 2–17.

Gonzalez-Jaranay, M., et al. (2007). Electron microprobe analysis in guided tissue regeneration: A case report. *European Journal of Dentistry, 1*, 40–44.

Guinier, A. (1937). Dispositif permettant d'obtenir des diagrammes de diffraction de poudres cristallines très intenses avec un rayonnement monochromatique. *Comptes Rendus Hebdomadaires des Séances de l'Académie des Sciences Paris, 204*, 1115.

Guinier, A. (1938a). Un nouveau type de diagrammes de rayons X. *Comptes Rendus Hebdomadaires des Séances de l'Académie des Sciences Paris, 206*, 1641.

Guinier, A. (1938b). Structure of age-hardened aluminium–copper alloys. *Nature, 142*, 569.

Henry, L. (1964). *Filtrage magnetique des vitesses en microscopie électronique* (Thèse, Orsay).

Huang, D., Hua, X., Xiu, G.-l., Zheng, Y.-j., Yu, X.-y., & Long, Y.-t. (2017). Secondary ion mass spectrometry: The application in the analysis of atmospheric particulate matter. *Analytica Chimica Acta, 989*, 1–14.

Jacquet, P. (1937). Sur le polissage électrolytique de l'aluminium. *Comptes Rendus Hebdomadaires des Séances de l'Académie des Sciences Paris, 205*, 1232.

Kumamoto, K., Warren, J. M., & Hauri, E. H. (2017). New SIMS reference materials for measuring water in upper mantle materials. *The American Mineralogist, 102*, 537–541.

Philibert, J., & Tixier, R. (1968). Electron penetration and atomic number correction in electron-probe microanalysis. *Journal of Physics. D, Applied Physics, 1*, 685.

Preston, G. D. (1938). Comment on Guinier's paper. *Nature, 142*, 570.

Rinaldi, R., & Llovet, X. (2015). Electron probe microanalysis: A review of the past, present and future. *Microscopy and Microanalysis, 21*, 1053–1069.

Suenaga, K., Tencé, M., Mory, C., Colliex, C., Kato, H., Okazaki, T., Shinohara, H., Hirahara, K., Bandow, S., & Iijima, S. (2000). Element-selective single atom imaging. *Science, 290*, 2280.

Yang, J., & Gilmore, I. (2015). Application of SIMS to biomaterials, proteins and cells: A concise review. *Materials Science and Technology, 31*, 131.

CHAPTER FIVE

Random recollections of the early days

V.E. Cosslett[a] **and Peter Hawkes (Afterword)**[b,*]
[a]Formerly Cavendish Laboratory, University of Cambridge, Cambridge, England, United Kingdom
[b]CEMES-CNRS, Toulouse, France
*Corresponding author. e-mail address: hawkes@cemes.fr

Contents

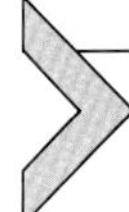

1. How I came into the subject

1.1 Berlin

I missed being in on the birth of the electron microscope (EM) by less than 10 km. In 1931, when von Borries and Ruska were getting the first low-magnification electron micrographs at one end of Berlin, I was studying gas reactions at the other end of the city in the Kaiser Wilhelm Institute (KWI) for Physical Chemistry, and was beginning to understand the political situation. The senior research staff were almost all Jewish: Bogdandy; Kalman; Ladenburg; Wigner; and Michael Polanyi, under whom I worked.

Advances in Imaging and Electron Physics, Volume 220
ISSN 1076-5670
https://doi.org/10.1016/bs.aiep.2021.08.005

I was disturbed by one of his Ph.D. students appearing occasionally in a Storm Trooper's uniform. My girlfriend was Jewish and was working with Lise Meitner in the next Institute. When this became clear to the landlord where I lodged, he made a transparent excuse to turn me out, though he belonged to the Stahlhelm and not to the Nazi Party.

From time to time there were colloquia at Hamack House, the center of the KWI complex (now the Max Planck Institutes), and there I first heard D. Gabor. Regularly I went to the colloquia at the university, in town, sitting in awe of the great men (Nernst, Bodenstein, Schrödinger, and Planck himself) and trying to keep up with rapid German. Once I went to hear Einstein in the Aula of the university, but his lecture was broken up by Nazi students. By the time I came home in the following year, Jewish shops and banks were being wrecked quite systematically and in early 1933 came the Nazi takeover.

1.2 University College, London

These events brought me, an unsophisticated provincial, up against the realities of the great depression and helped to form my political outlook. In England the economic situation was such as to change the whole direction of my scientific career. As a protégé of Imperial Chemical Industries (ICI), my future had seemed assured: I was under bond to accept a research job in one of their laboratories, if offered. To my dismay it was not offered, as the firm was having to retrench. In order to assist me in writing up my Ph.D., however, I was given a post as temporary assistant in the research group that ICI had set up in Donnan's laboratory at University College, London (UCL). My job was to build an apparatus for studying electron diffraction in gases, under Dr. H. G. de Laszlo, one of the sons of a famous portrait painter.

Thus, entirely fortuitously, I became apprenticed to electron beam technology. This was the first of a number of accidents that led me, in the end to electron microscopy in the Cavendish Laboratory, along what in retrospect might seem to have been a predetermined path. It was another 2 years before I learned about the EM and some 10 years later that I first laid eyes and hands on one, the Metropolitan Vickers EM1, the earliest British instrument. In between these years I had come into (written) contact with Ernst Ruska, but did not meet him until after the end of World War II, along with Boersch, von Borries, and other pioneers.

1.3 Bristol

After my year at UCL, I was fortunate enough in the continuing slump to obtain a research studentship back at Bristol, my alma mater, but now in the physics laboratory, not in physical chemistry. My brief was to set up an (improved) electron diffraction camera for a study of the structure and electrical conductivity of *in situ*-deposited thin-metal films. I found myself in a small group headed by E. T. S. Appleyard, a man of wide learning and left-wing sympathies, and including Bernard Lovell, who later went into radio astronomy. The department held regular colloquia, and one day in 1933 Professor A. M. Tyndall, knowing that I had some competence in German, asked me to report on the recent development of an electron microscope. He pointed me to the first papers by Knoll and Ruska (1932a, 1932b) giving a detailed account of the principles and construction of the magnetic EM. So I mugged them up into a 40-minute talk, of which I have little recollection apart from nervousness at the first occasion of addressing my new colleagues.

The nature and promise of the Berlin work made a great impression on me, though I did not grasp all the ideas at once nor the future possibilities. The use of an electron lens for imaging was revolutionary, in spite of the fact that we were already employing a simple ring coil to concentrate the electron beam in a diffraction camera (Cosslett, 1937). The perfection of the magnetic lens with polepieces (Ruska, 1934) and the demonstration of an image resolution beyond that of the optical microscope soon followed. In this rather accidental manner my attention was directed into electron microscopy by Professor Tyndall. He was a man of wide scientific interests and a rather laconic, humorous disposition. One of my abiding memories is of him standing on his head in the tea room to prove that gravity was not essential to the act of drinking.

1.4 Birkbeck College, London

When my time at Bristol was up, the economic situation was no better. In 1935 I was glad to take a part-time teaching job at Faraday House, in the middle of London, an electrical engineering college for the "sons of gentlemen." It had been established in the 1880s by the young electrical industry as a source of qualified staff, at a time when the subject was not respectable enough for any university. By 1935 the intake was to a large extent from overseas. After the war it had to accept ladies to keep afloat, but finally went under, as London University could not recognize its diploma toward a degree qualification.

My duties were light enough, teaching physics and chemistry (though I myself learned much about machine drawing and construction), and occupied not much more than half my time. Birkbeck College, largely an evening institution of London University, was less than a mile away, and Professor P. M. S. Blackett willingly took me in as a Ph.D. student in physics (my Bristol degrees were in physical chemistry). My task was to set up a β-ray spectrometer for 1- to 15-MV electrons. It was in effect an outsize electron lens, possibly the largest constructed up to that time. The maximum induction (1 T) was not unduly high, but it had to extend over a large volume between polepieces containing an axial stop. The six coils of outer diameter 40 cm were strip wound on a yoke 75 cm long originally designed by J. D. Cockcroft for a different purpose (Cosslett, 1940a).

The focusing problems of this apparatus impelled me to look into the electron optics and in particular the aberrations of magnetic lenses (Cosslett, 1940b). Only a few preliminary results were obtained before Blackett was appointed to the chair in Manchester in 1937, and the spectrometer went with him. His successor, J. D. Bernal, an X-ray diffraction expert, was interested in the instrumental as well as the interpretive side of his subject. We discussed the possibility of improving the intensity from an X-ray source by focusing the electron beam onto the target with a magnetic lens. A feasibility study proved promising, but showed that it would be a full-time job to design and construct an experimental tube, and to look into the associated problem of target cooling.

1.5 Oxford

So in 1938 I applied for a research studentship from the University of London, and was lucky enough to get it. But World War II broke out just when I was due to take it up in September, 1939. Bernal's group of crystallographers was evacuated to Oxford, to the laboratory of Dorothy Crowfoot-Hodgkin, where there was no room for me. After casting around in Reading as well as Oxford, I approached Professor J. S. E. Townsend at the Electrical Laboratory, Oxford, the old Physics Department, alongside the Clarendon Laboratory, of which Professor F. A. Lindeman (later Lord Cherwell) was head. The two were at daggers drawn, not only regarding space and finance but also over basic physics, since Townsend refused to accept the quantum theory. In the end the highest court of the university, the Visitatorial Board, had to remove him from his post for dereliction of duty, but at the end of the war he became Sir John Townsend, as a reward for his wartime efforts in training cadet officers in physics!

In 1939, however, he generously gave me space and facilities, since electron beams were respectable. My London grant continued—we all expected the war to be over quickly—and I set up a simple electron-optical bench to study focusing a beam into a small spot. The experimental work had to be largely abandoned when the war became serious, but it had brought home to me the role of spherical aberration in limiting the intensity in a focused spot. I continued my theoretical studies and, as a change from instructing officer cadets in elementary radio, aerodynamics, and navigation, I gave a lecture course on electron optics to the small class of physics students that continued. All graduate scientists had been enrolled on a central register shortly before the start of the war. I was in due course called for interview by the panel allocating jobs. It proved to consist of Lindeman himself, by that time Churchill's chief scientific adviser, and his assistant, James Tuck. The job they had in mind for me was to work on proximity fuses on a range out in the wilds of West Wales. They decided, quite rightly, that my knowledge of electronics was inadequate and let me carry on at Oxford with the cadet courses, which soon expanded to include Naval, Air Force, and Signaller cadets. Their initial education in physics had been quite varied. We had some amusement in correlating the Army radio textbook with the Navy's "Wireless Manual," which still treated capacity in terms of "jars," among other archaic concepts.

By the end of the war in 1945, Townsend was out and Lindeman was back in undisputed command of physics. Reviewing his plans for the future of physics in Oxford, he decided that there would be no place for electron optics in it. In his unemotional, remote way he informed me as much, but kindly offered to keep me on as a temporary lecturer for a year (this was summer, 1945) until I could find a post elsewhere.

Through my investigations of electron beams I had become increasingly interested in the electron microscope, in its applications as well as its fundamentals. On learning that a number of early RCA microscopes had been obtained from the United States under the lend–lease scheme, I had written in November, 1942, to the Department of Scientific and Industrial Research (DSIR) (forerunner of the Science Research Council) inquiring if one could be allocated to Oxford. From contacts with the physiologists there, as well as physical chemists and crystallographers, I stated the case for the Electrical Laboratory to be so favored. The Secretary of the DSIR, Sir Edward Appleton, replied to the effect that the destinations of the available instruments had already been settled, but that Oxford would be kept in mind if another should become available.

In 1944 his successor wrote me that if Oxford was still interested in getting an EM, I should approach the Metropolitan Vickers Company in Manchester, which was about to market an instrument that should be suitable. I wrote for details to Dr. T. E. Allibone, whose work was known to me, but, as he was away in the United States, a reply came from the head of the high-voltage research department, F. R. Perry. He confirmed that the EM1, built for Imperial College in 1937 and then used at the National Physical Laboratory (NPL), had been returned to Manchester, to serve as a basis for the development of an "improved type." The outline specification of this "austerity" model, later to become the EM2, was enclosed. It bore more resemblance to the RCA than to the Siemens instrument. The maximum voltage was to be 50 kV; the "microscope tube" (i.e., the column) would have three viewing windows and a single-plate camera, taking five 2-in. pictures per plate. The maximum magnification was limited to 10,000X in the interests of a relatively large image field combined with adequate brightness at top magnification. A full description was published in due course by Haine (1947). He gave a short account of it at the meeting of the newly formed Electron Microscopy Group of the Institute of Physics at Oxford in September, 1946 (Cosslett, 1947), seeking early publication because "the lens system is so embarrassingly similar to the Phillips [sic] one." He had returned in July from a visit to European laboratories, to catch up on postwar developments.

In this way my first contact was made with the EM activities at Metropolitan Vickers, which soon became part of Associated Electrical Industries (AEI), and with Dr. Haine. My association with the firm extended over 30 years and more, in a sort of love–hate relationship. The Manchester laboratory was notorious for its "NIH" attitude: if a device was "not invented here" it was unlikely to be of much value. From early times it had relied on in-house recruitment, from apprenticeship upward. It was many years before one of our Ph.D.s obtained a job there. Parallel development went on with Cambridge on both the scanning electron microscope and the microprobe analyzer. It was only when it came to the high-voltage microscope that some sort of cooperation was arranged, with the Ministry of Technology as intermediary.

We benefited in minor ways, however, from Manchester's expertise. Early in 19461 got a quotation for a magnetic objective lens "as used in the EM2," for what now seems the bargain price of £33.10.0. The idea was to use it to defocus the electron beam in the X-ray projection microscope that I was then starting to build at the Electrical Laboratory, Oxford.

The experimental setup was soon moved with me to the Cavendish later that year, together with David Taylor, who was my technical assistant. The only noteworthy point is that it was evacuated by means of a Holweck pump, an early form of turbomolecular pump, to avoid the possibility of contamination by mercury vapor or other diffusion pump liquid.

Before that development, however, we had held at Oxford the first formal symposium on electron microscopy. The story of how the EM Group grew out of early informal meetings set up to discuss the problems of the RCA microscopes imported from the United States has been told elsewhere (Cosslett, 1971). In September, 1946, at Oxford, we heard accounts of the work done in France by Dupouy (Toulouse), Brück and Grivet (CSF, Paris), and Lepine (Institut Pasteur), and in the Netherlands by Le Poole and van Dorsten. Gabor introduced a discussion on lens aberrations, on the basis of a manuscript sent to him by Scherzer. The contacts made at this meeting, and in the tours of European laboratories by Haine in 1946 and myself in 1948, were partly instrumental in generating the first International EM Conference in Delft in 1949.

In these ways I became familiar with the work going on in Europe on both the construction and applications of the microscope. We were aware of that in the United States through the contacts made over the acquisition of the RCA EMB instruments. So I was pretty well equipped with a theoretical knowledge of the subject by the time I was transferred to the "other place," but had hardly any practical experience, having only looked at an electron image once or twice and never having been responsible for an EM.

1.6 Cambridge

The chance to go to Cambridge was a great piece of luck, and was almost lost by an administrative muddle. Knowing that the Cavendish Laboratory had received one of the RCA batch, I wrote to Professor W. L. Bragg early in 1946 after it became clear that there was no future for me in Oxford. He advised me to apply for a research fellowship, under a scheme founded by ICI to assist mature scientists to get back into research after the interruption of the war. I did so at once, but after some weeks had had no reply. As I was short-listed for a lectureship elsewhere, I wrote inquiring when a decision would be made. Bragg wrote back in some distress, saying that the initial awards had been made already and that my application had been mislaid and therefore not considered. He asked me to come over for an

Figure 1 George Crowe at the RCA EMB microscope in the Cavendish Laboratory, Cambridge.

interview nevertheless. I did so, bringing with me the proofs of "Introduction to Electron Optics," which the Clarendon Press was publishing. To my surprise and delight he was able to persuade the electors to find an extra fellowship, for 3 years, from October, 1946.

It had always been my ambition to get to Cambridge, but an illness had prevented me from taking the scholarship examination 20 years before, and I went to Bristol University instead. I found it intriguing also that my research was once again financed by ICI, as it had been in 1930–1932.

At the Cavendish the RCA electron microscope had been installed some 4 years earlier. By 1946 it was well run in by George Crowe, a jovial man of infinite resource, who had been Rutherford's research assistant for many years (Fig. 1). The loss of two fingers from handling radioactive material did not prevent him from preparing specimens and skillfully manipulating the controls of an EM. He was also a keen sportsman, especially with the gun after pheasants and pigeons. The RCA instrument was accommodated in the new Austin wing of the laboratory, near the X-ray crystallographers. Soon after my arrival we were allocated by the enemy property custodians a Siemens microscope (Fig. 2) that had been liberated from the Krupps ar-

Figure 2 The early Siemens electron microscope at the Cavendish Laboratory, with (left) V. E. Cosslett and (right) R. W. Horne.

maments laboratory. This type of war booty was being divided up between the British and American technical field units. There was no difficulty over our machine, but a later model that went to the National Institute of Medical Research in Hampstead was subsequently returned to Germany after it was discovered to have been taken from some hospital there. A story also went the rounds that there was a skirmish between British and American marines over another Siemens EM, which the latter snatched from the former's store in Chelsea.

When the Siemens EM was delivered to Cambridge it was not possible to house it with the RCA microscope and so room was found in the Old Cavendish, across the yard. This room had been Aston's research room for many years and contained his second mass spectrometer. Rows of vials of gaseous specimens stood in mercury cups on the shelves. Most of this historic material went to the museum when we moved in at the end of 1946. Shortly afterward the neighboring room became vacant. It had been J. J. Thomson's research room in his old age, when he became Master of

Trinity College. His high desk, for writing standing up, had remained there, and we found in it some old papers relating to that wing of the Cavendish, built from the Nobel Prize awarded to Lord Rayleigh. I was amused to find that the wing was opened on the day I was born, June 16, 1908. The original Cavendish had been formally opened on June 16, 1874, and in due course the New Cavendish Laboratory was opened on June 16, 1974.

I came to the Cavendish with a 3-year appointment and have remained there for more than 10 times as long. How the Electron Microscope Section was developed during that time, with a succession of talented research students and a more permanent technical staff, has been related in some detail (Cosslett, 1981). Beginning as a service unit for researchers in the university generally, we progressed into the design and construction of electron beam instruments: the X-ray microscope and microprobe analyzer, a reflection EM, a high-voltage and later a high-resolution–high-voltage EM, as well as several experimental apparatus for investigating electron scattering and related phenomena. We also organized summer schools in electron microscopy (Fig. 3).

As time went on, most of the other laboratories in the University obtained their own EM, so that our service load gradually diminished, only to grow again with each new instrument that we constructed. At one moment there was a danger that we might be turned into a grand central service unit. Lord Rothschild, then an Assistant Director of Research in Zoology, proposed such a scheme as being more economical than for each department to have its own microscope, most likely underused. In fact, every researcher preferred having an instrument under his own control, ready when needed, rather than having to book ahead in a centralized service laboratory. So the scheme was dropped, luckily for me. In those days capital grants were not difficult to come by, but some years later the University Financial Board woke up to the fact that it was having to meet large bills for maintenance on the 40-odd EMs in the various faculties, and the prospect of a heavy cost for replacements in the future. By that time it was too late to put the clock back.

The Electron Microscope Section was thus able to carry on with developing its own lines of research in the Old Cavendish. More space became available as other sections, Metal Physics in particular, moved out to the new buildings in West Cambridge. By the time we should have followed, the money had run out and I was told that we would have to stay put.

Figure 3 Electron Microscopy Summer School, Cavendish Laboratory, Cambridge, 1950. Seated center, Dr. Cosslett, Dr. M. E. Haine (with pipe), and Dr. G. Drummond; far left, Dr. R. W. Horne; far right, Dr. J. W. Menter; standing behind him, Dr. J. Nutting.

This news did not upset me much: it would have meant a great upheaval to move the high-voltage microscope, apart from cutting us off from the many departments with which we were closely collaborating, especially Engineering and Metallurgy. On the other hand, we would lose the large laboratory planned to house a 600-kV high-resolution EM (HREM) on the new site. Luckily, however, it was found possible to fit it into a large room in the old building, originally earmarked by Maxwell for precision experiments in electricity and magnetism! Fears about disturbing vibrations in the center of Cambridge were overcome by the design and fabrication of a suspension system by our mechanical engineering colleagues. Other lines of research have been contracted or ended, but the HREM continues the practice of electron microscopy in the part of the Old Cavendish which it has occupied these 30-odd years.

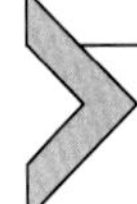

2. These I met along the way

2.1 British colleagues

2.1.1 Sir Lawrence Bragg

The younger Bragg received the Nobel Prize in 1915, along with his father (W.H.B.) for their elucidation of crystal structures by X-ray diffraction. In 1938 he was appointed to the Cavendish after the death of Rutherford, but only effectively took charge at the end of the war. Whereas the laboratory had been almost exclusively devoted to nuclear physics under Rutherford, it was Bragg's aim to diversify into several of the currently developing areas of physics. In addition to his own speciality of X-ray crystallography he encouraged the growth of work on radio astronomy, surface physics, aerodynamics, and metal physics. The atmosphere was already favorable for electron microscopy when I came along.

Bragg's philosophy was to give backing to individual initiative—not to plan a new field of research, but to allow freedom for it to grow in the hands of someone who gave proof of dedication to the job. The Cavendish thus developed a big section for the physics and chemistry of surfaces under Philip Bowden, while that for fluid dynamics (originally started by G. I. Taylor) remained small but distinguished under Alan Townsend. In X-ray crystallography, the "classical" aspects of which he left in the hands of W. H. Taylor, he was farsighted enough to support the infant study of molecular biology, at first in the persons of Max Perutz and John Kendrew, and later Francis Crick and James Watson. The former pair were down-to-earth, step-by-step workers as compared with the flights of fancy in which the latter indulged, although they finally had the good sense to pick a winner. I well remember them tossing about their ideas on protein structure at the tea table, Watson the puppy performing before the wise old dog Crick. Then they started building models of the double helix in the room in the Old Cavendish where we later placed the HREM, as no other space large enough was available to them.

By that time, the early 1950s, Bragg was approaching retirement and no longer actively conducting research. He was always keen to keep up with new developments, however, and would look in once a week or so whenever we were busy with some new experiments. Electron microscopy held for him the prospect of directly determining the structure of molecules, although I stressed that it would be a long time before we could do Perutz out of business. The "phase problem" was very much in Bragg's mind—how to work back from a diffraction pattern to the structure that gave rise to it. He had himself put forward and demonstrated an optical procedure for such a reconstruction, calling it "an X-ray microscope" (Bragg, 1939). It was not until much later that the EM Section was able to make a contribution to the solution in the form of the Gerchberg–Saxton algorithm (Gerchberg & Saxton, 1971), an iterative method relating the data in the electron image to those in the electron diffraction pattern.

In personality Bragg was warm and supportive. Although his father, W. H. Bragg, looked like a farmer, solidly built and whiskery (he came, in fact, from a Cumberland farming family), the son gave the impression of belonging to the landed gentry, spruce and interested in his garden as well as in science. He supported his staff unreservedly in applying for outside grants and in any difficulties that arose with the university administration. Proud of the achievements of the laboratory, he would frequently bring eminent visitors around to inspect new developments. I well remember him bringing Dr. Max von Laue, the cofounder of X-ray crystallography, to see the rather simple point-projection X-ray microscope that W. C. Nixon had set up in an unused photographic darkroom. The old man was clearly impressed with the possibilities of high-resolution imaging of wet specimens in air or a partial vacuum, a prospect that has still not been realized owing to lack of intensity and radiation damage. I was not sure whether his knowledge of English was up to understanding all the points of detail we were trying to press on him. We were certainly impressed by meeting the two fathers of X-ray diffraction.

Sir Lawrence continued his interest in our work after his move to the Royal Institution in London in 1953, dropping in now and then to look at what we were doing, asking for microradiographs for his lectures and for the Rontgen Museum in Remscheid (when he received the Rontgen medal). He invited me to give a Friday Evening Discourse on "Microscopy with X Rays" in 1959, and later the Children's Christmas Lectures (Cosslett, 1966) (Fig. 4).

2.1.2 Sir Charles Darwin

The man who was primarily responsible for getting the initial batch of RCA microscopes into the country, and thus starting up the subject in Cambridge and elsewhere, was a grandson of Charles Darwin, the father of evolution. Sir Charles Darwin grew up in Cambridge as one of his eminent descendants. A mathematical physicist of distinction, his first notable contribution was in the theory of X-ray diffraction. He extended the kinematical treatment of Bragg and von Laue of perfect crystals to the more

Figure 4 Children's Lectures, Royal Institution, London, 1960. Dr. Cosslett with model of an icosahedral virus.

realistic case of a mosaic of differently oriented blocks (Darwin, 1914). At the outbreak of World War II he was head of the National Physical Laboratory, to which the original Met-Vick EM1 had been transferred from Martin's laboratory in Imperial College, apparently for safety's sake before the expected enemy attacks on London. Possibly it was the arrival of this instrument that stimulated Darwin's interest in the subject. The head of the Metallurgy Department at that time, G. D. Preston, informed me of "a manuscript note by C. G. Darwin dated 24.7.40 dealing with electron optics from the Hamilton aspect—a very succinct piece of work in which I think he persuaded himself that an EM ought to work" (presumably with adequately high performance). "Anyway he went to the States and the RCA microscopes were delivered to NPL in 1941, I think."

In this way the NPL became the center of wisdom on electron microscopy. "We were put at the disposal of the recipients of the other instruments," in various government and industrial research laboratories. Many problems arose, since no one in England had any first-hand experience of the subject. So Darwin took the initiative in arranging a meeting of representatives of the seven active groups, which was held at the NPL in

November, 1943. It was followed by other meetings, which culminated in the formation in 1946 of the EM Group of the Institute of Physics (Cosslett, 1947, 1971). Darwin was elected chairman and myself secretary.

By that time I had made my peace with Sir Charles. He had ticked me off in 1944, both privately and in a letter to *Discovery*, for not mentioning the Met-Vick microscope in an article I wrote for it (Cosslett, 1943). "I think if we had the custom of behaving as journalists do in America your article would have been all about the Met-Vick machine, and it would have forgotten to mention that some of the photographs came from another instrument!" This was hardly fair, in fact, because the EM1 could not bear comparison with the RCA model B, in either design or performance. I remedied the matter in a later survey in *Discovery*, describing the EM3 in some detail (Cosslett, 1950a).

Initially, since he wrote me at my private address in Oxford, Darwin seemed to think I was a journalist! In due course all turned out right and we got on well in running the infant EM Group. Apart from my helping him with recruiting a successor to the NPL electron microscopist, F. W. Cuckow, we joined forces in persuading the Institute of Physics to found a Journal of Applied Physics, he acting for the Physical Society (before it fused with the Institute).

2.1.3 Sir George Thomson

The third member in the effort that got electron microscopes from the United States under the lend–lease agreement was G. P. Thomson (later Sir George). Son of J. J. Thomson, discoverer of the electron, he himself shared a Nobel Prize in 1937 for his part in the demonstration of electron diffraction, following de Broglie's theory of the wave nature of that "particle." Later he was Professor of Physics at Imperial College, London, and after World War II he became Master of Corpus Christi College, Cambridge, of which he had been a Fellow as a young man.

I believe he was a member of the Scientific Advisory Committee of the War Cabinet in Churchill's administration, along with Bragg and Darwin, which was responsible for the allocation of the RCA microscopes to various laboratories in 1941–1942. In personality the three differed greatly. They were all under the burden of being sons of famous fathers. Darwin seemed to take this lightly, a tall, dominating man who had taken a First Class in Mathematics at Cambridge, as had Thomson. I fancy that both the latter and Bragg suffered from the suspicion that their major work had been inspired by their parents. In Bragg's case this was patently not true, from the nature of his initial theory of diffraction. Yet his relations with his father remained difficult until old age; also, he was inclined to feel that his mathematics was inadequate. He thought in terms of physical models rather than in mathematical formulas, and (like myself) was always trying to prove himself to physicists (Phillips, 1979).

Thomson, on the other hand, was a physicist of the classical school, of experimental as well as mathematical ability, whereas Darwin was essentially a mathematical physicist. Slight in build and keen of eye, slow to speak but critical in attitude, G.P. (as he was always known), was a kindly man underneath it all; his great hobby was sailing, and later on building models of sailing ships. He was interested in the impact of science on society and was an early member of the Pugwash Conferences on nuclear control and disarmament.

I had come across him occasionally before he returned to Cambridge as Master of Corpus in 1952, initially in my electron diffraction days in London, where he had been appointed to the Physics chair at Imperial College in 1930 and was the first to introduce this new technique in England. He became interested in electron microscopy also, and through his connections with Metropolitan Vickers over high-voltage cathode-ray oscillographs was instrumental in arranging for the EM for L. C. Martin (mentioned earlier) to be financed and built (Moon, 1977).

In Cambridge he became aware of the lack of provision of living and social amenities for research students. Colleges had traditionally been devoted to the education of undergraduates, the purpose for which the great majority were founded. Doctoral and other research students had grown greatly in numbers since the end of the war but very little notice had been taken of their needs, aside from collecting their tuition fees. The university set up an inquiry, chaired by Lord Bridges, to look into the broader question of the relations between the university and the colleges, of which I was made a member. But before it reported, G.P., along with the senior tutor of Corpus, also a member of the inquiry, had devised a local solution. An existing Corpus property to the west of Cambridge, Leckhampton House, was extended and turned into an annex for research students and staff. As part of this extension of college facilities a number of new Fellows were elected, of which I was one. In this development, as in many others, G.P. showed himself to be concerned with the living problems as well as with the scientific activities of research students. He continued to be interested in the philosophy of science, the history of physics, and its applications in public affairs. In general he kept to the old rule of not talking "shop" in college, preferring to discuss (if anything) his personal enthusiasms, among which euthanasia figured strongly in his later days. He was always keen to salute successes by members of the college. I particularly prized a note from him when I was elected to the Royal Society in 1972: "[your] work has always followed the old Cavendish tradition of originality combined with simplicity. Your microscopes are indeed a long way from sealing wax and string but like Rutherford's and my father's your apparatus has always been simple, dependent for its success on having the right idea."

2.1.4 G. I. Finch

In the early days of electron diffraction, Imperial College, London, was the source of wisdom in this country: G. P. Thomson mainly on the theoretical side and G. I. Finch for experimental know-how. As de Laszlo had little experience of the subject and I still less, we relied greatly on the latter in our research at University College, London. It happened that de Laszlo knew Finch well, both having been educated partly in Switzerland.

Son of an Australian judge (and father of Peter Finch the actor), George Ingle Finch was originally trained as a chemical engineer at the Technical University in Zurich. Coming to England in 1912 he was first at the Royal Arsenal as a research chemist, then in World War I a captain in the artillery, and afterward a lecturer in electrochemistry at Imperial College. From studying the initiation of gaseous reactions by electrical discharges, he became interested in the catalytic effect of the cathode. X-Ray diffraction having given no clues, G. P. Thomson suggested the use of electron diffraction to examine the surface structure. Always a practical man, Finch set out to build a suitable apparatus; at that time there was none such manufactured in England or elsewhere.

The Finch camera, as it came to be called, became the standard electron diffraction instrument and remained so for many years, eventually being produced by Edwards High Vacuum Company in small numbers. At University College we largely copied it (Cosslett, 1934): a cold-cathode electron source, magnetic coil for focusing the beam, mercury vacuum pump, and a simple transformer–rectifier high-voltage supply for 55 kV. The set was in a wire-netting enclosure with a bare rod connecting the output to the cathode. I remember drawing a spark by pointing my finger at it once. The resistance of the human body is fortunately high, so I came to no harm. It was otherwise with de Laszlo one day when his welding torch was not properly insulated.

The insulator for the electron gun was a wine bottle with the base cut off. A Barsac bottle was preferred, according to G. D. Preston; Finch was a bon viveur and a collector of old wine bottles. He was also a famous mountaineer. He took part in the 1922 Everest expedition and for long held the altitude record for a climb without oxygen assistance. In celebration of his sixtieth birthday he climbed the Matterhorn for the sixteenth time.

Much of this experience with electron diffraction apparatus, joined with that of high-voltage oscilloscopes, formed the background to the later development of electron microscopes. It certainly stood me in good stead when it came to building electron beam instruments at Cambridge. That work led to a detailed interest in the properties of electron lenses and especially to possible means of overcoming spherical aberration. We received grants from the Paul Instrument Fund of the Royal Society for this research, and Finch was appointed assessor of the projects. To do so, he would drive over from Oxford two or three times a year. He had retired to Heyford, north of that city, in fox-hunting country, where he continued to ride to hounds until he was over 70. His other passion, apart from mountaineering, was sailing a yacht in the Channel, a sport he shared with G. P. Thomson. In old age he was still fit and erect. I do not remember anyone driving a car in reverse so fast (and safely) as he could do.

Our work on lens aberrations went on for many years, with the help of a series of research students, including the editor of this volume. We received encouragement and technical advice from Finch, but like other groups in this field we succeeded only in showing how difficult the problem was. The most immediately appealing solution, the use of quadrupole lenses, could be demonstrated on a bad magnetic objective, but full correction was hindered by adventitious effects, both mechanical and electrical (Hawkes, 1966). The alternative use of pulsed fields was similarly found to be feasible in principle but with severe limitations in practice (Vaidya, 1972).

2.1.5 J. D. Bernal

In my time at Birkbeck College I learned something of the properties of electron lenses, large and small, but much more about the strategy of scientific research. From Blackett I gained experimental confidence and the technique of setting out on a line of work. It was a real lesson to see how he would quickly decide on the essential parameters and make a rapid first-order calculation of the expected effects, and so of the type of apparatus needed to observe them. He was perhaps a little overeager for results once the project was under way, not realizing that inexpert hands were not so good at experimental work as his were.

His successor was a different type of man altogether: John Desmond Bernal, known as "Sage" because of the breadth and depth of his knowledge. Where Blackett had been commander of a destroyer in World War I, and carried himself as such, Bernal was everyone's idea of a scientist, short, with tousled hair, and entirely informal. He had a soft but penetrating look, and a disarming smile which made him a favorite with so many ladies: he seemed in need of mothering, I always thought. Blackett was leftish in politics, coming out after World War II in opposition to what he regarded as the "disastrous" policies of the British and American governments on atomic weapons (Blackett, 1948). Bernal was much farther left, taking part in many of the communist-inspired campaigns from his Cambridge days on. He became deeply concerned about the applications of science for the public good. His "The Social Function of Science" (Bernal, 1939) made a great impact on my generation.

Bernal was the quintessential polymath. During World War II he was advisor to the Chief of Combined Operations. His postwar "Science in History" ranged (in four volumes) all the way from the origin of life to the future of society (Bernal, 1954). Just before the war, in 1939, he was investigating the structure of tobacco mosaic virus by X-ray diffraction. On learning that electron micrographs of viruses had been obtained in Germany he began making plans to get such an instrument, and we had our eyes on one of the early Siemens machines when hostilities put a stop to them. As usual with him, once he got a new idea he went baldheaded after it. I recollect him telling me that the chief thing was to get started on a promising research as quickly as possible, leaving the financial aspects till later. A key point in getting a grant was to have some results, however provisional, to show that the approach was viable. Excellent advice when apparatus was simple and not difficult to rig up with the aid of a good workshop! It was another matter when it came to raising the funds to design and build a high-voltage electron microscope. There we would hardly have stood a chance of success if something similar had not already been constructed in France and in Japan, and in each case produced results of metallurgical interest.

2.1.6 Dennis Gabor

It is said of Edward Teller, father of the hydrogen bomb, that of the flood of ideas which poured from his brilliant mind, 9 out of 10 were doomed to fail in the cold light of practice. Dennis Gabor, a fellow Hungarian, had an equally fertile mind but a rather higher proportion of successes. He was by no means infallible, however. When I first heard him in 1931 in Berlin, he was speaking of his research into mitogenetic radiation, supposedly emitted from the root tips of actively growing plants. The results were indecisive, and "a mystery surrounds this work" (Allibone, 1980). Years later, I was external examiner for some of his research students working on a flat TV tube for hanging on the wall. This had remarkably complicated electron optics, especially for color, and most experimental physicists would have predicted that it would never perform satisfactorily. But, in this as in many other things, he may prove to have been ahead of his time, since a flat tube (of simpler design) is now announced by a manufacturer.

In between these flights of fancy, Gabor produced a mass of important work in communications theory, as well as holography, which justly earned him a Nobel Prize in 1971. He came to this country in 1934 and had a great effect on the development of the electron microscope as well as in many larger spheres of life, including social and philosophical problems. Here I first came across him toward the end of the war, through the Association of Scientific Workers, in which we were both active. He was then researching into electronic devices and communication theory at the British Thomson Houston (BTH) laboratory at Rugby (BTH, like Met-Vick, was part of AEI). I remember visiting him in the hut-annex specially built for him, outside their security fence, since he was initially regarded as an untrustworthy alien. He had already taken up interest in electron optics again, which he regretted not following up in his Berlin days (see Allibone, 1980).

During his leisure time he had worked on the application of Hamiltonian methods to electron optics, and to the correction of lens aberrations in particular. The results were first published in his short book, "The Electron Microscope" (Gabor, 1945), and were presented at the Oxford EM Conference in 1946. He had begun to attend the regular meetings of the Institute of Physics EM Group in January, 1946, and in this way we became more closely acquainted. He became deeply interested in the problems of defining and assessing the resolving power of an electron microscope. At the first International EM Conference (Delft, 1949) he was asked to organize a working party, "with the aim of reporting at the next Conference at Paris in 1950, on the theoretical and practical rating of the performance of electron microscopes." Like many subsequent attempts to pin down resolution, this initiative appears to have come to nothing. He prepared a short survey of the problems involved, which he circulated to the members of the Working Party (names not recorded), but there is no report in the published proceedings of the Paris meeting.

In November, 1948, he wrote telling me of his appointment as Reader in Electronics at Imperial College, London, and offering to send me the notes of his lectures on electron dynamics, in which I had expressed an interest. I replied with thanks and the news that I would not be moving to London myself after all, having turned down a Readership in J. T. Randall's laboratory at King's College, on the promise by Bragg of an "official lectureship from October next, and the assurance of adequate facilities to expand our work with the electron microscope."

Later, in 1959, Gabor put to us a problem suitable for our X-ray microscope or the scanning microprobe: the characterization of particles on a magnetic tape. The X-ray images from the latter instrument showed the proportions of Fe and Co overall, but it was impossible to get a reliable size distribution owing to the thickness of the coating. The CBS Laboratories in the United States, for which he was a consultant for many years, were not impressed.

By the time he left Rugby for London, Gabor had already "invented" holography. Although the validation of the principle was carried out at optical wavelengths (Gabor, 1948), the ultimate intention was to circumvent the effect of aberrations on electron microscope images. His plan was to take an electron micrograph by a method producing interference with a coherent background (a "hologram" because it contained the whole information including the phase) and then to reconstruct it in a light beam. The idea, which owed something both to Bragg's fly's eye method in X-ray diffraction and to Zernike's interference method of studying aberrations of optical lenses, had occurred to him on the tennis court on Easter, 1947. In the following year he published a detailed account of "microscopy by reconstructed wave-fronts" and the prospects for reaching a resolution of 1 Å in the EM (Gabor, 1949). The practical difficulties that prevented its realization in spite of careful experiments at the AEI laboratories (Aldermaston) are described by T. Mulvey later in this volume.

2.2 Contacts with European colleagues

Immediately after the end of World War II we began to develop contacts with electron microscopists abroad. There had been some interchange with the United States during the war, apart from the negotiations that brought

Figure 5 Siegbahn–Schönander electron microscope, Stockholm, about 1943.

the six RCA microscopes to this country. But we knew nothing of comparable developments at Siemens and elsewhere in Europe, apart from one isolated item of news. There was occasional scientific contact with Sweden, which was a neutral country. In 1943 Sir Lawrence Bragg was in communication with his old friend Manne Siegbahn, of Stockholm, a fellow pioneer of X-ray physics. He brought back news of an unusual electron microscope, designed by Siegbahn and manufactured by the firm of Schönander. It was unusual in being horizontal (Fig. 5). Although a number were produced, it failed to find favor in the face of the RCA and Siemens products.

More interesting, and revolutionary in its way, were details of a technique for cutting thin biological sections, devised by a young man named Sjöstrand. Some years earlier it had been "proved" by von Ardenne, on the basis of the mechanical properties of biological material, that it was impossible to cut sections thin enough (<1 μm) for imaging in an electron microscope (von Ardenne, 1939a, 1940), except as a wedge-shaped section. A handwritten note now described how it could be done by slicing a thicker section bent over the convex side of a watch glass. Images of muscle taken with the Siegbahn–Schönander microscope were shown in a later publication (Sjöstrand, 1943). In the following years several conventional microtomes were geared down to cut still thinner sections, with the specimen made more rigid by embedding it in some plastic material.

Aside from this window, the only light on European developments came from British and American teams that toured Germany and some of the occupied countries, and from a few leading German scientists brought over for interrogation. The FIAT Reports (Farrand, 1946; Hansell et al., 1946) contained a short mention of electron microscopy among mostly military technology. Much more information was contained in a British intelligence report that was put together in 1946–1947, but appeared only later (Sayer, 1948). It was the result of visits to German laboratories by a team of three, F. W. Cuckow (of NPL), M. E. Haine (of AEI), and L. J. Sayer (of the Admiralty Research Laboratories), and the interrogation of a number of electron microscopists brought over to London for this purpose.

Sayer's report included an extensive bibliography of the German literature on electron microscopy and related subjects from 1939 to 1946, over 400 entries in all, together with author and subject indexes. There was also a list of the Siemens EMs that could be traced, 27 out of the "over 40" which that firm had manufactured, and 6 of the AEG electrostatic type. At that time 6 Siemens instruments had been transferred to Britain and allocated to MRC (Hampstead), Metropolitan Vickers (Manchester), NPL (Teddington), Cavendish Laboratory (Cambridge), King's College (London), and the Chemical Defence Research Station (Porton).

2.2.1 Ernst Ruska

The original inventor of the electron microscope continued developing it through the war in makeshift laboratories in and near Siemensstadt, in Berlin. Fortunately I was able to get in touch with him soon after the Cavendish received a Siemens EM that had been brought over from the Krupps research laboratory. The output insulator on the high-tension transformer had been cracked in transit, so that we could not operate it adequately above 80 kV. Some projector polepieces were missing and there were no spare filament holders. The latter we got through the DSIR, which had allocated the EM to Cambridge; the former we made in our workshops.

An extended correspondence ensued to obtain from Siemens, if possible, a porcelain-pot insulator. I began it in October, 1946, soon after arriving in Cambridge. It was routed through the Sundry Materials Branch of the Board of Trade and the Economics Branch of the British Military Government in Berlin to Siemens-Erlangen. Not only was this a separate company from Siemens-Halske in Berlin, but it was in the American zone of occupied Germany. In June, 1947, I received a reply from the responsible British official enclosing a quotation for 85 Reichsmarks, signed "E. Ruska." I wrote accepting it, and was then informed that it was in the wrong currency: it should have been for £10.12.0. After our acceptance of this in July, I had a letter from Ruska to the effect that the quote had to be in dollars, which would amount to $44 (excluding packing, transport, and customs)! We finally received the insulator in August, 1948, after a great deal of correspondence. The invoiced price was £25.10.0, but included import duty of 50% as an *optical* instrument. On my pointing out to the Board of Trade that it was essentially *electronic* in design and operation, a duty-free license was issued.

Meanwhile I had received from Dr. Ruska detailed advice about installing the insulator in such a way as to avoid getting air into the transformer windings. Until recently I believed this was the first direct contact between us. At the European EM Conference in the Hague in 1980, however, he came to me with a letter of mine and asked exactly when it was sent to him. I had neglected to put the date on a request for his 1934 paper in *Zeitschift fur Physik* on magnetic lens design. The address showed that it was in fact sent in 1936. This incident was typical of Ernst Ruska's sense of order and attention to detail. Who else anywhere keeps a record of all reprint requests, and the precise documentation, over nearly 50 years? His office in Berlin-Dahlem houses, in fact, the most complete archive of electron microscopy that I have ever seen.

2.2.2 Bodo von Borries

As mentioned earlier, some of the leading German electron microscopists were brought over to England after the war for questioning about the state of the subject in their country. Among them were B. von Borries, H. Boersch, E. Kinder, H. Ruska, and E. Ruska. The last named I did not manage to see, although the DSIR wrote in October, 1946, that "Dr. Ruska will not be going back to Germany for a week or two and it will be possible for you to see him in two weeks time." He was interviewed "at the Admiralty" by Gabor, Sayer, and Haine, according to Sayer (who was employed at a Naval research laboratory).

Dr. von Borries and Dr. Boersch came to Cambridge later on, in August, 1947, and April, 1948, respectively. During his 3-day visit, von Borries was interviewed by Arthur Brown (research fellow), David Taylor (senior assistant), and myself. According to my later report to the Admirality, in whose care he was, he "elucidated several important matters concerning the construction and operation of the Siemens electron microscope.... Detailed discussion took place over the precautions necessary to obtain high resolution with the instrument, especially the establishment of optimum focusing conditions and illumination." He was anxious to arrange an exchange of scientific information and for the supply of spare parts for the six or seven Siemens EMs in the country. "A scheme of exchange of reprints was proposed and has been agreed by the Committee of the Electron Microscopy Group. As a first step Dr. von Borries has made available 150 papers published by the Siemens Laboratory and scientists using its machines." Subsequently he arranged for the import of spare parts through the London firm of H.C. North and Co. Ltd.

Dr. von Borries was very keen on the proper servicing and maintenance of the Siemens EMs in Britain. "He believed that they could produce results at least as good as the more recent RCA types." Efforts were made to arrange a contract with von Borries himself for the purpose. I wrote to the five other users known to me at the end of October, 1947, to inquire if at least three of them would contribute toward the cost (£100) of a visit of about four weeks, but unfortunately the scheme did not get the necessary support.

During his visit he mentioned that the Siemens EM laboratory was being reestablished under Dr. E. Ruska, but "was experiencing difficulty in re-erecting even one microscope." In fact it took several years for the post-war Elmiskop to emerge. The Cavendish Laboratory received one of the first of this new model in 1954, when Ruska himself came over to commission it, accompanied by his then assistant Dr. S. Leisegang. It immediately performed well, and continued to do so for many years.

This is not the place to enter into discussion of the relative parts played by von Borries and E. Ruska in the early development of the electron microscope. The story has been recounted in great detail by the latter (Ruska, 1980) and told again in a court action which refuted claims made by von Borries's widow long after his untimely death, an action made doubly painful in that she was Ruska's own sister. There is no doubt that the development of the magnetic lens was entirely the work of E. Ruska, guided no doubt by M. Knoll, his supervisor. The idea of a microscope was in the air at the time, the original patent being in the name of another man (Rüdenberg, 1932). To inquire into whether and how much von Borries contributed to the *idea* is an impossible task at this distance in time; his contribution to the construction and operation of the instrument is well documented. Suffice it to say that no undue claims are made in his first account of the work done (von Borries, 1949), in his later survey of the subject at the London Conference (von Borries, 1956), nor in the later short histories by his contemporaries (Gabor, 1957; Ruska, 1957); von Borries died in 1956.

We in England generally, and in Cambridge in particular, benefited greatly from the contacts built up with von Borries. Unlike his close collaborator, Ernst Ruska, he was a man of the world, outgoing, keen on international connections, and articulate in English. The foundation of the German EM Society was largely his doing: "Die Deutsche Gesellschaft für Elektronenmikroskopie ist ubrigens nicht in Mosbach gegründet, sondem bereits am 16 Febr. hier in Düsseldorf und zwar auf meine Initiative hin bei einem Zusammentreffen, des anlässlich der Einweihung meines Institutes Stattfand" [letter to me dated August 26, 1949]. His "Rheinisch-Westfälisches Institut fur Übermikroskopie" became the leading center for electron microscopy in Germany and probably in Europe. He took an active part in building up the International Federation, after abortive attempts with the aid of Ralph Wyckoff to set up a Joint Commission on the subject under the auspices of the International Council of Scientific Unions were frustrated by the "logic" of its French secretary (Professor Fleury) in interpreting the statutes of that august organization. He then became the first president of the IFEMS, with myself as secretary and a committee comprising T. F. Anderson, J. B. LePoole, M. Locquin, F. Sjöstrand, and M. Terada (Fig. 6). It was a personal blow to us all and a great loss to electron microscopy when he died of a brain tumor at the early age of 51 shortly before the First European EM Conference, held in Stockholm in September, 1956, which he had been so eager to promote.

2.2.3 W. Glaser

If von Borries was the most active proponent of the experimental side of electron microscopy in Europe, Walter Glaser was his theoretical counterpart, in his untimely death and in his influence. Glaser's were the earliest papers dealing with geometrical electron optics, with the theory of the electron microscope, and especially with its aberrations (Glaser, 1933a, 1933b). These were soon followed by a wave treatment of the subject, culminating in his monumental "Grundlagen der Elektronenoptik" (Springer, Vienna, 1952), which became the bible of the subject for theoreticians (Glaser, 1952).

Originally from Prague, he was, after the war, at the Technical University of Vienna. Having myself married a refugee from Hitler's Austria, I was anxious to take up his approach in the hope of also getting in touch with her parents in that city. He wrote me in November, 1947, about some

Figure 6 Committee of the International Federation at the Berlin Congress, 1958. From right: M. Terada (Japan); M. Locquin (France); E. Ruska (President); J. B. Le Poole (Netherlands); T. F. Anderson (United States); F. Sjöstrand (Sweden); V. E. Cosslett (secretary).

points in a paper of mine which had just been published (Cosslett, 1946a) and begging a copy of my "Introduction to Electron Optics" (Cosslett, 1946b), which he had begun to use in a seminar. Thus started an active correspondence, which later expanded as first Peter Sturrock and then Jack Burfoot took up the question of the perfect electron lens. But, in spite of attempts via the British Council's representative in Vienna, I failed to meet Glaser in person until 1950. The city was then as divided as Berlin still is, and it was not until the settlement of the "Staatsvertrag" in 1954 that movement was at all easy.

By that time I had twice met Walter Glaser, first at the Paris International Congress in 1950 and then at the Semicentennial meeting of the National Bureau of Standards in Washington in the following year. From 1954 to 1956 he was in New York as chief physicist of the Farrand Optical Company, who were at that time constructing an electrostatic electron microscope. He wrote apologizing for not being able to attend the London Conference in 1954, and I do not recall meeting him at the following EM Congress in Berlin in 1958. In 1960 he died at the early age of 54.

We had corresponded regularly and I translated a paper of his (Glaser, 1951) on the refractive index in electron optics, on which he had taken exception to a derivation by Ehrenberg and Siday. He was meticulously correct in dealing with scientific matters, as in personal behavior, but a difficult man to get to know. He was a dedicated scientist, a mathematical physicist of outstanding ability in the continental mold. I retain the impression of a modest and dignified personality, reserved to the point of aloofness. Ernst Ruska, who had worked with him before the war, wrote in *Optik* "W. Glaser war ein ausgeprägter Individualist und Humanist; Intoleranz war ihm fremd. Seine allem Dogmatismus abholde, kritische Denkweise, die durch grosse Vertrautheit mit der Physikgeschichte unterbaut war, suchte er auch auf seiner Schüler zu übertragen."

When the Deutsche Gesellschaft für Elektronenmikroskopie decided in 1975 to honor both him and Bodo von Borries with honorary membership, posthumously, it was a pleasure as well as a duty for me to present the encomium at its meeting in Berlin. Long before that time his magnum opus, "Grundlagen der Elektronenoptik," had become recognized as the classic treatment of the subject. I had been proud and honored when he sent me an autographed copy, which I still have on my bookshelves.

2.2.4 Otto Scherzer

The other leading German theoretician of electron optics, Otto Scherzer, I did not get to know until later. He went to the United States immediately after the war, and for several years was at a U.S. Army establishment. His early work with Brüche on electron optics had been one of the first formulations of the subject in mathematical terms (Brüche & Scherzer, 1934), but he was best known for his theoretical treatment of lens aberrations (Scherzer, 1936). In this and in a later paper (Scherzer, 1947) he showed in particular that spherical aberration could not be corrected in rotationally symmetrical electron lenses. Rather, as Glaser developed tenaciously the wave theory, Scherzer devoted the rest of his life to the practical task of finding means of getting around this prohibition and creating a corrected electron lens. He favored the method of introducing nonrotationally symmetrical elements, in the form of quadrupoles and octupoles. At the Paris Congress in 1950 he described the first practical attempts to do so, and by the time he retired 30-odd years later had at last demonstrated a degree of controlled correction.

Unlike Glaser, Otto Scherzer was a companionable man, fond of a joke and always accessible (Fig. 7). I have two abiding memories of later visits to his laboratory in Darmstadt, apart from his ready hospitality. First, I was struck by the small size of his workshop, and indeed of his department as a whole. He was not an organization man, and only late in life did he concern himself about the future scope of his chair in the Technical

Figure 7 O. Scherzer and J. B. Le Poole, with V. E. Cosslett (and Taffy), Cambridge, 1954.

University there. He complained that doctoral candidates did not apply themselves as they did in his young days—if only they would do so, instead of going off skiing, "we should have solved this problem of lens correction in short order." Second, he was famous for offhandedness in driving a car. I feared (and that is the right word) that he was normally controlling it with at most one hand, while he held forth on the beauties of the surrounding Weinviertel, or wherever we were.

2.2.5 H. Boersch

Standing apart from these contemporaries in his approach to the subject, though similar in some ways, was Hans Boersch. Rather like myself, he came to electron microscopy from electron diffraction, and indeed from experiments on diffraction in gases. I first came upon him when Bragg wrote about my recent appointment to the Cavendish in June, 1946, adding "I enclose a letter which has just come from Börsch. What about this? Is he so good that we ought to make some effort to get him over to this country?" This was a letter in which Boersch, writing from the University of Innsbruck, inquired about the possibility "to emigrate to England" and enclosed a list of his publications and his curriculum vitae. In support of his case he said that "my colleagues v. Ardenne, v. Borries, E. Ruska live now in Russia (Krim), so that probably I am the unique scientist in the western part of Europe, who took experimentally and theoretically part in the development of the electron-supermicroscope."

Apart from this misleading information about von Borries and Ruska (only von Ardenne had gone to Russia, and to Georgia not to the Crimea),

I learnt that Boersch had been studying the molecular structure of carbon monoxide and other gases in Mayer's laboratory in Vienna soon after I had been doing the same for carbon tetrachloride, in London and Bristol (Cosslett, 1934). He had in fact been working under Hans Mark in the physics department of Vienna University, while my future wife was studying physical chemistry next door (he was 3 years older). Not only that, but he had written to Bragg (he said) in 1939 "about the X-ray microscope," which had engaged my own attention at around the same time.

It seemed that we were on much the same wavelength, as was confirmed when he came to Cambridge under the displaced scientists scheme in April, 1948. My report on his visit states that he was interviewed by Arthur Brown and Peter Sturrock (research fellow) as well as myself, and "joined in a general discussion with Dr. J. B. Le Poole and Dr. A. M. Nieuwenhuys of the Netherlands Institute of Electron Microscopy who happened to be visiting the laboratory." Apart from his work in Vienna with a Siemens EM and an electrostatic microscope of his own construction, he told us of some preliminary research on an ion microscope (Boersch, 1942). He had been making theoretical studies of the interaction of electrons with matter and especially of Fresnel diffraction. An electrostatic EM from the AEG factory in Mosbach was about to be delivered to Vienna, "the first instrument commercially produced by AEG; previously they were prevented from marketing them owing to patent difficulties with Siemens." He himself was expecting to take up "a new post shortly in Brunswick in the newly reconstructed Reichsanstalt." This he did, and our correspondence continued from there until he moved back to West Berlin as professor in the Technical University at Charlottenberg in 1959.

Although it may appear from what has been quoted above that a certain degree of self-advertisement formed part of his makeup, this was far from the truth. Hans Boersch was a modest man, upright in every sense and if anything slightly reserved. He had had long experience (1935–1940) with the AEG research institute under Brüche, but had preferred to return to Vienna rather than to move with it to Mosbach. He was more interested in the basic physical problems of electron microscopy than in those of instrument design and production. His background in diffraction led him to produce the first ideas of selected-area diffraction in 1936 and, in a less clear way than Gabor, of holography (or two-wavelength microscopy) already in 1948 (Boersch, 1948). Being an individualist, never part of a team, his contributions to electron imaging took a long time to be appreciated.

My most striking recollection comes from much later. In 1971 I was invited by him to speak on the recent progress of our work at his laboratory in Charlottenberg. This was the period of widespread student unrest which had begun in France in 1968 and by that time was affecting strongly the German universities. I had come across it at the Free University of Berlin, but had not realized how deeply it penetrated the scientific–technical ranks until alerted by Boersch. He warned me that there would certainly be trouble at the meeting of the Physikalische Gesellschaft that I was due to address, because the students believed he was accepting money from American military sources in support of his research. My talk was brought forward and I was then taken off to a beer hall while the students staged their protest.

Professor Boersch did not suffer these ructions for long. He soon retired early and settled in Bavaria with his charming third wife. In England, fortunately, we were not so sorely tried.

2.2.6 Manfred von Ardenne

In 1939 I had begun to work with Bernal on the production of X rays and the application of electron optics to obtaining high intensity. At that time Birkbeck College was still situated in an old, converted warehouse, Breams Buildings, off Fetter Lane between the Strand and High Holborn. We made much use of the nearby Patent Office Library for foreign journals. In almost the last issue of *Naturwissenschaften* to reach us before the war broke out, I found a short paper by someone named von Ardenne (1939b) putting forward the idea of imaging with X rays, by projection from a point source. As it later proved it was not the first time such a proposal had been made (see Cosslett & Nixon, 1960, p. 49), but this was a definite suggestion that a small enough source could be produced by focusing electrons with a lens. The great attraction, of course, was that the specimen could remain in air, or at least partially hydrated if a transmission target could be used.

That was my first contact with the fertile mind of Baron Manfred von Ardenne, though it was nearly 20 years before I was to meet him in the flesh. It set me off on the search for an X-ray microscope, which led eventually to the microprobe analyzer. In the meantime all that we heard of the man was that he had gone off to Russia, while others had moved west, either to the Allied-occupied part of Germany or to the United States. The best-known case was that of Werner van Braun and his rocket team, who had set up shop in Texas and laid the basis for the later United States flight to the moon. It seems that von Ardenne had got into nuclear physics at his Berlin-Lichterfelde laboratory and so had been of interest to the Soviets, who had entered Berlin from that direction. The story is told in his autobiography, "Memoiren" (von Ardenne, 1972), though a pinch of salt should accompany it here and there (Ruska's part in the invention and development of the EM gets rather less than one page, for instance). The most remarkable thing is not that he and his family, with their furniture and heirlooms, settled on the Black Sea coast at Sucumi, but that all returned safely to Dresden, where he lived in a villa swapped for his Berlin property with the East German authorities. On later visiting him at Weisser Hirsch, the first thing that I saw was a polished suit of medieval armor, the second an impressive oil painting of Einstein, and the third a photograph of von Ardenne shaking hands with Walther Ulbricht, the head of state.

To resume this account from the start, what happened was that out of the blue I had a letter from von Ardenne in May, 1955, forwarded from Oxford, which was the latest address of mine in his files. It announced that he was back in Germany "after 10 years absence," proposed an exchange of reprints, and said that he intended to include some of my work in a forthcoming publication. This work followed shortly after—it was a two-volume compendium of theoretical and practical information concerning electron physics (von Ardenne, 1956). Begun as a benchtop collection of useful recipes, formulas, and references in his earlier days in Germany, he had put it into more logical order and amplified it during his years in the USSR, according to his Preface. The result comprised over 1350 pages, closely printed and with many footnotes. Nothing of the sort has come to my notice before or since. Slightly with tongue in cheek, I said as much in a review for *Nature:* "It may be thought unusual to say of a collection of data that it is hard to put down, that it is fascinating, intriguing and (occasionally) exhausting and infuriating. But this is an unusual book, and its author no ordinary man.... Only a German could have conceived of such a project, only a man of very wide knowledge and experience could have planned and carried it through, only a von Ardenne could have done so in the Soviet Union."

Writing to thank me for so warmly welcoming his work early in February, 1958, Professor von Ardenne expressed the hope that we might meet at the EM Congress in Berlin that September: "Da wird es mich besonders freuen, endlich Sie persönlich kennen zu lernen...." We did so, with some surprise on my part to encounter a jovial figure in what seemed to be a good English tweed suit and hat. We got on well, both scientifically and otherwise. An intermittent exchange of letters and reprints followed, and several visits to Dresden on my part. Some years later he visited our laboratory in Cambridge, along with his second wife, who was a champion tennis player. As he will be telling his own story in these pages, and in any case our acquaintance began in recent times, not in the early days of our subject, I refrain from saying more here, except to set down my pleasure at having met one of electron microscopy's earliest workers and longest survivors.

2.3 Other European colleagues

After the end of World War II we in England soon made connections with other colleagues who had been developing electron microscopy in France, Holland, and elsewhere in Europe. These contacts were founded on information gained from publications, from visiting scientists such as mentioned earlier, and from the British Council. They were frequently made at conferences and summer schools and, later, on a visit which I paid to a number of continental laboratories in 1948 under the auspices of the British Iron and Steel Research Association (see Cosslett, 1981). Earlier attempts to visit Boersch in the French-occupied zone of Germany and Glaser in Vienna were abortive, mainly owing to bureaucratic difficulties over postwar travel. Here I shall mention only the beginning of a number of fruitful scientific friendships; in each instance our opposite number is still alive and contributing to this volume, except for those in Czechoslovakia (Drahos, Rozsival, Wolf).

2.3.1 France

The development of electron microscopy had been carried on in Paris by Grivet and his team and in Toulouse by Dupouy. The former had developed a two-stage electrostatic microscope (Grivet, 1942), which later was the basis of the CSF commercial EM and of the first X-ray microprobe analyzer (Castaing, 1951). Although Grivet was at the Oxford EM meeting in 1946 to describe the microscope, I did not get to know him well until the International Congress in Paris in 1950. He showed me then not only his laboratory at the Ecole Normale but also, further afield, the invasion beaches of Normandy, introducing me on the way to the local cider and to the stronger drink, Calvados. Apart from an exchange of technical information, this initial contact led to reciprocal visits of personnel, to the English edition of Grivet's "Optique Electronique," and not least to the marriage of our editor.

The construction of a magnetic EM at Toulouse was described at the same Oxford conference of the EM Group by Gaston Dupouy. A short, dynamic man, he was insistent about speaking and writing in his mother

Figure 8 Toulouse, 1962. From right: E. Faure-Fremiet; unknown lady; G. Dupouy; Mme. Faure-Fremiet; V. E. Cosslett.

tongue, although his command of English was perfectly adequate. Out of those early experiments grew his first high-voltage electron microscope (Dupouy et al., 1960), which paved the way for our Cambridge HVEM work. In preparing the application to the Paul Instrument Fund for the latter, I visited the Toulouse laboratory in 1962. There I was most hospitably received and rapidly cemented a rapport with Dupouy, aided by a common interest in rugby football and MG motor cars. Among other delicacies I was introduced to quail and to Armagnac, the local version of brandy (Fig. 8).

Dupouy has always had a deep interest in the history of experimental science. Subsequent to his visit to the Cavendish Laboratory in 1947, he requested photographs of some of its historical apparatus and we sent him prints of the instruments built by Aston, Rutherford, J. J. Thomson, and Wilson. He was already planning at that time to build a 250-kV electron microscope, so he wrote. The target ultimately escalated to 3 MV, as we now know.

2.3.2 Holland

At the Manchester meeting of the EM Group in January, 1947, a young man stood up to ask a question. His English was almost perfect, but not quite. Over lunch I found that he was from Delft, and had learned his English by listening to the BBC broadcasts clandestinely during the war. In conditions of near secrecy he had been building a revolutionary electron microscope. Jan Le Poole tells in this volume in his own words how it was concealed from the Germans. Here, I only mention his warm personality and his integrity. Thus began a friendship which has long continued, with frequent visits in each direction (Fig. 7). In case he does not mention it in his own contribution to this volume, it is right to pay tribute here to his patient and loving support of his first wife through all her affliction. I shall long remember his care of her during the Grenoble Congress in 1972, when she was so enfeebled that he had to turn for her the pages of the book she was reading.

It was, however, at the Oxford meeting in September, 1946, that I had first come in contact with Jan Le Poole's handiwork, in the form of the commercial EM brought out by the Philips Company (van Dorsten et al., 1948). With some effort, one of the first of the production batch had been flown over to Oxford, in the care of Adrian van Dorsten, but in spite of many attempts it refused to produce an image. It was only after its return to Eindhoven that the cause was discovered—a misplaced aperture.

This mishap, however vexatious to our Dutch colleagues, was the start of a long acquaintance with Dr. van Dorsten and others at the Philips laboratories. They were, and still are, in the forefront of EM development. Not only the finished version of Le Poole's microscope, with its innovative diffraction lens and inclined column, but also the first high-voltage instrument (above 200 kV) was due to them (van Dorsten et al., 1947). For many years they supported our Cambridge work on X-ray microscopy.

2.3.3 Czechoslovakia

In 1947 one of the technical groups then swanning around Europe on other matters found that there was an EM of unusual design in the University of Prague. Dr. Charles Smith, a wartime radar expert who was converting to studies of radiation damage in biology, brought back some micrographs taken with it. This news was followed by a letter from the head of the Physics Department, Professor Petrzilka, giving the name and address of the scientist in charge of the EM (Rozsival, 1947). In this way we made contact with Dr. Rozsival and other Czech electron microscopists.

It turned out that his main interest was in oxide replicas of metals, with which Jack Nutting was occupied in our laboratory, while one of his colleagues (Professor Jan Wolf) was an anatomist developing a method for replicating wet surfaces (Wolf, 1944). Rozsival was using one of the Swiss EMs made by Trüb-Taüber, of which more later, and Wolf was using an RCA microscope. The latter visited us in Cambridge after the Stockholm

Conference in 1956, and in a subsequent visit to Prague with my family we made the acquaintance of his (much younger) wife, Darina. A motor trip through Czechoslovakia followed, in my first MG, taking in Marianske Lazne (Marienbad), where we were shown one of Edward VII's lodges, near the still-operating golf course. The culminating event was a symposium at the castle of Smolenice which belonged to the Slovak Academy of Sciences. It was made memorable by a barbecue in the woods, greatly enlivened by Professor Spivak from Moscow University with Russian songs and dances.

On the way to Austria we visited the EM factory at Brno, where we met Dr. Drahos, the designer of the first Czech microscope (Delong et al., 1961). He was a lively and ingenious man, as much a theoretician as a practical man, who remained a friend until his untimely death in 1972. Part of the story of the development of that microscope and related apparatus is told below by his close collaborator, Dr. Delong.

2.3.4 Switzerland

The "unusual" electron microscope mentioned earlier was indeed unusual in design and construction. It was an unorthodox assemblage of magnetic and electrostatic lenses, produced in Zurich by the firm of Trüb-Täuber and designed by a Swiss–Italian engineer named Induni. His intention was to avoid the effects of chromatic aberration by using an electrostatic objective, with a magnetic condenser and projector for flexibility of operation (Induni, 1946). Like some electron diffraction instruments it had a cold cathode in the form of an aluminum rod (Bas, 1955). A high-speed rotary molecular pump of the Holweck type ensured an excellent vacuum while minimizing specimen contamination.

After learning of this unique instrument, I included a visit to Zurich in my forthcoming tour of European laboratories in 1948. I was cordially received by Dr. Induni, who proved to be an electrical engineer with little previous experience in microscopy. The microscope was very solidly built, rather in the old-fashioned German style, but worked well up to 15,000X. Although only a limited number were produced, some of its features have been retained in apparatus produced by the Liechtenstein firm of Balzers.

On the same trip I visited the designer of the rod cathode, and, later, of the accelerator for the 1.5-MV set for the Toulouse HVEM. Dr. Bas was working on electron-optical problems in the Technical University, the ETH. Of Turkish origin, he was very much the brisk, young scientist, involved in both theoretical and practical problems. He had developed the rod cathode originally for the Swiss television system, the Eidophor. It produced a very stable electron beam and, when it began to be pitted, could readily be rotated to provide a new point of emission. Later, when we spent a half-sabbatical in the ETH (1955), we saw much more of him and his work.

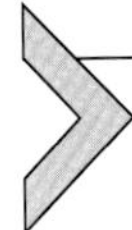

3. Aid from America

3.1 RCA

The maintenance of the six RCA microscopes that came to Britain during the war was initially undertaken on an ad hoc basis. Operating and servicing information (and personnel) were exchanged between the several laboratories concerned, at first in an informal way and then by means of the regular meetings organized by Sir Charles Darwin, mentioned earlier. A great deal of progress was made, as shown by the proceedings of the EM Group, but after the end of the war it was soon evident that we needed the help and experience that were available in the United States. There work had proceeded on both the construction and the applications of the EM, as evidenced by many publications and especially by the comprehensive treatment (765 pages) that came to be known as "We, the People" (Zworykin et al., 1945).

As secretary of the EM Group it fell to me to make the arrangements. I had already been in touch with Jim Hillier of RCA before the end of the war, requesting some micrographs for a forthcoming book of my own (Cosslett, 1946b). Through him I was put in touch with RCA International Division in New York, and by them with their London agents, RCA Photophone, as well as with the development laboratory in Camden, New Jersey. A detailed and active correspondence followed with Perry Smith of the latter office and with Holmes Halma and Dr. L. Garten of the New York office. In London Mr. S. Johnson was also involved, on behalf of the Electron Microscope Committee of the Department of Scientific and Industrial Research. The purpose was to arrange terms for Mr. Halma to come to England to service the 12 (as the number now was) RCA instruments. Three of them had been inspected by Mr. Garten on a visit in early 1947, who said in his letter to Mr. Johnson "It was a pleasant surprise for me to learn how well everyone managed despite numerous difficulties.... [Nevertheless] the people of NPL explained to me that during the first few years all of the instruments were carefully maintained, but have been allowed to depreciate gradually since."

Negotiations for a general agreement failed, however, despite RCA reducing their charges from $3800 to $1000 per month for Mr. Halma's services, in the interests of "our fullest cooperation" and "to have the instruments in first class condition." It was left to individual laboratories to make their own arrangements with Mr. Halma for his visit to them: $1000 per month seemed to be too large a sum for the Ministry of Supply to meet. So our EMB at the Cavendish came to be cared for by him, as part of his European responsibilities, from October, 1947. An energetic, tubby man, he was always optimistic and cheerful. I imagine that he was not daunted by instrument breakdowns because he had seen it all in his travels, whereas we were still rather in awe of the instrument and reluctant to make any major modifications on our own initiative, still less to bore a hole in the column. That revolutionary step had to await the arrival of Peter Duncomb, who converted the EMB into our first microprobe analyzer a few years later. By that time we had gone over to AEI electron microscopes and Holmes Halma had returned to the United States. It was a pleasure to be welcomed by him and his charming wife Lily when many years later I visited Fritjiof Sjöstrand's laboratory in Los Angeles, where he had become Fritjiof's personal assistant.

3.2 Jim Hillier

In the course of these interchanges, I had come into contact (by correspondence) with Jim Hillier, who had been responsible in part for the prototype of the EMB when he was at Toronto University and then continued to develop it at the RCA Camden laboratories. In addition to sending technical details of new improvements, and later of the EMU model, he told me of the foundation of the Electron Microscope Society of America (EMSA), of which he was a founder member. I joined a year or so later (1946), when it became possible to transfer the membership fee of $5. But in spite of various plans we did not meet until he came to Europe in 1949 to attend the Delft International Conference. The following year he came to visit the Cavendish, during which he gave us a colloquium, the title of which (surprisingly enough) was "The Cutting of Thin Sections for the Electron Microscope." Subsequently, on several occasions, I visited him in Princeton and made the acquaintance of many people with whom I had been in correspondence. It was a pleasure to meet him again at the Toronto EM Congress in 1978, where he received an honorary degree from the university, along with two other pioneers of electron microscopy in America, Cecil Hall and Alfred Prebus.

3.3 Bill Marton

I had already been acquainted with the pioneer work of Dr. Marton before the war, when he was still in Brussels. Of Hungarian origin, of which his speech (and especially that of his wife) never lost a strong trace, he emigrated to the United States in 1938. Initially he was at the RCA laboratories in Princeton, where he built the Type A microscope. RCA preferred the Type B model for production, however, so just before the United States joined the war he moved to Stanford University. There he constructed a three-stage microscope, briefly described (along with his earlier work) in his autobiography (Marton, 1968). In it he mentions incidentally "The need was felt for a bibliography of electron microscopy. Mrs. Marton undertook the task in collaboration with Samuel Sass." That is the topic on which I first made contact with him.

In England we had independently started, some years later, a collection of reprints and an abstracting system for the EM Group, for which I was responsible (Cosslett, 1950b). Partly on this account Marton visited us in Cambridge in 1948, along with his wife, Claire. By that time he was back on the East Coast, as head of the Electron Physics Section of the National Bureau of Standards in Washington. There I met him again in 1951, when he invited me to attend a symposium in connection with the Bureau's Semicentennial Celebrations. That occasion introduced me to many American electron microscopists, with some of whom I had already corresponded, and also to the Marton's spacious home, open-plan in the California style. We became good friends, as did our wives, who renewed the historical Austrian–Hungarian relationship. Our common interests helped to cement this relationship and in particular eased relations when I began (with Robert Barer) to edit a series of *Advances in Optical and Electron Microscopy* a few years later. We consulted each other frequently about the respective coverage of this and the series which he had already started in 1948, in which this article is appearing.

An account of his earlier work as well as of his later years at the Bureau of Standards is given in this volume by Dr. Süsskind, and he is referred to in the article by Dr. Hall, so it is not necessary to mention here his many other contributions to our subject. In personality he was a jovial, welcoming host, always au fait with his subject and full of stories about it. His sociable nature was reflected in his name: universally known as Bill, when hardly anyone knew that it was in fact Ladislaus.

4. Final word

I am very conscious of failing to deal with many other pioneers in electron microscopy, especially Mike Haine and Gerhard Liebmann of AEI; Gordon Drummond of the Shirley Institute, Manchester (and Sydney University, Australia); and W. T. Astbury of Leeds. My apologies to those of them still with us; some inadequate reference to their early work is to be found in the cooperative bibliography provided by Cosslett (1950b).

5. Afterword by Peter Hawkes

Ellis Cosslett was born on 16 June 1908 and died in Cambridge on 21 November 1990. Further information can be found in the entry by Tom Mulvey in the *Biographical Memoirs of Fellows of the Royal Society* (**40**, 1994, 61–84) and in a long-two-part article by Cosslett in *Contemporary Physics* **22** (1981) 3–36 and 147–182. Obituaries appeared in the *Journal of Microscopy* (**161**, 1991, ii) and *Ultramicroscopy* (**35**, 1991, 169–170).

References

Allibone, T. E. (1980). *Biographical Memoirs of Fellows of the Royal Society*, *26*, 107.

Bas, E. B. (1955). *Zeitschrift für Angewandte Physik*, 7, 337.

Bernal, J. D. (1939). *The social function of science*. London: Routledge & Kegan Paul.

Bernal, J. D. (1954). *Science in history*. London: Watts.

Blackett, P. M. S. (1948). *Military and political consequences of atomic energy*. London: Turnstile Press.

Boersch, H. (1942). *Naturwissenschaften*, *30*, 711.

Boersch, H. (1948). *Monatshefte für Chemie*, *78*, 163.

Bragg, W. L. (1939). *Nature (London)*, *143*, 678.

Brüche, E., & Scherzer, O. (1934). *Geometrische Elektronenoptik*. Berlin and New York: Springer-Verlag.

Castaing, R. (1951). Ph.D. Thesis, University of Paris.

Cosslett, V. E. (1934). *Transactions of the Faraday Society*, *30*, 981.

Cosslett, V. E. (1937). *Proceedings of the Physical Society of London*, *49*, 121.

Cosslett, V. E. (1940a). *Journal of Scientific Instrumests*, *17*, 259.

Cosslett, V. E. (1940b). *Proceedings of the Physical Society of London*, *52*, 511.

Cosslett, V. E. (1943). *Discovery*, *4*, 311.

Cosslett, V. E. (1946a). *Proceedings of the Physical Society of London*, *58*, 443.

Cosslett, V. E. (1946b). *Introduction to electron optics*. London and New York: Oxford Univ. Press (Clarendon).

Cosslett, V. E. (1947). *Journal of Scientific Instrumests*, *24*, 113.

Cosslett, V. E. (1950a). *Discovery*, *11*, 245.

Cosslett, V. E. (Ed.). (1950b). *Bibliography of electron microscopy*. London: Arnold.

Cosslett, V. E. (1966). *Modern microscopy*. London: Bell.

Cosslett, V. E. (1971). *Physics Bulletin*, *22*, 339.

Cosslett, V. E. (1981). *Contemporary Physics*, *22*(3), 147.

Cosslett, V. E., & Nixon, W. C. (1960). *X-ray Microscopy*. London and New York: Cambridge Univ. Press.
Darwin, C. G. (1914). *Philosophical Magazine [6]*, *27*, 315, 675.
Delong, A., Drahos, V., & Zobac, L. (1961). *Delft, 1960* Vol. I, p. 89.
Dupouy, G., Durrieu, L., & Perrier, F. (1960). *Comptes Rendus Hebdomadaires des Séances de l'Académie des Sciences*, *251*, 2836.
Farrand, C. L. (1946). Report No. 765, Washington, D.C.: Field Inf. Agency (FIAT).
Gabor, D. (1945). *The electron microscope*. London: Hulton Press.
Gabor, D. (1948). *Nature (London)*, *161*, 777.
Gabor, D. (1949). *Proceedings of the Royal Society of London. Series A*, *197*, 454.
Gabor, D. (1957). *Elektrotechnische Zeitschrift. Ausgabe A*, *78*, 522.
Gerchberg, R. W., & Saxton, W. O. (1971). *Optik*, *34*, 275.
Glaser, W. (1933a). *Zeitschrift für Physik*, *80*, 451.
Glaser, W. (1933b). *Zeitschrift für Physik*, *83*, 104.
Glaser, W. (1951). *Proceedings of the Physical Society, London, Series B*, *64*, 114.
Glaser, W. (1952). *Grundlagen der Elektronenoptik*. Berlin and New York: Springer-Verlag.
Grivet, P. (1942). *Revue Générale d'Électronique*, *51*, 473.
Haine, M. E. (1947). *Journal of the Institution of Electrical Engineers*, *94*, 447.
Hansell, C. W., Banca, M. C., & Barnes, R. B. (1946). FIAT Rep. No. 769, Washington, D.C.: Joint Intelligence Objectives Agency.
Hawkes, P. W. (1966). *Quadrupole optics*. Berlin and New York: Springer-Verlag.
Induni, G. (1946). *Helvetica Physica Acta*, *19*, 231.
Knoll, M., & Ruska, E. (1932a). *Annalen der Physik (Leipzig) [5]*, *12*, 607.
Knoll, M., & Ruska, E. (1932b). *Zeitschrift für Physik*, *78*, 318.
Marton, L. (1968). *Early history of the electron microscope*. San Francisco, California: San Francisco Press.
Moon, P. B. (1977). *Biographical Memoirs of Fellows of the Royal Society*, *23*, 529.
Phillips, D. C. (1979). *Biographical Memoirs of Fellows of the Royal Society*, *25*, 75.
Rozsival, M. (1947). *Fysika v Technice*, *2*, 129.
Rüdenberg, R. (1932). *Naturwissenschaften*, *20*, 522.
Ruska, E. (1934). *Zeitschrift für Physik*, *89*, 90.
Ruska, E. (1957). *Elektrotechnische Zeitschrift. Ausgabe A*, *78*, 531.
Ruska, E. (1980). *The early development of electron lenses and electron microscopy*. Stuttgart: Hirzel Verlag.
Sayer, L. J. (1948). *British intelligence objectives subcommittee*. Rep. No. 1671, London.
Scherzer, O. (1936). *Zeitschrift für Physik*, *101*, 593.
Scherzer, O. (1947). *Optik*, *2*, 114.
Sjöstrand, F. (1943). *Arkiv för Zoologi*, *35A*, 1.
Vaidya, N. C. (1972). *Proceedings of the IEEE*, *60*, 245.
van Dorsten, A. C., Oosterkamp, W. J., & Le Poole, J. B. (1947). *Philips Technical Review*, *9*, 193.
van Dorsten, A. C., Le Poole, J. B., & Verhoeff, A. (1948). *Journal of Applied Physics*, *19*, 1190.
von Ardenne, M. (1939a). *Zeitschrift für Wissenschaftliche Mikroskopie und Mikroskopische Technik*, *56*, 8.
von Ardenne, M. (1939b). *Naturwissenschaften*, *27*, 485.
von Ardenne, M. (1940). *Zeitschrift für Wissenschaftliche Mikroskopie und Mikroskopische Technik*, *57*, 291.
von Ardenne, M. (1956). *Tabellen der Elektronen Physik, Ionenphysik und Übermikroskopie*. Berlin: Dtsch. Verlag Wiss.
von Ardenne, M. (1972). *Memoiren*. Munich: Kindler Verlag.
von Borries, B. (1949). *Die Übermikroskopie*. Berlin: Saenger-Verlag.
von Borries, B. (1956). *London, 1954*, pp. 4–25.
Wolf, J. (1944). *Zeitschrift für Wissenschaftliche Mikroskopie und Mikroskopische Technik*, *59*, 240.
Zworykin, V. K., Morton, G. A., Ramberg, E. G., Hillier, J., & Vance, A. W. (1945). *Electron optics and the electron microscope*. New York: Wiley.

CHAPTER SIX

Early history of electron microscopy in Czechoslovakia

Armin Delong[a] **and Bohumila Lencová (Afterword)**[b,*]
[a]Formerly Institute of Scientific Instruments of the Czechoslovak Academy of Sciences, Brno, Czechoslovakia
[b]Tescan Brno, Brno, Czech Republic
*Corresponding author. e-mail address: lencova@cmail.cz

Contents

1. Introduction

Czechoskovakia, with its 15 million inhabitants, ranges among the relatively small countries of Central Europe. On the average, there are some 0.36 ha. of soil, 0.32 ha. of woods, 0.15 automobiles, and not more than 0.000034 electron microscopes per citizen in the country. In other words, one electron microscope is shared by 30,000 people. Some 90% of available electron microscopes are of home production. This means that besides atomic power plants, locomotives, and automobiles, electron microscopes are also produced on a commercial scale in Czechoslovakia. In 30 years as many as 1800 electron microscopes of different types, from the two-stage transmission electron microscope of the year 1950 to the ultrahigh-vacuum scanning electron microscope using a field emission cathode of 1977, have been built. Most of them were exported to socialist countries. With its modest share, Czechoslovakia belongs to the few countries that have contributed to the development of electron microscopy. Thus, many laboratories in different fields of science and industry both at home and abroad

Advances in Imaging and Electron Physics, Volume 220
ISSN 1076-5670
https://doi.org/10.1016/bs.aiep.2021.08.006

could be equipped with commercially produced electron microscopes that opened unexpected possibilities of studying microscopic structures of an organic and inorganic nature. The author of this paper finds it very pleasant to meet unknown colleagues abroad who assure him that, thanks to the device at whose birth he stood, they entered the large community of electron microscopists, having experienced the same fascinating moments on first meeting the fantastic world of the microcosm as did, years ago, the author himself.

The beginnings of electron microscopy in Czechoslovakia date back to a more recent time than do the beginnings in Germany, Great Britain, the United States, and some other countries, and so there might seem to be nothing worth writing about. The first electron microscope was designed as late as 1949, by which year RCA had already produced some hundreds of instruments, 10 years after the successful construction of the first commercial electron microscope by Siemens. Yet the time of construction of the first electron microscope goes back to the period after World War II when the liberty of the Czech and Slovak nations was restored and when all inventive forces were roused to an unusual extent in the whole society. Nothing seemed impossible or unrealizable!

After their forced closing by Nazis in 1939, technical and other universities were opened again. Their lecture halls and laboratories attracted keen young novices of science. The Nazi occupation had had fatal consequences for the staff. Many professors did not live to see the end of the war. New ones had to be appointed to prepare new lectures, scientific and research work had to be reestablished, and new laboratories had to be built. All that took some time.

In this connection the earliest activity of the Institute of Theoretical and Experimental Electrotechnology of the Technical University in Brno should be mentioned. It was Professor Aleš Bláha who became head of this institute after World War II. Professor Bláha, with experience as lecturer at a technical university in France before the war, soon recognized that the era of electronics was approaching. Even though he was engaged in problems of high-tension lines at first, he soon occupied himself with the idea of designing a continuously evacuated oscillograph for measurements of transient voltage in networks. During the time when he was working at ŠKODA Works, several variants were built under his guidance. The last of them, dating from 1938, is shown in Fig. 1. The work that provided Professor Bláha with the experience of designing electron-optical devices was interrupted by World War II and could be continued only after the

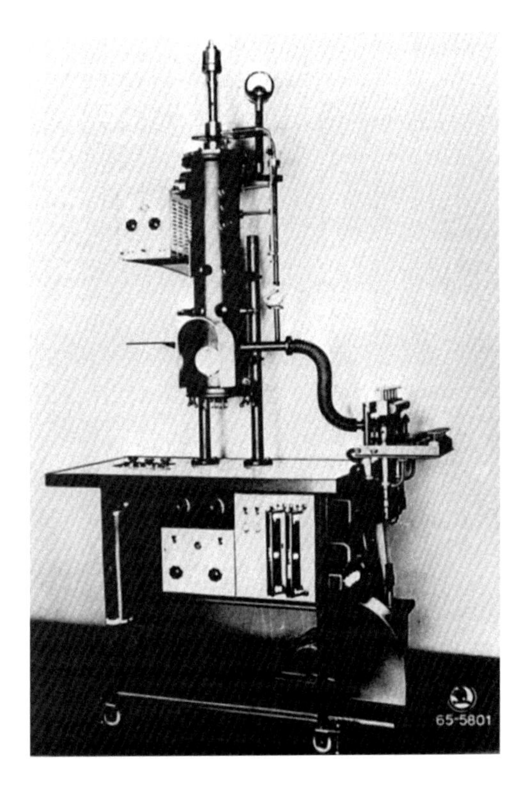

Figure 1 Continuously evacuated oscillograph designed by Professor Bláha in 1938.

war. As there were very few graduate workers, Professor Bláha chose all his collaborators from among the students. He did so in the course of examinations, because he could not know all the students attending his lectures. When I was asked if I wanted to work at the laboratories of the Institute I agreed enthusiastically. These facts may seem of ephemeral interest, but they elucidate the atmosphere of that time. We were working till late at night at the laboratory, having attended seminars and laboratory courses in the daytime hours. Some of us even graduated a year late since the research work was very time consuming. First we started designing a simple continuously evacuated oscillograph with a cold cathode. The task Professor Bláha entrusted us with was a well-thought-out one. We had to gain experience first of all. The idea of constructing the Czechoslovak electron microscope was accepted with great enthusiasm. Professor Bláha wanted us to get acquainted with a source of electrons other than Induni's cold cathode and with image-forming electron lenses other than the focusing

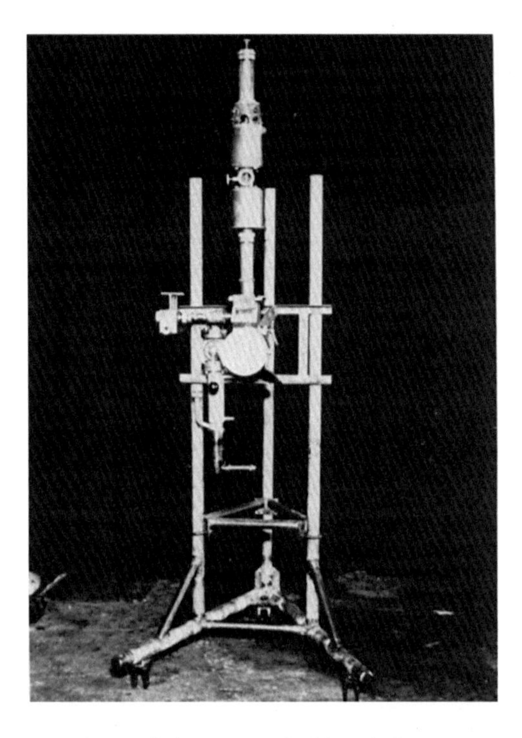

Figure 2 Two-lens experimental electron-optical bench from 1947.

ones used in cathode-ray oscillographs. In this way the well-known tripod (Fig. 2) was constructed that, in fact, enabled us to repeat Ruska's experiments. Then nothing stood in our way and we could start designing an actual microscope.

At that time several electron microscopes were already working in Czechoslovakia. The first electron microscopes imported to Czechoslovakia after the war were two RCA EMU 2 instruments, both of which were obtained within the scope of the action of the United Nation Relief and Rehabilitation Administration (UNRRA). One of them was destined for Professor Wolf at the Medical Faculty of the Charles University in Prague (Professor Wolf introduced the replica method for wet biological tissues) and one for Professor Herčík at the Medical Faculty of the Masaryk University in Brno. It was Professor Wyckoff who personally took part in the installation of this microscope. Later on, electron microscopes from the French firm CSF, the Swiss Trüb-Taüber, and the Swedish Siegbahn-Schönander were supplied, but this was at the time when the work on the first Czechoslovak electron microscope was already in progress.

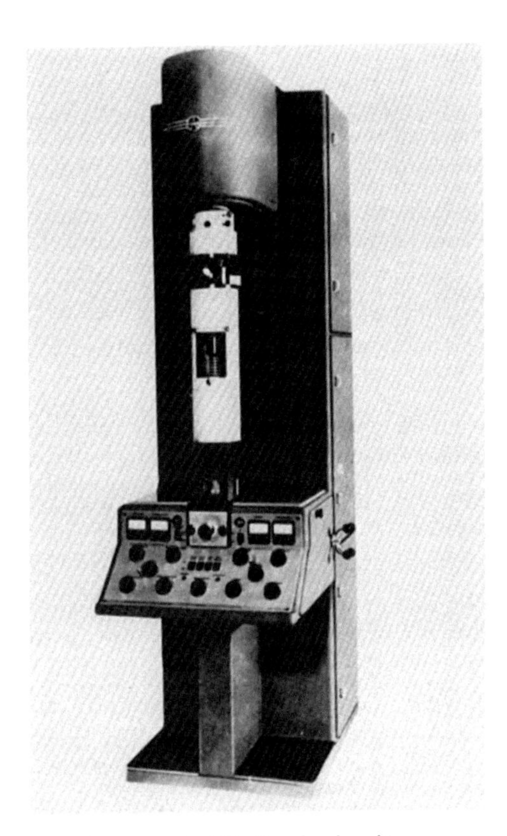

Figure 3 First electron microscope built in Czechoslovakia.

The book "Elektronen-Übermikroskopie" by Professor M. von Ardenne from which we gathered information made attractive and thrilling reading for us. It was really an adventure to overcome all problems step by step, and finally the first image appeared on the luminiscent screen of our electron microscope in 1950. This device already possessed all the requisites of the standard electron microscope: thermionic cathode, condenser lens, objective lens, and projector lens (Fig. 3). It was the RCA EMU 2A electron microscope—the most easily accessible yet the most progressive device of that time—that we used as a model to a certain extent.

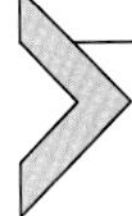

2. Construction of the first electron microscope in Czechoslovakia

There was a unique atmosphere about the construction of the first electron microscope at the Institute of Theoretical and Experimental Electrotechnology. There were several reasons for this. In 1949 the electron microscope was considered a sophisticated device with a number of technological secrets that required special materials and component parts. This seemed to pose an insoluble problem in Czechoslovakia after the war, and insoluble tasks are necessarily given to students and new graduates who do not know what is considered insoluble in general. But in view of the élan and courage of the young though inexperienced students under the sensitive leadership of Professor Bláha, who were given an unusual degree of independence in deciding both essential approaches and details, it was to be hoped that the effort would not fail.

The team which started to work on the first Czechoslovak electron microscope consisted of our members: three students [A. Delong (age 23), V. Drahoš (23), and L. Zobač (22)] and one designer [J. Speciálný (26)]. We had at our disposal the workshops of the Institute, with their skilled craftsmen from the engineering factories in Brno. They produced not only all we had designed but also helped us with their advice and experience. It seemed that the production of the electron microscope proper would not meet with any great difficulties. We succeeded in getting the necessary materials. After all, the demands were not very exacting at that time: soft iron, brass, copper, and aluminum. The polepieces of the objective lens made of Poldi-Armco steel were carefully lapped so that even the first images showed a very low astigmatism. A certain problem was presented by the luminiscent screen. It was very difficult to get a suitable luminiscent material. We therefore used the phosphor of the transmission screens of X-ray devices, though it was not entirely suitable. Later we used phosphor belonging to the accessories of the EMU 2A electron microscope. The vacuum system was based on the existing commerically available parts. The firm Fysma produced diffusion pumps of some tens of liters per second. The pump used paraffin as the pumping medium. The designer of this pump was Professor Dolejšek of the Charles University in Prague. Rotational oil pumps were produced in Czechoslovakia as well. Thus it was necessary to design the vacuum pipe lines and to add vacuum-meters, which were developed by another team of students concerned with problems of achieving and measuring high vacuum.

The construction of power supplies, especially of the source of accelerating voltage, was a more serious problem. Czechoslovakia was the traditional producer of X-ray diagnostic devices. Therefore, the first source of an accelerating voltage of 50 kV was built from materials and parts we had at our disposal. But this source was not sufficiently stable and in spite of using a relatively large filter capacitor the filtration of the dc high voltage was not sufficient either. The chromatic aberration caused by insufficient filtration disappeared after disconnecting the power supply, but only as long as the filter capacitor was able to supply the electron gun. There was an urgent need for a new high-frequency source of accelerating voltage, similar to those used in the RCA microscopes. The producer of capacitors supplied us with the necessary high-voltage capacitors, and the producer of X-ray tubes provided us with rectifier vacuum diodes. A high-frequency generator to supply the high-frequency power source was built using power tubes originally produced to meet the needs of the German Army. As a matter of fact, the German Army left behind many materials and electronics parts that could be utilized with advantage—a negligible boon that would not have been necessary had there been no war. In this way we succeeded in designing an electronic current stabilizer using the same power tubes as for the generation of accelerating voltage. In 1949 the electron microscope was ready for tests. The first pictures were made. It soon became apparent that the resolution was limited by the axial astigmatism of the objective lens. Therefore, we attempted to use Hillier's method of correction using eight screws of soft iron between the polepieces of the objective lens. This method was, however, very laborious and that is why we decided to employ a stigmator consisting of four coils below the objective lens that had to be mechanically rotated to adjust the correct orientation of the correction field. This electron microscope became a production model, 25 of which were produced to cover the needs of some research laboratories in Czechoslovakia (Delong & Drahoš, 1951).

3. Desk transmission electron microscope BS 242

The construction of the desk electron microscope belongs to the successful achievements that contributed to the development of electron microscopy in Czechoslovakia. The idea of building a desk electron microscope originated in the Chair of the Institute of Theoretical and Experimental Electrotechnology as early as 1951. The work on it started, however, 2 years later in 1953. The aim was to build an electron microscope of as simple construction as possible using available materials that do not require specialized treatment—a microscope that would not make too great demands on production. On the other hand, it was to provide the user with maximum operating possibilities. The designers of the desk microscope gained experience with building the two-stage electron microscope (based

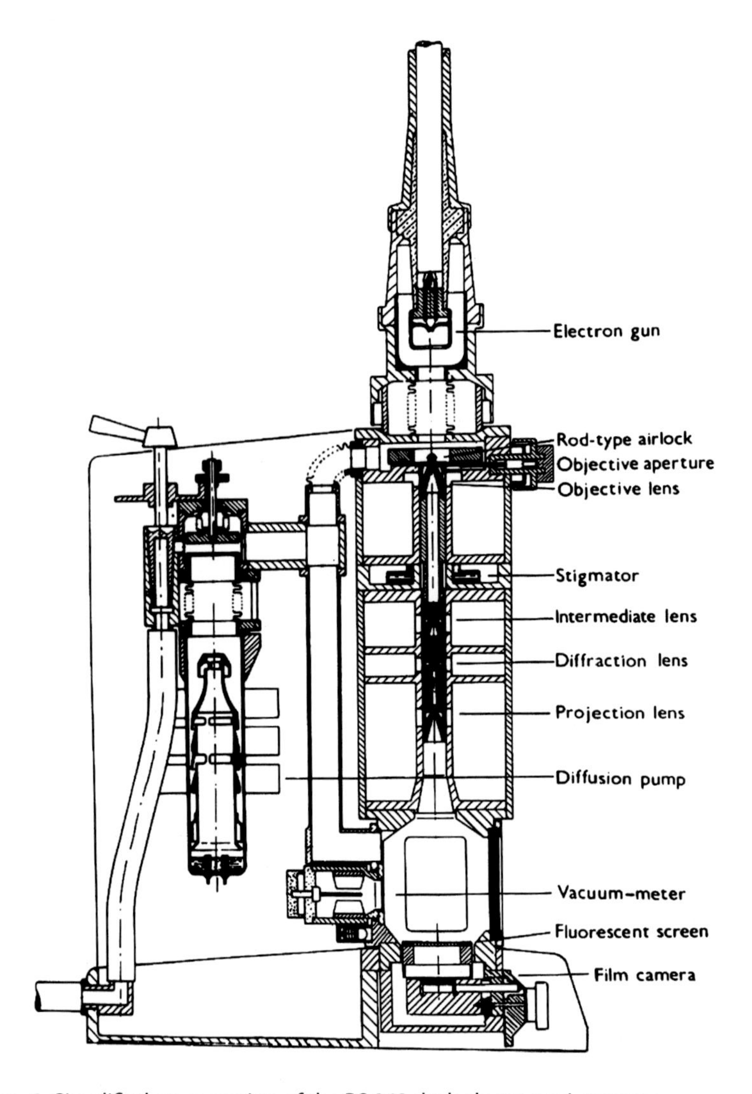

Figure 4 Simplified cross section of the BS 242 desk electron microscope.

on the RCA EMU 2A). Thus they were able to design some parts with much greater safety. A small team of young engineers and technicians completed a prototype in 1954. Its cross section is shown in Fig. 4. The general view can be seen in Fig. 5.

The cross section in Fig. 4 shows that the desk electron microscope was of a relatively ambitious arrangement as compared to the original aim. It is true that the illuminating system consisted only of an electron gun

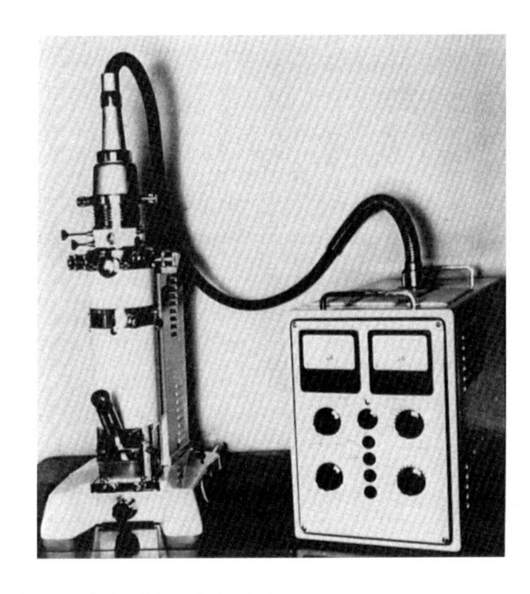

Figure 5 General view of the BS 242 desk electron microscope.

with the "telefocus" as designed by Steigerwald (1949); it thus provided a relatively narrow range of current densities and aperture angles of illumination at the object. However, the image-forming system was composed of four magnetic lenses—objective lens, intermediate lens, diffraction lens, and projection lens—and permitted not only a wide magnification range but also selected-area diffraction. The vacuum system consisted of a rotary oil pump and glass diffusion oil pump located behind the column and cooled by air convection only. A simple valve system was placed above the diffusion pump. The microscope was opened to the air only during the exchange of the photographic material (35-mm film). The exchange of the specimen was carried out through the rod-type air lock. Therefore, the objective lens was provided with a flat upper polepiece to facilitate the task of placing the object at a sufficient distance from the objective lens. The rod-type air lock consisted of two parts. The part carrying the object holder was inserted in the x-y stage, ensuring the motion of the specimen in directions perpendicular to the optic axis. The other part, when screwed together with the first one, protected the specimen when inserting the rod into vacuum. After unscrewing, it sealed the object chamber. This simple principle proved successful and was used without any modification for many years. The same principle was used for the construction of the rod-type air lock, which facilitated the automatic lowering of the object into the bore of the upper polepiece. The axial astigmatism was compensated by a stigmator which consisted of four coils located outside the vacuum. They were thus easy to turn without any vacuum feed-through. The three-lens projection system consisted of mechanically centered polepieces, to be inserted into the magnetic circuits. The electron-optical system comprised three diaphragms that could be centered from outside: a diaphragm limiting the illuminated area, the diaphragm of the objective lens, and a diaphragm for selected-area diffraction. The chambers for image observation and the film camera were made by turning and milling. The column of the microscope was mounted on a stand provided with operating elements on both sides—specimen displacement and focusing. For electron acceleration, a high-frequency power supply of 60 kV with oil insulation was designed. Because of its size it could be positioned beside the column on the desk. The accumulators originally used to supply the windings exciting the lenses were soon replaced by electronic stabilizers placed together with the rotary pump under the desk (Delong & Drahoš, 1955).

The resolution of the microscope was 25 Å, and later on even 15 Å. The operation was very simple and the handicap of using 35-mm film as photographic material was solved by many users by drying the film by evacuation.

More than 800 of the microscopes were made and exported to 20 countries. Production was stopped only after nearly 15 years. During this time no part of the microscope was substantially changed. If we compare this device with the modern transmission electron microscope there are great differences in resolving power and versatility of application, but on the other hand also in complexity, ease of operation, and price, of course. Has a device of this category quite lost its sense? What use could be made of a device whose parameters have been surpassed many times over? In connection with these questions, many others come to mind. Let us stop for a moment and consider some of the features that we readily gave up in favor of progress.

One of the properties the device was to possess was a simple construction and consequently simple operation. It stood on a desk, which facilitated its dismounting. An operator of average technical training did not need to be afraid of mounting and dismounting it and he could get well acquainted with the functions of all its parts. It is apparent that this fact is of great importance for pedagogical purposes. It was really possible to mount and dismount the device without any problems. Modern electron microscopes do not admit this possibility. Yet the device comprised all the important modes of operation of the transmission electron microscope: electron-optical imaging with a resolution that exceeded the resolution of the best light microscope by two orders of magnitude. It was easy to demonstrate the effect of the objective-lens diaphragm on the contrast and thus the principle of contrast and image formation in the bright- and dark-field modes. The diffraction lens made it possible to demonstrate electron diffraction at the crystal lattice, and using the selected-area diaphragm it was even possible to assign the diffraction image to the optical image of the part of the object investigated. It is obvious that this simple device could not approach the properties of a light microscope, which works reliably without any service for many years, though this would naturally be desirable. The present transmission electron microscopes are designed with the aim of reaching the theoretical resolving power. To maintain such a complex device at optimum performance level is not possible without service. Perhaps it would be worth considering how to exploit the present progress in technology for designing a device that would be optimized from quite different points of view, a guaranteed resolution being reliably achieved with minimum effort. It is the scanning electron microscope that comes closest to satisfying all these considerations, especially in a simplified version which permits simple preparation of the objects investigated. There are, however, objects that are not imaged with satisfactory information content when using a scanning electron microscope. The TEM field lacks a simple device—a counterpart to the simplified SEM. The question is, to what degree would it be possible to combine both devices into one without affecting the design principles: simple construction, easy operation, and low price.

There is one more interesting aspect in connection with the desk electron microscope: a simple and cheap production and low price. A sophisticated TEM is a very complicated device with regard to materials, technology, and production, if the limiting performance is to be achieved. If we accept a resolution several times lower than the attainable limit, the demands are accordingly reduced as well. The material costs and production times of the desk electron microscope were so low that the device could be sold for some thousands of dollars only. The precision machining concentrated on the production of polepieces; the production of other parts, including the simple vacuum system, was not so difficult. The construction of the power supply unit was not complicated either. We used available standard elements that complied with relatively low requirements of short-term and especially long-term overall stability. During the whole period the production did not meet with serious problems, and the number of parts unusable owing to defective material was minimal.

At the beginning of the 1950s the construction of the desk electron microscope proved very successful. The resolution limit of other electron microscopes produced at that time was not so much better that a microscope of this category became uninteresting for most applications. The reverse seemed to be the case. A device of such simple construction, easy operation, and low price met the requirements of many biological laboratories. Surface replicas, which were often investigated at that time, also conformed to the possibilities of the device. Using relatively simple means the microscope was adapted for reflection microscopy and even for thermoemission electron microscopy (Delong et al., 1956, 1957).

It is an interesting question whether it is possible to return to the idea of constructing a simple device that would find its way into the many fields of human activity to which electron microscopy cannot be applied because of the very high price of modern, though, we must admit, sophisticated devices.

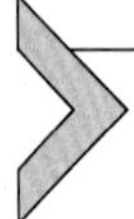

4. Production of electron microscopes in Czechoslovakia

The development of the first Czechoslovak electron microscope was financially supported by industry with the aim of introducing its industrial production. Therefore, preparations were made at the beginning of the 1950s to start the production of electron microscopes together with other devices to cover the needs of the research and development work in Czechoslovakia. In Brno a Scientific Workshop was founded as an industrial enterprise that manufactured 25 transmission electron microscopes on the basis of documentation compiled during the development of the first Czechoslovak electron microscope. The Scientific Workshop soon became a part of the enterprise Tesla Brno, which started production of the desk electron microscope in 1957. This production was gradually increasing up to the maximum number—more than 100 instruments a year. Then it began to decrease in consequence of increasing demands on the transmission electron microscope. This was at the time when transmission electron microscopes with a substantially higher performance were coming on the market. Nevertheless, in addition to the more perfect and more complex devices, the desk electron microscope was produced until the beginning of the 1970s. The attempt to place the microscope on a special table and

Table 1 Production of electron microscopes in Czechoslovakia.

Type	Production		Number of instruments[a]	Developed
	Started	Ended		
TEM BS	1952	1953	25	Institute of Theoretical and Experimental Electrotechnology
TEM BS 242 (desk type)	1957	1973	827	Czechoslovak Academy of Sciences
TEM BS 413 (high resolution)	1964	1973	358	Czechoslovak Academy of Sciences
TEM BS 500 (540)	1973	–	399	Tesla Brno
SEM BS 300	1976	–	142	Tesla Brno
SEM BS 350	1977	–	18	Czechoslovak Academy of Sciences

[a] Of a total of 1769.

to reconstruct some of its functional parts failed. It led to an increase of price and resulted in a decrease of interest. The boom era of the device was irrevocably gone. The production of 827 electron microscopes of this type during 15 years can be regarded as an undisputable success, confirmed by a number of awards, among others the gold medal of the World Exhibition in Brussels in the year 1958, but above all by the appreciation of many users.

It is beyond all dispute that the production of desk electron microscopes in the national enterprise Tesla Brno had a positive influence on the initiation of production of more advanced electron microscopes. Though the stimulation and documentation for production came from a team which has been working under the author's leadership at the Czechoslovak Academy of Sciences (an institution associating research institutes of different orientation, such as the Academy of Sciences of the USSR or the French CNRS) since 1954, in Tesla Brno a team of experienced workers, engineers, designers, and craftsmen was successively formed. The production of electron microscopes also had a great influence on the initiation of production of other scientific devices (NMR spectrometers), and also on the quality of electronic measuring instruments—the main production program of Tesla Brno. In the laboratories of Tesla Brno two types of electron microscopes of their own design were built. Table 1 gives a survey of electron microscope production in Czechoslovakia.

5. Conclusion

The scientific workers in Czechoslovakia became acquainted with the electron microscope only after the end of World War II. It was not simply a passive meeting. The fascinating impression of the device that opened never dreamt of possibilities of investigation of the submicroscopic world became an inspiration for the attempt to participate in the ever-increasing worldwide tendency to pursue further development of the device of the century. Even though it was a modest contribution it enlarged the family of electron microscopes, and not only in Czechoslovakia. Just as in other countries of the world, in Czechoslovakia the electron microscope led to the rapprochement of microscopists from different fields of science and technology. On the initiative of Professor Herčík the first conference of Czechoslovak microscopists took place as early as 1952. The increasing activity of the community of electron microscopists led to the organization of a conference in Smolenice in the year 1959 with the participation of a great many well-known specialists from many countries of the world.

The culmination of Czechoslovak efforts to improve contacts between East and West also in the field of electron microscopy was the Third European Regional Conference on Electron Microscopy, held in Prague in 1964, in which many prominent scientific workers of East and West took part. This conference became a milestone for the engagement of the microscopists of the socialist countries in the development of electron microscopy—work that has always served the purposes of peace.

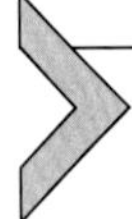

6. Appendix A: bibliography of related publications

Delong, A., and Drahoš, V. (1964). L'observation de la structure de l'aluminium et de ses alliages à l'aide de l'émission électronique froide. *J. Microsc. (Paris)* **4**, 711–714.

Delong, A., and Drahoš, V. (1965). Ein kombiniertes Emissions-Elektronenmikroskop. *Prague, 1964* Vol. I, pp. 25–26.

Delong, A., and Drahoš, V. (1966). Emissions-Elektronenmikroskopie ebener Oberflächen mittels Tunnelelektronen. *Mikroskopie* **21**, 151.

Delong, A., and Drahoš, V. (1967). Emissions-Elektronenmikroskopie ebener Oberflächen mittels Feldelektronen. *Trans. Czech. Conf. Electron., 3rd, 1965* pp. 633–644.

Delong, A., and Drahoš, V. (1968a). On some features of a new electron emission microscope. *J. Sci. Instrum.* [2] **1**, 397–400.

Delong, A., and Drahoš, V. (1968b). Investigation of electron tunnel emission in the emission microscope. *Proc. Czech. Conf. Electr. Vac. Phys., 4th, 1968* pp. 375–380.

Delong, A., and Drahoš, V. (1968c). On some results of electron tunnel-emission microscopy. *Rome, 1968* Vol. I, pp. 113–114.

Delong, A., and Drahoš, V. (1971a). On the project of a versatile emission electron microscopy. *Grenoble, 1970* Vol I, pp. 197–198.

Delong, A., and Drahoš, V. (1971b). Low energy electron diffraction in an emission microscope system. *Nature (London), Phys. Sci.* **230**, 196–197.

Delong, A., and Drahoš, V. (1972). Auger spectroscopy in an emission microscope. *Manchester, 1972* pp. 158–159.

Delong, A., Drahoš, V., and Zobač, L. (1961). An experimental high performance electron microscope. *Delft, 1960* Vol. I, pp. 89–91.

Delong, A., Drahoš, V., Kroupa, J., and Speciálný, J. (1968). Experimental electron transmission microscope of very high performance. *Rome, 1968* Vol. I, pp. 199–200.

Delong, A., Kolařík, V., and Štěpánková, J. (1973). Low energy electron diffraction on the planes non parallel to the surface. *Czech. J. Phys., Sect. B* **23**, 918–922.

Drahoš, V., and Delong, A. (1958). A small universal electron microscope. *Br. J. Appl. Phys.* **9**, 306.

Drahoš, V., and Delong, A. (1964). The source width and its influence on interference phenomena in a Fresnel's electron biprism. *Opt. Acta* **11**, 173–182.

Drahoš, V., and Delong, A. (1965). Observation of charges on specimens in a transmission electron microscope. *Czech. J. Phys., Sect. B* **15**, 760–765.

Drahoš, V., and Delong, A. (1966a). Resolution in low-angle electron diffraction. *Nature (London)* **209**, 801–802.

Drahoš, V., and Delong, A. (1966b). Einige Bemerkungen zur Verwendung von Schattenprojektion und Kleinwinkelbeugung in einem Durchstrahlungs-Elektronenmikroskop. *Mikroskopie* **21**, 149.

Drahoš, V., and Delong, A. (1967). The use of shadow projection in a transmission electron microscope. *Trans. Czech. Conf. Electron., 3rd, 1965* pp. 655–664.

Drahoš, V., and Delong, A. (1971a). Low-angle electron diffraction from defined specimen area. *Grenoble, 1970* Vol. II, pp. 147–148.

Drahoš, V., and Delong, A. (1971b). On a new technique of electron small angle diffraction. *Czech. J. Phys., Sect. B* **21**, 604–613.

Drahoš, V., and Delong, A. (1972). Selected area LEED in an emission electron microscope. *Manchester, 1972* pp. 156–157.

Komrska, J., Drahoš, V., and Delong, A. (1964a). Fresnel diffraction of electrons by a filament. *Opt. Acta* **11**, 145–157.

Komrska, J., Drahoš, V., and Delong, A. (1964b). Application of Fresnel diffraction to the determination of the local filament diameter in an electron biprism. *Czech. J. Phys., Sect. B* **14**, 753–756.

Komrska, J., Drahoš, V., and Delong, A. (1965). Interpretation of interference phenomena in Fresnel's electron biprism. *Prague, 1964* Vol. I, pp. 1–2.

Komrska, J., Drahoš, V., and Delong, A. (1967). Intensity distribution in electron interference phenomena produced by an electrostatic biprism. *Opt. Acta* **14**, 147–167.

Books

Bartl, P., Delong, A., Drahoš, V., Hrivňák, I., and Rossenberg, M. (1964). "Methods of Electron Microscopy" (in Czech). Nakladatelství ČSAV, Praha.

Delong, A., and Drahoš, V. (1958). "Practical Electron Microscopy" (in Czech). Nakladatelství ČSAV, Praha.

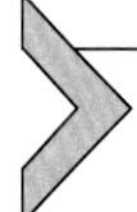

7. Appendix B: publications from the Institute of Scientific Instruments of the Czechoslovak Academy of Science in Brno (in Czech)

Bauer, M. (1967). Magnetické pole v objektivu elektronového mikroskopu při nasycení pólových nástavců (The magnetic field in the objective of an electron microscope at the saturation of pole pieces). *Elektrotech. Cas.* **18**, 10–26.

Delong, A. (1966). Ultravakuová aparatura pro emisní elektronovou mikroskopii (An ultra-high vacuum apparatus for emission electron microscopy). *Jemna Mech. Opt.* **11**, 244–245.

Delong, A., and Drahoš, V. (1951). Československý elektronový mikroskop (Czechoslovak electron microscope). *Sb. VST Brno*, **20**, 334–348.

Delong, A., and Drahoš, V. (1955a). Použití československého stolního elektronového mikroskopu (The use of Czechoslovak table model electron microscope). *Slaboproudy Obz.* **15**, P31–P32.

Delong, A., and Drahoš, V. (1955b). Československý stolní elektronový mikroskop (Czechoslovak table model electron microscope). *Slaboproudy Obz.* **15**, 358–366.

Delong, A., and Drahoš, V. (1965). Kombinovaný emisní elektronový mikroskop (A combined emission electron microscope). *Slaboproudy Obz.* **26**, 214–220.

Delong, A., and Drahoš, V. (1972). Nová metoda difrakce pomalých elektronú (A new method of low energy electron diffraction). *Cesk. Cas. Fyz.* **A22**, 202–203.

Delong, A., and Drahoš, V. (1975). Aktuální tendence v prozařovací elektronové mikroskopii (Trends in transmission electron microscopy). *Cesk. Cas. Fyz.* **A25**, 14–27.

Delong, A., Drahoš, V., Bezděk, L., and Růžička, D. (1956). Možnosti použití odrazové elektronové mikroskopie při studiu struktur kovů (Possibilities to use reflection electron microscopy for the examination of metal structures). *Hutn. Listy* **11**, 355–363.

Delong, A., Drahoš, V., Bezděk, L., and Růžička, D. (1957). Možnosti použití emisní elektronové mikroskopie při studiu struktur kovů (Possibilities of using emission electron microscopy in the investigation of metal structures). *Hutn. Listy* **12**, 206–215.

Delong, A., Drahoš, V., Speciálný, J., and Zobač, L. (1960). Pokusný elektronový mikroskop s vysokou rozlišovací schopností (An experimental high resolving power electron microscope). *Slaboproudy Obz.* **21**, 195–206.

Delong, A., Drahoš, V., and Kroupa, J. (1962a). Rychlostní analyzátor pro měření stability urychlovacího napětí elektronového mikroskopu (A velocity analyzer for measuring the stability of the accelerating voltage in an electron microscope). *Slaboproudy Obz.* **23**, 311–316.

Delong, A., Drahoš, V., and Zobač, L. (1962b). Elektronový mikroskop s vysokou rozlišovací schopností (A high resolving power electron microscope). *Cesk. Cas. Fyz.* **A12**, 471–478.

Delong, A., Drahoš, V., and Speciálný, J. (1964). Optická soustava elektronového mikroskopu s vysokou rozlišovací schopností (The optical system of an electron microscope with a high resolving power). *Slaboproudy Obz.* **25**, 509–515.

Delong, A., Drahoš, V., and Speciálný, J. (1968). Emisní elektronový mikroskop s ultravakuem v komoře pro preparát (An emission electron microscope with ultrahigh vacuum in the object chamber). *Slaboproudy Obz.* **29**, 65–70.

Delong, A., Drahoš, V., Kolařík, V., Lenc, M., Hladil, K., and Šálek, R. (1978a). Rastrovací elektronový mikroskop s autoemisní tryskou (A scanning electron microscope with field emission gun). *Slaboproudy Obz.* **39**, 443–450.

Delong, A., Kolařík, V., Lenc, M., and Vašina, P. (1978b). Mikroanalýza pevných látek v rastrovacím elektronovém mikroskopu (The microanalysis of solid matter in a scanning electron microscope). *Slaboproudy Obz.* **39**, 451–455.

Drahoš, V., and Delong, A. (1962). Elektronová interferometrie a fázový kontrast (Electron interferometry and phase contrast). *Pokroky Mat. Fyz. Astron.* **5**, 80–90.

Drahoš, V., and Delong, A. (1963). Úprava prozařovacího elektronového mikroskopu pro interferenční elektronovou mikroskopii (Adaptation of a transmission electron microscope for interference electron microscopy). *Cesk. Cas. Fyz.* **A13**, 278–286.

Drahoš, V., and Delong, A. (1964). Optická soustava interferenčního elektronového mikroskopu (The optical system of an interference type electron microscope). *Slaboproudy Obz.* **25**, 523–527.

Drahoš, V., and Delong, A. (1965a). Komplexní metoda měření tlouštky a vnitřního potenciálu v interferenčním elektronovém mikroskopu (A complex method of thickness and internal potential measurement in an electron microscope). *Cesk. Cas. Fyz.* **A15**, 476–483.

Drahoš, V., and Delong, A. (1965b). Přehledové zvětšení v prozařovacím elektronovém mikroskopu použitím stínové projekce (Survey magnification in a transmission electron microscope using shadow microscopy). *Slaboproudy Obz.* **26**, 643–649.

Drahoš, V., and Delong, A. (1966). Difrakce pod malými úhly v prozařovacím elektronovém mikroskopu (Low-angle diffraction in a transmission electron microscope). *Slaboproudy Obz.* **27**, 494–500.

Drahoš, V., and Komrska, J. (1962). Úhlová apertura osvětlovacího svazku a osvětlená oblast preparátu v elektronovém mikroskopu (The angular aperture of the illuminating beam and the illuminated specimen area in an electron microscope). *Cesk. Cas. Fyz.* **A12**, (1962), 479–488.

Drahoš, V., Delong, A., and Komrska, J. (1963). Interferenční elektronová mikroskopie (Interference electron microscopy). *Jemna Mech. Opt.* **8**, 242–246.

Frank, L. (1982). Rastrovací Augerova mikroskopie (Scanning Auger microscopy). *Cesk. Cas. Fyz.* **A32**, 134–147.

Frank, L., and Kolařík, V. (1975). Energiově selekční elektronový mikroskop s dvojitým magnetickým hranolem (Energy selecting electron microscope with a double magnetic prism). *Cesk. Cas. Fyz.* **A25**, 48–52.

Komrska, J. (1975). Ohyb a interference elektronů (Diffraction and interference of electrons). *Cesk. Cas. Fyz.* **A25**, 1–13.

Komrska, J., and Drahoš, V. (1964). Měření mezirovinných vzdáleností metodou selekční difrakce v elektronovém mikroskopu (Measurement of interplanar spacings by selected area diffraction in electron microscope). *Cesk. Cas. Fyz.* **A14**, 435–442.

Lenc, M. (1975). Perspektivy mikroskopie s odrazem pomalých elektronú (Prospects of low energy electron reflection microscopy). *Cesk. Cas. Fyz.* **A25**, 28–37.

Lencová, B. (1975). Metoda konečných prvkú v elektronové optice (Finite element method in electron optics). *Cesk. Cas. Fyz.* **A25**, 57–61.

Pavlík, K. (1973). Supravodivé magnetické čočky elektronového mikroskopu (Supraconductive lenses in an electron microscope). *Slaboproudy Obz.* **34**, 405–417.

Podbrdský, J. (1983). Použití miničoček v prozařovacím elektronovém mikroskopu (The use of mini-lenses in the transmission electron microscope). *Cesk. Cas. Fyz.* **A33**, 237–250.

Studeník, J. (1973). Objektiv elektronového mikroskopu se supravodivým vinutím (Objective lens for an electron microscope, using supraconductive winding). *Slaboproudy Obz.* **34**, 417–420.

Studeník, J., and Pavlík, K. (1975). Nová supravodičová magnetická čočka (A new superconductive magnetic lens). *Cesk. Cas. Fyz.* **A25**, 154–155.

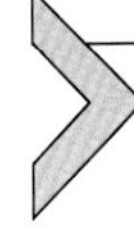

8. Afterword by B. Lencová: Remembering Delong, Drahoš and Zobač

This year we celebrate 70th anniversary of the production of the first Czechoslovak electron microscope in Brno. The early years are characterized by Armin Delong in the "Early days....", which ends at the late 1950s and mentions the 1964 European Electron Microscopy Conference in Prague plus a list of several papers from the 1970s.

The three people responsible for making the Brno electron microscope are shown in the photograph taken around 1953 when they were awarded one of the highest orders of that time, the "Order of Work, three young men aged just 28 (Fig. 6). The two technicians involved were not included, they were not that young. Nor was Professor Aleš Bláha, the supervisor who was the head of department where the three graduated, included. As Ladislav Zobač wrote in his Memoirs, "We were not sure that we really deserved the Order, but the Military Academy pushed this to make themselves more visible"; the Academy was where they were teaching at that

Figure 6 From left to right: Armin Delong, Vladimir Drahoš and Ladislav Zobač.

time and Delong and Vladimir Drahoš were just starting their postgraduate study. The Military Academy took over parts of the Brno University of Technology in 1951 and forced Professor Bláha to leave Brno and cancelled his professor title; he moved to Bratislava where he could continue teaching and his research. Staying at the Military Academy or working for a PhD was a safe way to avoid military service, at least for Zobač, who managed this at the last moment. They were working mostly in the Scientific Workshop established by Bláha, later taken over as Development Center of Tesla Elektronik; it later became the Development Workshop of the CSAV (Czechoslovak Academy of Sciences), and in 1957 became part of the Institute of Scientific Instruments (ISI) of CSAV. Let us at this point mention academician Ferdinand Herčík, who got the first commercial microscope from RCA. He showed this to the three of them and so they managed to make an original electron microscope with similar quality to that of the RCA microscope. This was not just a copy of the RCA device; in additions they designed a small tabletop microscope. Herčík took care that two academy institutes, Biophysics and ISI, were built close to one another on Královopolská street, and just 1 km south of newly built Tesla factory. He was also active in the United Nations and UNESCO.

Let us briefly mention also the "boundary" or "initial" conditions of that time. Czechoslovakia was formed only in 1918 after WW1 from former Bohemia, Moravia, and Slovakia (part of the Hungarian part of the Habsburg monarchy) and Carpathian Ukraine, annexed later by the USSR, and lasted only twenty years, with interesting economic growth. The German occupation of Czechia and Moravia by the Nazis was preceded by the annexation of the Sudetenland in autumn 1938, inhabited mostly by close to 3 million German population; after the war they were massively forced to leave the country for Germany. In March 1939 first Slovakia split, making their own fascist state, and the rest of Bohemia and Moravia were occupied. The holocaust reduced the Jewish population of 80000 to 10%; many professors were of Jewish origin, and many other Czech professors perished in prisons and concentration camps. The universities were closed from November 1939 throughout the war. The war also forced many young people to enter the communist party, because most of the country was liberated by the Soviet army. The economic situation was not good and started with a record drought in 1947, but the help of the Marshall plan was refused by the communists in the post-war government and exchanged for much less grain from the Soviet Union; this enabled a communist coup in February 1948, making the country a satellite state of the Soviet Union. Later in the same year the number of party members was increased by forced unification with social democrats. The government, factories and even universities were run mostly by the communist party candidates and under close supervision by secret police and Russian advisors. In 1968, with improving economy, people expected that the country would change, hoping for "socialism with a human face", which was terminated by Russian tanks, and mass emigration resulted. This was followed by a "normalization" period, when many of the reformist communists were not only removed from the party but often also from their jobs, preventing their children from getting higher education.

Prof. Armin Delong was the most visible of the three people who made the first Czechoslovak microscope. He was certainly an exceptional person. Born in January 1925, he started to study at the VTU (Technical University in Brno) in the first post-war year; all universities had been closed for almost six years. In 1957 he (and Drahoš) obtained the title CSc (Candidate of Sciences, equivalent to PhD); in 1969 he obtained his DrSc. He was also the most adventurous of all three, and the only party member. In 1961 he became the director of the institute. This established a tradition that the director of the institute is from the largest department, electron optics. He remained in this position until the next revolution, the velvet one. Other departments of ISI included NMR (Dadok emigrated in 1968 to US) with a cryo-NMR department, and lasers for measurement. In 1973 he became corresponding member and in 1981 full member of the CSAV. The biggest department, electron optics, attracted many talented engineers and physicists, and had several diploma and doctoral students, who mostly remained in ISI. They continued with developing further microscopes and cooperated with Tesla. The only bad decision was the "tsar microscope" that never worked owing to mechanical and electrical instabilities. The ambitious aim was to get as close as possible to 1 Angstrom resolution 50 m away from tram and trolleybus lines.

At the end of the 1960s interest shifted towards emission electron microscopy (EEM) with papers in *Nature* (1971) and the *Journal de Microscopie* and conference presentations. The EEM had a UHV sample chamber, two magnetic lenses and a gun below the screen with a hole in the middle. Many techniques were included: ion bombardment, sample heating, electron spectrometry. The UHV environment allowed the observation of LEED patterns, which have the same dimensions independent of the energy of primary electrons; this was reported in *Nature* and rediscovered at the end of 1990s in a paper in *Optik*.

Other direction of interest for Delong became ion beam implantation and hemispherical LEED (and Auger analyzer) in the UHV specimen chamber. In 1969 Delong became external head of the Department of Solid-state Physics of the Faculty of Natural Sciences for three years. Delong´s activities were reduced when he started to build a nice house for his family of two children. Most of his coworkers started as students of Delong, most of the diploma theses were devoted to EEM and surface physics. In the late 1980s he made a prototype of 5 keV FEG miniTEM.

The career of Drahoš continued more smoothly, he was teaching at the Faculty of Electrical Engineering as well as instrument technology at BUT (Brno University of Technology as it was now called), published two books and several student texts in Czech. In 1964 he was awarded his DrSc. and in 1968 he became professor of BUT. In ISI he was the head of the electron optics department and Deputy director of ISI. He pursued the development of X13 series of high resolution TEMs, produced by Tesla, which developed their own BS 500 and 540 routine TEMs. His closest coworkers were Jiří Komrska, whose diploma students included Michal Lenc and Josef Podbrdský. Komrska became one of the victims of normalization and was thus unable to supervise diploma students, although we published a paper in *Optica Acta* explaining why model and mathematics of electron biprism work well. Thus he suggested that I make my diploma under the supervision of Delong. After graduation I started in the group of Drahoš, who at the 1972 European Microscopy meeting in Manchester gave an invited talk and talked there to Tom Mulvey and Eric Munro. As a result, he asked me to try to write a Finite-Element Method program. This I managed in six months, and my first task was to improve the behavior of the BS500 microscope, which should have worked as well as the BS540 but did not. I soon found that the wrong shape of the lower polepiece was causing a significant parasitic field. We also started with the design of a new TEM, but this was never realized. Electron interference, diffractography, holography, reflection diffraction system, were Drahoš's work. He was very pleasant person, caring about his coworkers. Both he and Delong could speak French, English, German and Russian.

In 1968 an extension to the main building was added, because the number of people had grown too much, up to 260 people. In the first days of the Soviet occupation, Radio Brno broadcast for a few days from this building, as disclosed only 30 years later This event was called brotherly help, which even included by mistake the GDR army in its plans, but they withdrew within the first 24 hours.

In 1974 it was recommended from higher places that Tesla should start to make a scanning electron microscope. The work proceeded in parallel at Tesla where a system with a thermionic emission gun was built, and at ISI attempts started with a cold FEG, inspired by the CwickScan of Crewe. The chief designer was Kolařík, a former student of Delong, who produced a very skillful design with two magnetic lenses, FEG immersed in the upper lens, and UHV specimen chamber that also contained Auger and EDAX spectrometers. Traditional plastic material of SE and BSE detectors could not be used in UHV, so it was replaced with a new type of scintillation material, YAG Ce-doped single crystal, also used later for TEM screens (Rudolf Autrata).

In 1978 Drahoš was seriously ill with flu in winter, and later in spring he was diagnosed with cancer and died in June. Delong, evicted from Brno university, started to give lectures in Olomouc, and later all five of us who had no chance to obtain a PhD were in 1978 were awarded a funny title RNDr.

A little later, another request came from above that the Eastern Block needs to improve its semiconductor industry and use electron beam lithography. A system had been developed in Jena in GDR, excessively big and oversized. Delong and Kolařík participated in the 1978 International Congress on Electron Microscopy (ICEM) in Toronto, where they got inspired by more stable Schottky gun, and only in 1982, for the first time, did a bigger group of five people go to the ICEM in Hamburg, where the FEG SEM was exhibited. Before that the only contact with abroad was via Tom Mulvey, who regularly visited ISI and shared with us his ideas and

Figure 7 Cold FEG SEM BS 350 produced by Tesla company, but developed at ISI.

proceedings volumes. As an exception Jiří Komrska spent several months at Aston University in 1968 and Podbrdský spent several months in the University of Arizona in Tempe in the late 1970s. As a reciprocal gesture of the Royal Society Exchange scheme that financed most of Mulvey's travels, Podbrdský spent a month in the UK in the early 1980s, and even I was able to travel there in 1987.

The EBL machine was completed in 1984, and several pieces were sent to the USSR. It allowed a field of view of 6 × 6 mm, shaped beam up to 6.4 × 6.4 μm and a resolution 0.1 μm, with 1 μA in a 15 keV beam. The part of Tesla that was built to produce the EBL machine now houses a Museum of Technology, with a nicely organized and complete show of Brno microscopes. This includes also the SEM in Fig. 7.

Zobač was also working in the institute, and he was engaged in introducing special technologies such as electron beam welding and an UHV Titanium Orbitron pump, needed for the UHV toys of Delong, but he was also interested in medical devices and cryotechnology. Later he married a nice lady from ISI and had a son, who is now continuing his work in ISI. Ladislav Zobač died a year after Delong, who passed away in 2017.

The velvet revolution inspired many people to criticize ISI, who even tried to get rid of Delong; however early in 1990 he became a deputy prime minister of Czechoslovakia for science. This career did not last too long, because the country split into Czechia and Slovakia (the Slovaks had their own fascist state only during WW2, historically they were a part of Hungary). Also, the funding was significantly reduced, and the number of people in the institute, 260, had to be reduced almost by half. Many people discovered their entrepreneur genes, trying even to take over ISI, and Tesla lost most of its markets, was privatized and collapsed. By some miracle I was abroad at that time, spending three months at Imperial College in 1987; a year later I was invited to Imperial College for half a year but got support for three months only, and for the summer months I was invited to Delft and for an invited talk at EUREM in York. At TU Delft I spent three happy years until autumn 1991.

In the post-revolutionary turmoil in 1990 Jiří Komrska became ISI director, but he resigned in autumn and left for BUT. He was followed by Josef Jelínek from the NMR department, who selected Michal Lenc as his deputy, and became seriously ill less than six months later. In 1992 Lenc left for the Department of Theoretical Physics. After a one-year intermediate director, Professor Autrata became director of ISI until his death in 2006. Several professors Sklenář, Kasal, Komrska, Lenc made an academic career after they left ISI, the same number as those who stayed at ISI. In 1994 I started to teach part time at Faculty of Mechanical Engineering of BUT at the same department as Komrska, had 8 diploma and 8 PhD students, and in 2006 I became professor as well; after completing my last grant project I quit ISI and started to work at Tescan. The next two directors were Luděk Frank followed by Ilona Mullerová.

Was this the end of electron microscopy in Brno with no Tesla company and reduced number of employees at ISI? In 1990 three companies emerged, with people from Tesla and ISI. First of all the SEM part of Tesla was taken over by Tescan, which grew from the six original people almost hundred times. Another group of about 20 people established a company Delmi and started to produce a routine TEM called Morgagni; Delmi was then taken over by Philips EO/FEI company, and in 2015 FEI was taken over by Thermo Fisher Scientific. And in the same year, 1990, the company Delong Instruments was established by Kolařík and colleagues; some transmission electron microscopes were built, operating at 5 keV and from 2014, at 25 keV, see Fig. 8. It supplied many other companies and institutions. In 2000 EUREM was held in Brno, and in 2014 ICEM was organized in

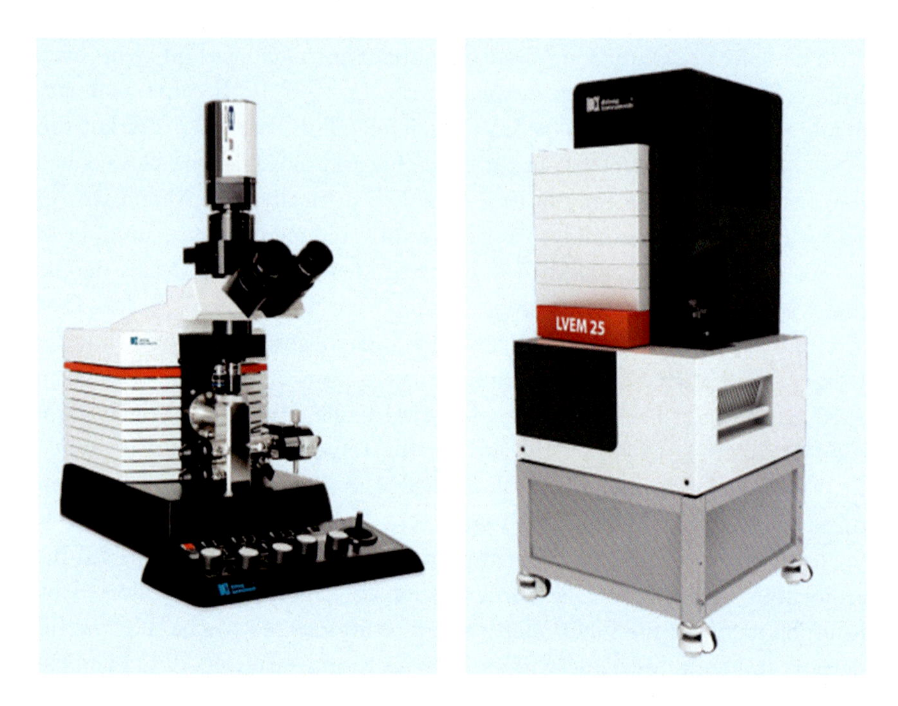

Figure 8 LVE5 and LVEM25, two low voltage TEMs produced by Delong Instruments.

Prague, and a claim was raised there that some 30% of world production of microscopes happens in Brno, making it a Microscopy Valley.

References

Delong, A., & Drahoš, V. (1951). *Sbornik VST Brno*, *3–4*, 335.
Delong, A., & Drahoš, V. (1955). *Slaboproudy Obzor*, *15*, 358.
Delong, A., Drahoš, V., Bezděk, L., & Růžička, D. (1956). *Hutnické Listy*, *11*, 355.
Delong, A., Drahoš, V., Bezděk, L., & Růžička, D. (1957). *Hutnické Listy*, *12*, 206.
Steigerwald, K. H. (1949). *Optik*, *5*, 469.

CHAPTER SEVEN

Personal reminiscences of early days in electron microscopy☆,☆☆

D.G. Drummond[a]
Formerly 45 Albert Drive, West Killara, NSW, Australia

Contents

1. Preamble

I first became involved in electron microscopy in 1942 at the Shirley Institute in Manchester during World War II. I am now retired and live in

☆ D.G. Drummend's original text is completed by a tribute to him by Kyle Ratinac and further reminiscences covering his Australian years. These form Section 17 of this chapter.

☆☆ Corresponding author: Kyle Ratinac. e-mail address: kyle.ratinac@mq.edu.au.

[a] In charge of the Electron Microscope Unit (EMU), University of Sydney, from its creation in 1958 until 1973.

Advances in Imaging and Electron Physics, Volume 220
ISSN 1076-5670
https://doi.org/10.1016/bs.aiep.2021.08.007

Sydney, Australia, and no longer have available the records of those early years, which would have helped greatly in the preparation of these reminiscences. Memories of names, places, and dates have become dimmed with the lapse of years and I have to apologize for inadequacies on these counts. The modern scene in electron microscopy is so different, however, that it may be of some interest to try to recall the flavor of those times and the difficulties which then confronted electron microscopists in Britain, arising from two causes—first, the primitive state of the science of electron microscopy itself and the lack of any general body of experience in its practice, and, in the second place, the shortages, restrictions, and difficulties of communication arising in the wartime conditions, which compounded the effects of the first cause.

To the electron microscopist of today the kind of specimens we could tackle, the magnification and resolving power we could attain, and the convenience and, above all, the speed, of working would seem extremely modest, if not trivial, but we could already outstrip the achievements of the optical microscope, in some directions at least, by a very substantial factor, and provide results of immediate practical importance.

2. Beginnings in Britain

In 1936 an English optical microscopist, Professor L. C. Martin of Imperial College, London, was moved to try to emulate the work of the pioneers, von Borries and Ruska and others in Germany, and to build an electron microscope (Martin et al., 1937). But he did not have much success, being defeated by the electrical requirements of the instrument, which lay outside his main field of expertise. He went back to the optical microscope, and in a letter to me much later remarked "a cobbler should stick to his last."

When war broke out in 1939 there was no functioning electron microscope in England. In Germany, on the other hand, the firm of Siemens and Halske was just beginning to produce a commercial model, based on the work of von Borries and Ruska, and their first instrument was installed at I. G. Farben Werke only a month or two after the war began. This was the "bathtub" model, so called (by the irreverent English) from the open-topped tank which surmounted the column and housed those, otherwise unguarded, items which were at a high voltage to earth, including the accumulator battery used to provide a stable power supply to the gun filament.

Martin's efforts had, however, done something to alert the authorities in England, and his instrument was sent to the National Physical Laboratory (NPL) in an attempt to have it developed urgently into a working instrument. Metropolitan Vickers in Manchester (Met-Vick) were also involved, as Martin had commissioned them to prepare the working drawings and to make some of the components for his instrument, though they had not till then been concerned with its overall design.

But when Sir Charles G. Darwin, the director of the NPL, visited the United States in the early part of the war, he found that the electron microscope currently being developed by the Radio Corporation of America (RCA Type B) had also reached the stage of commercial production, and realized that instruments of this model could clearly be got into action in England, to meet wartime exigencies, much more quickly than a new model developed virtually from scratch. So he ordered six instruments to be delivered under the lend-lease arrangement.[1]

These six RCA microscopes were brought across, one by one, during 1942 and taken, in the first instance, to the NPL, from whence they were dispersed to various laboratories up and down the country. It was rumored that one instrument had been lost to submarine action but certainly six arrived; maybe a replacement was sent. The laboratories chosen were (1) The National Physical Laboratory, Teddington; (2) The Cavendish Laboratory, Cambridge; (3) The British Cotton Industry Research Association, Shirley Institute, Manchester; (4) Department of Biomolecular Structure, Leeds University, Leeds; (5) The National Institute for Medical Research, Hampstead; and (6) Rothamsted Experimental Station, Harpenden.

3. The microscope at the Shirley Institute

I was employed at the Shirley Institute as a spectroscopist, and, knowing nothing of the new branch of science (an ignorance which I shared with the majority of my contemporaries), I was selected by the then Director, Dr. (later Sir) Robert Pickard, to install and operate the "machine"—as he invariably called it. In this task I had the assistance of J. B. Todd, especially with the electronics.

[1] The Australians would also have liked to have electron microscopes available but all lend-lease material for that part of the world required the personal approval of General MacArthur and this was not forthcoming. Presumably the General did not consider the electron microscope offensive enough!

In due course, in, I think, August of 1942, seven large crates were delivered, containing unfamiliar items of equipment from which we had to assemble, without any instruction book or working drawings, an instrument which none of us had ever seen, nor, as I remember, even heard of, until very shortly before that time. No direct technical assistance from RCA was feasible and we were "on our own," though Cuckow and Trotter at the NPL were a little ahead of us, since they got the first RCA instrument to arrive and had also had some experience with the Martin instrument, so were able to advise on some points. However, we unpacked[2] and began to put the parts together where it seemed right that they should go and where the flanges fitted! Fortunately, the diagrams and instruction books then began to arrive, section by section, though it was several months before all the information was assembled. Meanwhile we had found that the microscope was designed for an American standard electricity supply of 120 V, 60 cycles, and we had only the local 240 V, 50 cycles available. In the hard-pressed conditions of wartime there seemed at first to be no possibility of finding a suitable transformer (ignoring the frequency discrepancy), but our neighbors in Manchester, Metropolitan Vickers—soon to become our very good friends, came to the rescue with an offer to make a transformer for us, an offer which was gratefully accepted. Metropolitan Vickers were already being drawn into the electron microscope scene by their interest in the Martin instrument and were anxious to see the RCA model in action. M. E. Haine, who was our principal contact at Met-Vick, became the leader of the team responsible for the development of their first commercial model electron microscope.

It was still a couple of months before our transformer could be delivered, so we filled in the time by testing every component we could and by making sketch drawings, for future reference, of the less familiar mechanical and vacuum sealing arrangements (the latter mainly using flat gaskets and sylphons; "O" rings had not then come into popular use, and, indeed, the Siemens "bathtub" model employed greased, ground-metal joints, though we did not know this until later). These drawings were much appreciated by the Leeds University team when, shortly, they came to assemble their instrument. The delay over the transformer, though frustrating, proved a

[2] We were incidentally much entertained by the American newspapers used for wrappings, which, for instance, included an account of a dear old lady school teacher who, in her alter ego, had turned out to be the notorious "two-gun Bessie" who had lately been terrorizing her neighborhood! At least it made a change from the war news in our own papers.

blessing in disguise, for our tests showed that at least a dozen of the ceramic-coated, wire-wound resistors in the circuitry had suffered corrosion damage during their wartime transit through arctic waters,[3] and we were able to obtain (not without some difficulty) and install replacements before power was ever switched on to the equipment. We had no trouble thereafter on this score. At the Cavendish Laboratory they had had the apparent advantage of having a suitable transformer immediately on hand, but when they switched on, with undetected faulty resistors still in circuit, a chain reaction of further damage took place which greatly delayed the successful commissioning of their instrument.

Another corrosion problem occurred in the air locks to the camera and specimen chambers. The mechanism of these air locks involved a thrust rod driven by a cam in the operating handle, and it was clear that the length of this rod was of vital importance to the vacuum sealing. The design seemed unnecessarily critical but we took much trouble to adjust the length of the rod to get satisfactory operation in each case. It was not until after the war, when we first met an RCA technician, Holmes Halma, that we learned that any inaccuracy in the length should have been taken up by a concealed spring loading, which, in both air locks, had been rusted solid and was still quite unyielding several years later. Surprisingly, our vacuum locks had remained effective all that time.

There was other, potentially more serious, corrosion damage. Our projector lens had been shipped with the polepiece in place in its socket, and when it reached us the two were firmly rusted together. We managed to drive out the polepiece without further damage, and judiciously cleaned it and the lens socket, but, since the fitting was cylindrical and not conical this must have opened up a slight gap in the magnetic circuit of the assembly. We had to operate with the equipment in this condition for some considerable time. Luckily the central bore of the polepiece had not been affected. I do not think that we, at that time, fully appreciated the importance of close magnetic continuity in the lens–polepiece assembly, but we did, at first, suspect that this misfortune with the polepiece was responsible for an effect which worried us considerably at the higher magnifications. This was a unilateral distortion of the image which caused a field of spherical particles to appear as an array of elliptical images all oriented with the major axis in the same direction throughout the field. The maximum magnification

[3] Interestingly, our crates were marked as having been shipped in the "Samuel Bakker," the ship which was later used for a leading role in a film on the Northwestern approaches.

attainable was nominally 20,000×, and at this level the difference between the major and minor axes of these ellipses was about 20%, a very substantial distortion. The distortion diminished with diminishing magnification, and below 5000× it virtually disappeared, so for a long time we confined ourselves to magnifications of 5000× or less, though pincushion distortion was then more in evidence. Investigation showed that the direction of this unilateral distortion and its magnitude were unaffected by removing and replacing the projector polepiece or by rotating it in its socket, though the construction of the instrument was such that rotations of the lens body were impracticable. However, on traversing a mini-search-coil along the projector lens axis we found that, when the excitation exceeded that required to give a magnification of 5000×, a secondary peak of magnetic field appeared a few centimeters below the level of the polepiece gap. In effect this added to the lens a weak second element which was optically imperfect and evidently had a cylindrical component. Apparently the iron wall of the lens bore became so thin at this level that it saturated at the higher excitations and the field spilled out onto the lens axis.

We put this problem to G. Liebmann of Met-Vick, and he was able to design for us a new polepiece of higher magnetic reluctance which operated at a lower flux, thereby avoiding saturating the iron in the lens shroud. We had this polepiece machined, from a billet of Swedish soft iron, in our general laboratory workshop, without the benefit of specialized tooling, and, though we could no longer expect to reach a magnification of 20,000×, we were then able to get magnifications up to, I think, 13,000× without the unilateral distortion that had troubled us.

4. Inconveniences of early electron microscopes

The early electron microscopes were quite primitive compared with our modern instruments and, as well as tending to have design faults such as that detailed above, they left a good deal to be desired in convenience of operation. Usually the high-tension cable was connected externally to the gun, in which the cathode assembly was insulated from the body of the microscope by an exposed glass cylinder and subject to humidity problems, and it was also necessary to provide corona rings to prevent aerial discharges from the exposed top electrode. To make adjustments to the gun one had to switch off the high tension and use a grounding rod to discharge the circuits before climbing up to make a tentative alteration.[4] This process had to be reiterated until a satisfactory gun setting was arrived at, a very tedious, not to say risky, business. Trotter, of the NPL, told me he was once saved only by his hair rising as he climbed up, having forgotten, at the *n*th reiteration, to switch off the high tension.

The objective lens too was inconvenient and was often used with no aperture since there was a feeling abroad that the spherical aberration was so severe anyway that it automatically stopped down the lens. This was fallacious, of course, but in the RCA Type B instrument there was no provision for insertion, interchange, or centering of apertures except by breaking the vacuum, removing the polepiece and its attached aperture tube—which could carry a single aperture only, fitting or interchanging the aperture as required, and adjusting its centering relative to the polepiece (using the jig provided) before reinserting the assembly into the microscope. After this the instrument had to be reevacuated before one could observe the result of the operation. One was lucky to get any beam through at all at the first attempt, and could be forgiven for balking at the use of an aperture. But, at the modest magnifications then available, and when examining high-contrast specimens such as inorganic pigment particles, the lack of an objective aperture hardly mattered.

The RCA Type B microscope had no built-in lifting device to allow one to open the column at intermediate points without dismantling from the top downward. Thus, any work on the internal arrangements at the projector lens or viewing screen was a major operation. At the Shirley Institute we circumvented this by installing a rail, above the microscope, carrying a runner with a chain block, and made a lifting frame which could be clamped round the column to hoist it in sections, or as a whole, when necessary. The lifting tackle could also be used to lift out the high-tension tank, which was awkwardly situated at the top of the cabinet, close behind the gun.

5. Group contacts in Britain during the war

After the six RCA microscopes in Britain had all been installed and operating for some months, Sir Charles Darwin ordained a meeting of

[4] We found, after the war, that the Siemens "bathtub" model had an automated discharge rod, driven by a motor which whirred into action on switching off the high tension. The whirr culminated in a loud crash as the spark leaped to earth; quite unnerving the first time one heard it!

workers from the six laboratories to present and exchange their experiences with the equipment and their methods of handling specimens. This meeting was held in November, 1943, in the Royal Society's rooms in Burlington House, London, and was the first full meeting of the group that later developed into the Electron Microscope Group of the Institute of Physics. It was decided that speakers should be called in the alphabetical order of the institutions which they represented, and since I was from the *B*ritish Cotton Industry Research Association, I had, fortuitously, the honor of delivering the first paper.

Similar meetings were called at irregular intervals—about one per year—and were held in various centers, and many friendships developed as a result. It was at the meeting of June, 1945, at the National Institute for Medical Research, then at Hampstead, that I first met Dr. V. E. Cosslett, who was, I think, just about to move from the Clarendon Laboratory at Oxford to the Cavendish Laboratory at Cambridge. (I do not know why the name of the pub where we lunched should have stuck in my mind—it was the "King of Bohemia.")

It so happened that just before a later meeting, which had been scheduled to take place in Manchester and which I and my colleagues from the Shirley Institute and Haine and others from Met-Vick had the responsibility of organizing, the war came to an end and our now somewhat enlarged but still cosy little group of English electron microscopists was suddenly exposed to a rush of people from the continent who had been starved of conference facilities for so long and who insisted in coming to our meeting. This strained our resources and produced logistic problems—food rationing was still in force—but I do not think we had to turn anyone away and many fruitful contacts began at that meeting. Le Poole, van Dorsten, and Gabor are among the foreign guests whose names spring to mind, but there were many more. The Germans were not able to come, mainly for political reasons, but they were also beset by other problems of the aftermath of war; in a letter which I received from E. Ruska shortly afterward he wrote about the difficulties in his work caused by "hunger, cold and darkness."

6. Immediate postwar conferences in Europe

Following on from that first postwar meeting in Manchester, the Dutch arranged an ad hoc conference in Delft in 1949, the first which could be described as overtly "international." Among those attending were large contingents from England, France, Sweden, and, of course, Holland, as well as representatives from half a dozen other European countries and even from Uruguay and India. Only four Americans were listed, and three of those were from RCA. I suppose there had been more opportunities for meetings and discussion, during the war years, in the United States than there had been on our side of the Atlantic. There were still no Germans or Japanese.

Quite apart from the official business and the very fruitful contacts and exchanges of ideas that were engendered at Delft, the occasion was memorable for the hospitality shown and the friendly informality on the surface, which concealed a superb organizing efficiency. I had perforce had some small experience previously at Manchester and so could well appreciate what lay behind the Dutch effort. The trip organized for the midweek break must have covered nearly all the sights of Holland, and, though it got more and more behind schedule as the day went on, the apparently effortless way in which a lavish late dinner, complete with impromptu speeches, was conjured up for, I suppose, about 100 people, at short notice in Haarlem, miles from the home base, could only excite admiration. The official conference dinner at Scheveningen the following night was almost an anticlimax. So far as organization was concerned, we in Manchester had rather had "greatness thrust upon us"; in Delft the Dutch certainly "achieved greatness." Of course, in comparison with modern conferences, with their vast numbers of participants and multiple parallel sessions, ours were small affairs, but they had an intimacy which I think added to their coherence and value, as well as to their enjoyment.

In Paris in 1950, the French ran a conference, already growing bigger, and described this as the "First International Conference in Electron Microscopy," but those of us who had been at Delft perversely persisted in referring to that as the "First" conference, so that when the next international meeting was held in London in 1954, it became the Third International Conference in Electron Microscopy. There never was an official "Second" conference.

However, the Paris conference was truly international and now included Germans (who in a sane world had more right than anyone else to be there), Japanese, and Italians, as well as the other Europeans and Americans. At the official dinner in Paris my wife and I were sitting almost opposite von Borries, who, when he heard we were English, held out his hand and said, "Where are the photographs? The English always carry photographs of the children!" On my right was Professor Tani of Japan, who surprised me by telling me that there were no less than five firms manufacturing electron microscopes in Japan at that time: Hitachi, Toshiba, Shimadzu, Japan Electron Optics Laboratory, and Denshi Kagaru Kenkyusho. Across the table was an Italian (whose name I do not remember), who at one stage stood behind his chair and called for "More speeches"—"Plus de discours"— whereupon Tani whispered to me, "Now we get a speech from Giulio Cesare!"

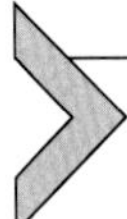

7. Specimen preparation and "the practice of electron microscopy"

It was earlier, at our Hampstead meeting of June, 1945, that at a morning tea break I incautiously suggested that we should collect our experiences in specimen preparation into a book, perhaps in the form of an updateable loose-leaf symposium, for the benefit of future workers. Sir Charles Darwin pounced on the idea as an excellent one and, as might have been foreseen, charged me with the task of implementing it. "Only," he said severely, "don't call it a 'symposium.' That's what we're doing now—drinking together!" We were standing round drinking cups of tea!

Specimen preparation was then at a very elementary level and success was all but restricted to the examination of small particles and randomly disintegrated materials, and some thin films and evaporated layers. Successful sectioning of biological material on a routine basis was far in the future and metallurgical specimens could only be handled by uncertain replica methods. However, I solicited information from the original six laboratories and from a few others that were coming into the field on the working details of their methods, and in due course received manuscripts which almost all began by describing how to prepare a supporting film and mount it on a grid, and, in general, overlapped so very considerably that there was no alternative but to rewrite the bulk of the material, condensing it to eliminate redundancies. This considerably delayed completion of the project and produced a sense of outrage among some of the contributors, especially those in more senior positions, at my cavalier treatment of their manuscripts, but I had to stick to my editorial guns, though it meant that the work did not finally appear until 1950. It was then published as a complete number of the *Journal of the Royal Microscopical Society* under the title "The Practice of Electron Microscopy" (Drummond, 1950). The idea of a loose-leaf format had been abandoned along the way, but a cyclostyled preliminary version was prepared in 1948 for use in the Summer School in Electron Microscopy, organized by Dr. V. E. Cosslett at the Cavendish Laboratory in that year and repeated in two or three subsequent years. Experience at these schools, and discussion with fellow demonstrators, helped greatly to improve the final printed version, as did information which was by then becoming available from a wider field of laboratories, both in England and in the postwar world generally.

Two thousand copies were printed, in addition to the normal circulation of the Microscopical Society Journal, a figure which at the time seemed optimistic, but this was the only publication of its kind to date and the edition was quickly bought up and went out of print. It was also, of course, going rapidly out of *date*, since the subject was developing fast, and it was not thought desirable to reissue it without considerable revision and rewriting, which circumstances precluded me from undertaking.

8. Grids

Incidentally, I feel I can claim to have established "grid" as the English term used to describe the mount for an electron microscope specimen. In those days a variety of other terms was being used, such as "screen," "mesh," "mount"—none of which seemed satisfactory. "Screen" tended, in writing, to be confused with the phosphorescent viewing screen, "mesh" implied a woven gauze structure (which *was* tried but proved too unstable dimensionally, and did not give a sufficiently flat area of support even after heavy rolling), and "mount" was not sufficiently specific to electron microscopy; but my manuscript demanded a consistent usage and I deliberately selected "grid" as the preferred term, and it has since been generally adopted.

Of course, at the time, grids were not available as individual predesigned items but were punched as disks from sheets of the perforated metal—usually nickel—which, so I believe, were originally produced as material for air filters to keep desert sand out of tank carburetors. I personally had some qualms about the effect of using a magnetic material such as nickel so close to the observed specimen in the lens field, but I never, in fact, noticed any adverse effect on the image—at our level of performance.

The maximum open area of these grids was never more than 25%, and though the holes were nominally square their edges were very irregular. Occasionally one received a batch in which the deposition of nickel had been overdone such that the holes were almost closed up, though they were still spaced at 200 to the inch. In such cases it became quite troublesome even to find a grid aperture to view the specimen and, as the holes no longer bore any resemblance to squares, the direction in which to move to find another aperture was also problematical. One could only hope that the next batch of grids, when they eventually came, would be more satisfactory. The modern, individually made grids, clean and regular, with a much greater open area, initiated by Smethhurst Highlight, came as a great boon to electron microscopists everywhere, but it was several years before production could match the demand. For this reason procedures were tried in many laboratories to clean and reuse grids, but no one was happy with this; frequently the work in hand required the comparison of specimens that differed only slightly from one another, and there was always a worry that some of a previous specimen would turn up, unrecognized as such, on a reused grid.

9. A problem with photographic plates

As the electron microscope must operate in a high vacuum and the photographic recording material has to go into the vacuum chamber, there was at first some doubt as to whether film-based material could be used or whether glass plates were essential. A bigger choice of emulsions and speeds was available on film but some, at least, of the films were too "gassy" for vacuum work (probably this depended on the plasticizer used in the base), and glass plates were generally favored. The RCA Type B microscope used a 10 × 2-in. plate which could be racked along to take several exposures along the length of the plate. This size may sound unusual but it was an available size, generally supplied for spectrographic work. An unexpected problem arose, however, since 10 × 2-in. plates were coated on a 10 × 8-in. size and cut up into 2-in. strips. In coating a 10 × 8-in. plate, the emulsion was poured onto the center of the glass and made to flow to the edges by judicious tilting, a very skilled job, I should imagine. However, though the center of the plate was quite uniform, viscous and capillary forces caused variations in coating thickness in an area about $\frac{3}{8}$ in. wide all round the margin of the plate, and in this region the sensitivity (to electrons at least) was changed. Hence on two out of every four 10 × 2-in. plates (i.e., 50% of all plates) the electron microscope images showed a band of excessive darkness all along one edge. This effect was very mysterious and we sought explanations in several directions before we came to suspect that the fault lay in the plates themselves. In spectroscopic work, spectra would usually only be recorded on the central strip of the plate and the varying sensitivity near the edge would not be noticed, but, in the electron microscope, images had to be recorded right to the edge. The solution to the problem was to order plates with "no coated edges," where 1 in. at each side of the 10 × 8-in. plates was cut to waste, and only three 10 × 2-in. strips were obtained from each 10 × 8 in. This was proportionately more expensive, of course, but at least we got 100% of satisfactory plates instead of only 50%. The coating effect was still present on the 2-in. ends of our plates but this region could easily be avoided.

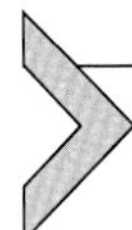

10. Capabilities and use of electron microscopes during the war

During the war the six lend-lease RCA electron microscopes in Britain were expected to be available for any investigations arising from wartime requirements—not only in their host institutions—and we had to attempt to cope with queries from any official, or semiofficial, or industrial source. There was, not surprisingly, a great lack of understanding among our clients as to what the instrument and the techniques were capable of, and often, also, a considerable lack of clarity as to the questions they wished to have answered. In many cases we were unable to help.

One textile manufacturer brought in a sample of cellulose monofilament at least a millimeter in diameter and as tough as horn, with the request "just to have a look at it in the electron microscope to see if there's anything interesting." That sample was hung on a peg and stayed there!

A more reasonable request was to examine a piece of armor plate taken from the back of the pilot's seat in the plane that Rudolph Hess had landed in Scotland. But we were essentially a textile research laboratory and had neither the knowledge nor the equipment for metallurgical investigations and had to pass that problem on elsewhere.

In the early days the favorite specimens used to demonstrate the electron microscope to interested visitors were small particles, particularly those of metal oxide smokes from burning magnesium or zinc. A grid with, or even without, a supporting film could be simply waved in the smoke to provide an "instant" specimen to show to an admiring audience. Such smoke particles are of high contrast and have striking forms.

Most of the specimens which we were able to help with at first resolved themselves into "small-particle" problems. Pigments of various sorts, including, for instance, those for mineral khaki, camouflage paints, etc., kept arriving, and since the information required would usually be the particle size and size distribution, the preparative methods used were generally based on the "rubbing-out" methods traditionally used by paint manufacturers to disperse the particles, but modified to enable a representative field of the particles so dispersed to be displayed on a supporting film on a grid. For carbon blacks, required as rubber fillers, a similar principle was employed but the rubbing-out medium was an uncompounded crepe rubber into which the sample was milled. This, it was argued, should give the sort of dispersion likely to be significant for the practical application of the material. The crepe rubber had of course to be washed away after the material had been mounted on a grid. However, if the state of aggregation of the particles within a rubber sample was of interest such a method would not do. We had no means of cutting sections, but we found that a surface layer thin enough for electron microscope transmission purposes, and containing filler particles *in situ*, could be stripped from a sample of prepared rubber by pulling away a pellicle of gelatine which had been dried onto the surface. The gelatine was then removed by dissolving it in water. In this way we could demonstrate, for instance, the conducting paths formed by the chaining together of carbon black particles in "conducting rubber." This was needed for aircraft tires to facilitate discharge of static electricity from a plane after a landing.

Substitute rubber fillers also came our way, e.g., such materials as finely divided calcium carbonates and calcium silicates. Carbon blacks were normally made by burning oil products, and as these were in short supply and needed for other purposes, anything which could reduce the demand for carbon black was to be investigated.

Abrasives as fine powders formed another class of small-particle specimens which we had to examine, and another class comprised the magnetic particles from recording tapes—our own and the German products.

One project which reached us, surrounded by a great air of mystery and what then seemed exaggerated security, was concerned with the corrosion produced by a gas named "HEX." We never were told what it was all about, but with postwar hindsight it seems evident that the gas was uranium hexa-fluoride, and one guesses that corrosion was interfering with ^{235}U enrichment by diffusion. We were able to show only something of the external form of the crystalline corrosion particles, information which at the time did not seem commensurate with the effect involved, but our clients seemed happy with the results.

The magnification and resolution at which we were able to examine specimens of this sort were at least an order of magnitude better than could be achieved with the optical microscope, and as such gave important information, but often the particles were thick and heavy by electron microscope standards and quite opaque to the electron beam. This gave rise to the unfortunate impression, in some quarters, that the electron microscope image was no more than a silhouette of the object, and did not and could not reproduce internal details in a range of tone values. This idea took some time to dissipate, but gradually other types of specimen came into evidence in which differences in image density were clearly representing internal details.

At the Shirley Institute, for instance, fragments of disintegrated textile fibers showed the fine fibrillar nature of the material which theory had postulated but which we had not seen before at this level, and differences were demonstrated between highly oriented fibers such as ramie, and the less oriented ones such as cotton, in images which were certainly not mere silhouettes. In the conventional biological field, although structures of tissues as a whole could not be shown, fragments of proteinaceous material, and especially the fibrous protein collagen, showed convincing periodical striations which could be related to the chemical makeup.

11. Shadowcasting

The introduction of shadowcasting methods by Williams and Wyckoff (1944), following on Müller's work (1942), was a revelation in many ways—not the least of these being cosmetic. The shadows, of course, gave additional information on the dimensions and distribution of the specimen material in the direction normal to the supporting surface, but, at the same time, the improved definition and contrast resulting from the presence of the heavy metal produced a most attractive, sharp picture, with a vivid impression of solidity and depth, which reproduced well in print and was good for public relations!

Moreover, all kinds of stratagems could be devised, using shadowcasting, to derive more information about the specimen. For instance, by shadowing in two directions at right angles, once before and once after carrying out some treatment on the specimen, the states before and after the experiment could be distinguished.

Replicas also were much improved in clarity and contrast by shadowing. There was some discussion as to how flat the outer surface, as distinct from the working side, of a replica film might be, and here again shadowing on the outer surface as well as the working surface could help to give the answer.

12. Section cutting

Shadowcasting has had relatively little place in biological electron microscopy since the development of section-cutting methods (where the specimen is ostensibly flat on both sides—not always a justifiable assumption, as shadowing might reveal), but it was some time before whole tissues as distinct from randomly disintegrated tissues could be studied in this way.

Attempts to develop sectioning techniques went on everywhere and it is perhaps salutary to consider how success was often hampered by preconceived ideas derived from the techniques used for optical microscopy. Biologists trained in optical microscopy, for instance, often insisted on carrying specimen preparation right through to the staining of their sections with the conventional dyes—which subjected their material to the hazards of an additional process, and was irrelevant. Also, the natural predisposition was to try to cut the thinner sections over an area suitable for optical microscopy, whereas the field of the electron microscope covered so much smaller an area that this was quite unnecessary and caused unnecessary difficulties (Porter & Blum, 1953). But one had to start somewhere and the basic new requirement was that sections should be, say, 50 times thinner than before.

Some early attempts were made, using standard microtomes, to cut sections with a tapered edge, such as by cutting a flat face on a block of embedded material and cutting again on a plane at a slight angle to the first cut. Small regions near the wedge-shaped edge might then be thin enough for examination, but this was a "hit-or-miss" business, and the mainstream of development required improvements to the microtome.

Modifications both mechanical and thermal to the feed mechanisms of standard microtomes were devised with varying success, but it was the Porter–Blum microtome (manual model) (Porter & Blum, 1953), specially designed for the job, that made the breakthrough. By itself, though, this would have been of little use without the other developments in technique that had been going on in parallel.

The optical microscopist's steel microtome knife was, naturally, the immediate choice for cutting sections, but this was less than satisfactory on several counts. The extremely thin and fragile sections cut "dry" onto the back of the knife were virtually impossible to remove and handle, and were in any case badly distorted, and the idea of cutting wet so that the sections floated away from the knife as they were cut (Gettner & Hillier, 1950) was not very practical, since the aqueous bath had to wet up to the extreme cutting edge of the steel and caused blunting by rusting. Also, it was considered (with what justification I cannot say) that the granular crystalline texture of the steel itself prevented formation of an edge fine enough for the purpose in hand. There was, briefly, a suggestion that knives might be made of beryllium-copper, which might have helped on both these counts, but this idea was killed at birth by the advent of the glass knife (Latta & Hartmann, 1950). This appears to have been an event of sheer serendipity, when a sheet of plate glass was shattered accidentally in an electron microscope laboratory and the fragments were seized upon as perhaps having edges which might make suitable knives. This indeed proved to be the case, and, glass being practically a noncorroding material, it now permitted the use of a water bath behind the cutting edge, onto which the sections could be floated as cut, with much less compression damage, and much less hazard in the subsequent handling, than in any "dry" cutting process.

But there were still other difficulties.

The conventional embedding media, such as paraffin wax, suitable for sections of 5 μm or so, did not give sufficient support to material cut one or two orders of magnitude thinner than this; here again alternatives had to be found. Many suggestions were made, including resort to harder waxes such as carnauba, but the first embedding material to emerge as in any way satisfactory was butyl methacrylate (Newman et al., 1949), which held the field until ousted by Araldite and other epoxys.

Further, although the fixatives favored for optical work preserved the structures visible at that level, they did not always work for detail in the electron microscope size range—fine coagulation of proteins could occur, for instance. But osmium tetroxide, buffered onto the alkaline side, proved to be satisfactory (Palade, 1952).

When all these modifications were brought together in the early 1950s (that is, the use of buffered osmium as fixative and butyl methacrylate for embedding, cutting onto a water bath with a glass knife on a special microtome such as the Porter–Blum, and bearing in mind that only very small areas need be cut) the production of satisfactory sections of biological material on something approaching a routine basis became possible at last, and this resulted in an explosion of knowledge of biological fine structures which would be difficult to parallel.

As an afterthought one should add that the supremacy of osmium tetroxide as the preferred fixative was perhaps due as much to its heavy-metal staining qualities, which made the fixed material visible in the electron microscope, as to its undoubted success as a preservative of structure.

Formaldehyde, which was discarded early on, turned out later to be a better fixative than had been thought, but this had not been recognized because it failed to produce material which could be seen easily in the electron microscope unless supplementary staining with a heavy element was applied. Since the 1950s there have, of course, been many improvements in specimen preparative equipment and techniques in the field of biology, but it was at that point that the subject "took off."

It may be interesting to recall that the diamond knife was first developed in Caracas, Venezuela, under the aegis of H. Fernández-Morán (1956), who was a wealthy man, reputed to be a millionaire. He traveled about the world carrying a supply of diamond knives with him and presented them free of charge to several electron microscope laboratories to try out, an activity which certainly promoted the well-deserved popularity of the diamond knife. But he was too prominent a citizen of Venezuela and, following a political upset in that country, had to go into exile, and the supply of diamond knives came to an abrupt halt. Fortunately his associates and workpeople remaining in Caracas were able, in a year or two, to get production going again and to put diamond knives onto the commercial market—but there were no more free handouts!

However, the adoption of one very useful technique, avoiding much of the damage surface tension forces can do to specimens, lagged far behind. "Critical-point drying" was described (Anderson, 1953) at the Paris conference in 1950 but it was not generally taken up until the 1970s. This long time lag seems surprising in view of the value of the method and the speed with which advances were being made in other directions. Maybe the need to set up the specialized equipment was a deterrent and perhaps we were a little nervous about moving into the unfamiliar field of high pressures (in excess of 73 atm with CO_2 as first proposed) with makeshift equipment. The requirements of scanning microscopy have finally popularized the method, and suitable pressure vessels and ancillary items are now available off the shelf.

13. Replicas and metallurgy

The other major field, besides biology, in which great efforts were made from the beginning to develop useful techniques for electron microscopy was metallurgy. Here sectioning methods could not be used—and even when, many years later, diamond knives made it possible to cut thin films from some metals, the stresses involved, and the consequent distortion of lattice structure, vitiated any observations made. Other methods of thinning metals have now been developed, but in the early days the only possible approach was by way of thin-film replicas of etched surfaces. The methods of polishing and etching used by optical microscopists formed the starting point but here also modification of the preconceived requirements were necessary. In particular, the etches had to be much lighter; the "key" provided by a conventional depth of etch made it very difficult to remove a replica film from a surface without tearing or distortion, and this applied even in a two-stage process wherein the final, or positive, replica was taken from a more robust intermediate plastic negative. Indeed, even using a lighter etch, the final replica necessarily consisted of a thin, fragile film, weakened by the valleys and ridges impressed by the surface being replicated and very subject to the hazards imposed by the surface tension forces to which it was inevitably exposed, whether it was a negative, floated directly from the metal, or a positive, released by a solvent from an intermediate negative. In general these thin films were difficult to handle and the success rate in producing satisfactory replicas was really too low to sustain an ongoing research program, though the occasional spectacular result might maintain interest.

A breakthrough came with the introduction of the "dry stripping" technique, enthusiastically exploited by J. Nutting. It was found that a thin Formvar film cast over a lightly etched metal could be lifted off with a piece of Scotch tape which had been pressed over it (Barrett, 1943); moreover, if a grid was interposed between the Scotch tape and the Formvar before they were pressed into contact, the area of the Formvar replica immediately below the grid would come away at the same time as the surrounding area which was in direct contact with the adhesive tape (Nutting & Cosslett, 1950). This produced a replica already mounted on a grid, without having been subjected to any surface tension hazards, and for the first time we had a very simple replica method which, though not leading to the highest resolution of which the microscope was capable, was sufficiently quick and reliable to sustain a routine research program. But there was a trick to it—it was necessary to breathe heavily on the Formvar-coated specimen before pressing down the Scotch tape. No one was quite sure why this was so, but it worked, and it was essential to success, though the practitioners came in for some chaffing as to what they found necessary to drink at lunch time!

At first, the pundits of optical microscopical metallurgy refused to recognize the validity of results obtained from replicas in the electron microscope, and so long as replica methods remained less than reliable there was some justification for their attitude, but F. W. Cuckow (1947) showed that the drystripped Formvar replicas gave the same story (though with much better resolution) as the reflection phase-contrast optical method which he had developed for metallography. Both these methods were able to differentiate, directly and reliably, between areas of different depths of etch and, therefore, between different metallurgical phases, whereas the conventional optical methods showed phase boundaries only. This helped greatly to establish confidence in, and acceptance of, the electron-microscopic results.

14. Optics of the electron microscope

The early commercial electron microscopes were strongly influenced by the traditional lens system of the optical microscope, and had condenser, objective, and eyepiece, the latter being a projection eyepiece, so that the image could be made visible on a phosphorescent screen. This system, with two stages of magnification, is entirely suitable for the optical microscope and is able to take that instrument to its theoretical limits of resolving power and magnification, largely because sufficient variables are at the disposal of the optical lens designer to enable him to correct for the many aberrations. The designer of an electron lens does not have this resource. Electron lenses always have strong positive spherical aberration, and no compensating component with negative spherical aberration is possible.

When a single projector lens is used, therefore, the final image shows strong pincushion distortion which is a direct consequence of the spherical aberration in the projector. To minimize this distortion it was desirable to utilize only a paraxial region of the projector aperture, and this dictated a change from the traditional balance of the stages of magnification in the optical microscope, in which the objective lens produced the main magnification and the eyepiece provided a relatively small factor, whereas, even in the earliest electron microscopes, it was the projector lens which largely determined the overall magnification. But this changed balance of the stages was only a palliative, and pincushion distortion remained an inevitable defect in the final image, somewhat less in evidence at the higher magnifications, but always present.

This was a great disadvantage, not only for the correct assessment of the shapes and sizes of objects in different parts of the field—of considerable importance where particle size distributions were in question—but also for investigations involving counts of particles, or other features, where areas of the field had to be referred back to unit areas in the specimen.

It was only with the introduction of the intermediate lens and further stages of magnification (not required in optical work) that ways were devised to overcome the pincushion distortion and to give us pictures which were true both qualitatively and quantitatively. The additional lenses also conferred great flexibility on the optical system of the electron microscope, increasing the useful range of magnifications (downward as well as up) and facilitating ancillary applications such as electron diffraction work.

15. Later instrumental developments

The development of the transmission electron microscope and the methods of preparing specimens for it have, as we have seen, all stemmed from the established methods of optical microscopy, but, in every direction, a decided deadweight of our own preconceived ideas has had to be overcome and drastic modifications have been necessary. Indeed, going back to the very beginning, Ruska has spoken of the "genuine reservations and strong psychological inhibitions" concerning the acceptance of the use of electrons at all, instead of light, in microscopy. Transmission electron microscopy was, essentially, created by physicists who strove to develop it into a convenient, reliable routine tool for the benefit of biologists, metallurgists, industrialists, and others, and this development included the working out of specimen preparative procedures (in collaboration), no less than the designing of the microscope. But the opportunities for direct application of the techniques to the problems of the science of physics itself are limited, and the interest of physicists turned to the extraction of more of the information inherently present in the interaction of an electron beam with a specimen than is made use of in the transmission electron microscope. Hence we now have the scanning electron microscope (SEM), the electron probe X-ray microanalyzer, and a range of cognate instruments which may no longer be recognizable as derived from the optical microscope, but which have greatly extended the range and form of the specimens that can be examined, and the "functional" as well as "structural" information that can be gleaned from them.

16. Conclusion

I would like to finish with a picture which has stayed with me over the years. At a conference at St. Andrews, in 1951, a few of us were taking a late evening stroll past the famous golf course and over the dunes onto the beach, and the talk turned to recent thoughts on the optics of the electron microscope. Despairing of getting the ideas across purely verbally, some of the group (Le Poole and Haine were certainly among them; I do not remember who else) fell to tracing the ray diagrams with a stick on the smooth sand. One could not help but recall the lines from Walter Savage Landor (1863):

Well I remember how you smiled
To see me write your name upon
The soft sea sand ...

. . . .

I have since written what no tide
Shall ever wash away ...

17. Gordon Drummond in Australia

David Gordon Drummond confines his fascinating article to his years in England, where he participated in the beginnings of the subject. But this was only a small part of his career. We reproduce two "Moments" from *50 Great Moments. Celebrating the Golden Jubilee of the University of Sydney's Electron Microscope Unit*, edited by Kyle R. Ratinac (Sydney University Press, Sydney 2008).

17.1 *Great Moment 2*: 'The First Director, Dr D. Gordon Drummond' by Kyle Ratinac[5]

Dr David Gordon Drummond, known as Gordon or 'Doc Drummond' to his colleagues and friends, was the first director of the the University of Sydney's Electron Microscope Unit (EMU). He led the unit from its earliest days in 1958, and so he faced the considerable challenges of installing the first TEM in the University of Sydney, educating the community about the new technology, and helping the University's researchers to use the instrument. His success in meeting the electron microscopy needs of researchers from the University of Sydney, and elsewhere, was evident in the rapid growth of the EMU. By the time of Dr Drummond's retirement in 1973, the unit had four TEMs, one SEM and a variety of equipment for specimen preparation, all supported by a 14-person staff; not bad for a unit that started with just two staff and one column some 15 years earlier (see Great Moment 1).

[5] Reprinted from 50 Great Moments, pp. 13–21, courtesy of the author and Sydney University Press.

Dr D. Gordon Drummond during the EMU era.

To understand how Dr Drummond came to lead the EMU through its early years, we need to travel halfway around the globe and back to 1929. After receiving a Bachelor of Science in physics that year, the young Drummond began a Master of Science in physics in 1930 at Armstrong College, Newcastle-on-Tyne, which was part of the University of Durham. He studied the optical properties of quartz and produced a thesis entitled "A search for weak absorption bands in fused quartz between 0.44 μ and 2.7 μ". He then substantially expanded the scope of his research to undertake a doctorate at Armstrong College, and in 1934 he submitted a PhD thesis entitled "Infra-red absorption spectra of silica". A particular aspect of both of these studies was the use of large pieces of silica – lengths of up to 306 cm of fused silica and 69 cm of quartz – in contrast to work by earlier researchers in which only small lengths of silica were used. Dr Drummond wrote a number of short letters in *Nature* from 1932 to 1936 about his measurements on the optical properties of the various forms of silica. His letter of 1932 examined the differences in absorption of infrared radiation by large thicknesses of quartz when the radiation travels parallel or perpendicular to the crystal axis (i.e. the optical axis) (Drummond, 1932). In 1934, one letter presented new results on the infrared-absorption spectra of various forms of silica (Drummond, 1934a) and another presented revised refractive indices of quartz for infrared radiation (Drummond, 1934b). His letter of 1936 discussed the possible origins of the 2.73 μm absorption band in fused silica, which he considered most likely to be due to absorption of CO_2 gas (Drummond, 1936a). He also published two extended articles in the *Proceedings of the Royal Society of London* during that year (Drummond, 1936b, 1936c).

In 1934, Dr Drummond joined the Shirley Institute in Didsbury, Manchester, which was headquarters of the British Cotton Industry Research Association (Monteith, 1986). He was employed as a spectroscopist in the Rayon Department of the institute and his work included studies of the optical anisotropy of cellulose, and the attachment of dye molecules to cellulose and subsequent fading of colors. While employed at the Shirley Institute, Dr Drummond continued to publish manuscripts from his doctorate (see above) and from his new duties (Drummond, 1940). He might very well have gone on working in this spectroscopic field for the rest of his career, and he might have been quite content to do so. However, the Fates had other plans for Dr Drummond, for the recent developments in the field of electron microscopy in Germany – associated with names like Ruska, Knoll, von Borries, Brüche, and von Ardenne – were starting to be noticed in England. As explained by Dr Drummond (Drummond, 1985), when the Second World War erupted in 1939, the British leadership realized Germany's technological advantage in this new form of microscopy. They sought to remedy the situation by asking the National Physical Laboratory to recommission an early electron microscope from Imperial College, which had been built by the Metropolitan Vickers Company three years earlier (Mulvey, 1985), but had not performed well (see Great Moment 4). Shortly thereafter, the Director of the National Physical Laboratory, Sir Charles Darwin, was in the USA and learned that Radio Corporation of America (RCA) had a microscope in commercial production, the RCA Type B. Recognizing that it would be more expeditious to order these commercial instruments than to redesign the instrument from Imperial College, Sir Charles placed an order for seven microscopes to be delivered to Britain under the wartime lend-lease arrangement (Reed, 1985).[6]

These microscopes were shipped to England, one at a time, during 1942 and were allocated to laboratories such as the National Physical Laboratory;

[6] Dr Drummond's personal reminiscences (Drummond, 1985) mention only six microscopes; it seems he forgot about the seventh instrument that Reed (Reed, 1985) states was allocated to "an undisclosed Scientific Research and Development Establishment run by the Ministry of Supply". The fate of this instrument is unknown.

the Cavendish Laboratory, Cambridge; the National Institute for Medical Research; and, to many people's surprise, the Shirley Institute in Didsbury. Some time prior to the arrival of the microscope – which appeared at the Shirley Institute in seven bulky crates in late 1942 – the institute's director, Dr Robert Pickard, chose Dr Drummond to assemble and operate the new 'machine'. Displaying his keen wit, Dr Drummond later attributed his newfound role to his complete ignorance of the science of electron microscopy at the time (Drummond, 1985). As to the allocation of the microscope to the Shirley Institute in Manchester, Dr Drummond "always maintained in a marked Northumbrian accent that it was given by the Southerners as a sop to the rough tribesmen living north of Watford" (Reed, 1985). As a more practical reason, Reed suggests that the instrument was sent to the Shirley Institute because at least some of its technical research, for example, in polymer impregnation of fiber assemblies, was important for the war effort.

Aided by Mr J. B. Todd, Dr Drummond faced considerable challenges in assembling the new microscope and the whole process was fraught with delays (and frustrations, I suspect) (Drummond, 1985). One difficulty was the initial absence of an instruction manual or assembly diagrams. Although these gradually arrived piecemeal over the course of several months, it seems that the early stages of construction resembled the assembly of some monstrous three-dimensional jigsaw puzzle, where finding matching flanges and the general look of the instrument were the only guides. Another challenge was plugging the thing in; naturally enough, the American-made instruments were designed for an electricity supply of 120 V and 60 Hz, rather than the 240 V and 50 Hz standard for Britain. The Manchester engineering firm of Metropolitan Vickers came to Dr Drummond's aid by making a suitable transformer to deal with the voltage difference, although the wait for this further delayed the installation. Corrosion of components during the microscope's long sea voyage from the USA also caused many headaches. For instance, despite their protective ceramic coating, numerous wound-wire resistors were corroded and needed replacement, which posed another difficulty during wartime. The ongoing operation of the airlocks for the specimen stage and camera was complicated by rusted spring mechanisms, and the projector lens suffered from rusting of the magnetic polepiece, necessitating some impromptu repair work to this crucial and sensitive component.

Once the instrument was operational, the fledgling microscopists rapidly learned that the RCA Type B left something to be desired in terms of ease of use. Adjustments to the electron gun brought the risk of electrocution, while inserting, changing or aligning objective apertures necessitated complete disassembly of the column. Nevertheless, with persistence and the help of Mr Todd and, later, Mr S. F. Ward, Dr Drummond began to make inroads on the use of the instrument, learning its peculiarities and developing suitable approaches to preparation of specimens. Many of the early samples studied were from government departments and industry involved in the war effort, because each microscope came with the stipulation that the instruments were to be made available for such research for the duration of the war. The most common war-related samples were particulate materials, such as pigments for camouflage paint, carbon blacks, and pulverized minerals to be used as fillers for rubber, and abrasive particles. Although Dr Drummond (Drummond, 1985) offered no insight into whether these external jobs were seen as a pleasant distraction or merely a burden, it seems reasonable to conclude that running an electron microscope on behalf of researchers from other organizations provided him with valuable experience for his future role in managing the service-oriented EMU at the University of Sydney.

A year or so after the arrival of the electron microscope at the Shirley Institute (either November 1943 (Drummond, 1985) or April 1944 (Reed, 1985)), Sir Charles Darwin of the National Physical Laboratory convened a meeting of the scientists working with the seven microscopes, as well as various defence staff, at the Royal Society's premises in London. There the microscopists had the opportunity to discuss their early results – Dr Drummond gave the first presentation – and to meet each other socially. Given the success of the first meeting, regular meetings of this type were held once or twice a year thereafter, with attendance gradually growing in size and diversity, especially as scientists from Europe began to participate after the end of the war. At the third such meeting in June 1945, Dr Drummond rashly suggested that it might be worthwhile to compile a booklet on the art of specimen preparation from the group's collective experiences to provide a resource for other researchers. The merit of this idea was obvious to the group and, of course, Sir Charles asked Dr Drummond to act as editor for the work and to collect contributions from each laboratory, which he duly did. When he found that there was substantial overlap among the contributions, Dr Drummond proceeded to revise and update the submissions, which was a lengthy process that delayed the publication of the book until 1950 and incurred the displeasure of some of his counterparts at his editorial treatment of their contributions.

Nevertheless, when the magnum opus finally emerged as a single volume of the *Journal of the Royal Microscopical Society* (Drummond, 1950), the value the editor's uncompromising attitude quickly became evident, and soon even the extra 2000 prints of the special-edition journal sold out. The book was comprehensive and highly practical in nature, covering methods for making and mounting films on grids, ways of preparing powders and airborne particles for viewing, methods for forming surface replicas, approaches to preparing micro-organisms and biological tissues, routes to calibrating magnification, and issues pertaining to photography of electron micrographs. The impact of Dr Drummond's work was profound. In the foreword to a subsequent book on electron microscopy techniques published some 11 years later (Kay, 1961), Dr V. E. Cosslett – a respected electron microscopist from the Cavendish Laboratory in Cambridge – highlighted the importance of Drummond's book: "It is not too much to say that the rapid growth of electron microscopy in Great Britain during the next decade was mainly based on 'Drummond'. It found much use in other parts of the world also."

Dr Drummond made other contributions to the developing British electron microscopy community in the 1940s and 1950s. For instance, he was responsible for the local organization of the electron microscopy conference held in Manchester in March 1946, which had over 100 attendees, and he still managed to find time to present results on gold-shadowed dust particles collected from cotton mills (Reed, 1985). During the following conference, held in Oxford in September of the same year, Dr Drummond presented a detailed review of specimen preparation methods (Cosslett, 1947). He presented papers on electron microscopy of fragments of various textiles in 1947 (Sayer et al., 1948); on the design of a new polepiece for their RCA microscope to solve a unilateral distortion at high magnifications in 1952 (Cosslett et al., 1953); and on techniques for preparing surface replicas of fibers in 1953 and 1956 (Challice, 1954, 1957). Of course, he was a regular attendee at most of the electron microscopy conferences, even when not speaking, and he engaged in what must have been lively and enjoyable discussions at these conferences (Drummond, 1953; Drummond & Liebmann, 1952; Sayer et al., 1948). He took a formal leadership role in the community, too; when Dr Cosslett retired from the chair of the Electron Microscopy Group at the annual general meeting in July 1955, Dr Drummond was elected as the new chairman. Dr C. E. Challice of the Wright-Fleming Institute of Microbiology, London, then the honorary secretary of the Electron Microscopy Group, had precipitated this move by

Dr Drummond and some other leading European electron microscopists at the Conference on Electron Microscopy, St Andrews, June 1951. From left: J. B. Le Poole, V. E. Cosslett, D. G. Drummond, G. Liebmann and M. E. Haine. (Image is courtesy Elsevier. It appeared as Figure 9 in: A. W. Agar, "The story of European commercial electron microscopes", Advances in Imaging and Electron Physics, 96: 415–584, copyright Academic Press (1996). Sincere thanks to Dr Alan Agar for providing the original photograph.)

writing a brief letter to Dr Drummond in June 1955, in which he stated: "if you would allow me, I should have great pleasure in putting your name forward [as new chairman] at the Meeting". After first clarifying that it was appropriate for him to be chairman of the group while continuing as secretary of the Manchester branch of the Institute of Physics, and that he could do so without excessive additional administrative burdens, Dr Drummond had agreed to be nominated. He remained as chairman of the group until 11 September 1957, just prior his move to Australia.

During the 15 years between the arrival of the crated RCA microscope and his departure for Australia, Dr Drummond continually honed his knowledge of the art of electron microscopy and built a strong reputation in the British electron microscopy community, in which he provided leadership and enjoyed an obvious sense of camaraderie. Throughout that time, he was employed by the British Cotton Industry Research Association at the Shirley Institute where his day-to-day investigations were determined by his superiors. After a number of years, these investigations became increasingly at odds with his growing interest in electron microscopy. Dr Drummond's studies in the card-rooms of cotton spinning mills, in particular, gradually drew him away from the regular use of electron microscopy as he explained in an undated, unaddressed and self-titled memorandum (from archival documents in the EMU):

> *In the early days with the electron microscope I was naturally fully occupied in finding out how to use it and to prepare simple specimens. I was then, at a quite early stage, diverted into the card-room dust field and later into that of the card-room bacteria, which prevented much progress in the other applications of the electron microscope and ultimately took me away from this instrument entirely.*

Without doubt, the work on the card-room dust and bacteria was important. The high levels of dust and bacteria in the card-rooms of cotton mills were responsible for the frequent cases of byssinosis – a chronic and often fatal respiratory disease – in laborers in British mills (e.g. Drummond & Hamlin, 1952). Nevertheless, these investigations were clearly not as dear to Dr Drummond as research on the electron microscope and its applications. With the closure of this period of work, he was anticipating the return to electron microscopy, as detailed in the memorandum:

> *I have been looking forward to returning to the EM work proper, with the advantage of additional personal experience and a knowledge of many new developments which have taken place in [electron microscopy] techniques, in order to tackle some of the textile problems other than card-room dust.*

Though the memorandum is not dated, from the details it contains, we can conclude that Dr Drummond wrote it in late 1952 or early 1953. The impetus for this document was that his "future course at the Shirley Institute appear[ed] to be undecided". Presumably this indecision coincided with the conclusion of the extensive investigation on card-room dust and bacteria and it obviously caused some disquiet for Dr Drummond, because he felt the need to write suggestions on his future work in the institute. His preferred future direction was to make detailed investigations of cotton fiber with the electron microscope. However, he also noted that the suggestion had been made to him that he "drop electron microscopy and take over various items of optical work in the Physics Division", a move that would have been, to some extent, a return to his field of research prior to the arrival of the electron microscope in 1942. Dr Drummond then proceeded to provide a list of optical problems in textiles that, on the basis of discussions with other staff, needed investigation, after which he still managed to propose that the electron microscope might be of use to at least one of those areas. Evidently, it had even been intimated to Dr Drummond that the microscope might need to be moved to another department, which would not have been a trivial task and further highlights the extent of the ambiguity around his future directions.

So what was the outcome of all this uncertainty? Two studies from Dr Drummond's work at the Shirley Institute published in the *Journal of the Textile Institute* shortly after his move to Australia incorporated substantial use of the electron microscope (Cumberbirch et al., 1959; Drummond & Warwicker, 1959). These manuscripts confirm that Dr Drummond continued to use the microscope after the period detailed in the memorandum, even though it is unclear whether this completely satisfied his scientific interests and removed the doubt over his "personal standing in the Institute". In any case, we can speculate that this uncertainty, and the soul searching that usually goes with it, prompted Dr Drummond to reassess his future career directions and consider other possible avenues of employment, given that some four years later he applied for the newly created role of electron microscopist at the University of Sydney.

At the same time that Dr Drummond was struggling with his future course at the Shirley Institute, discussions were underway in the University of Sydney about acquiring an electron microscope, primarily to meet research needs in the biological and medical areas. During 1956, the University committed to the bold step of providing this microscope as part of a centralized service, supported by a dedicated electron microscopist who, to ensure impartiality, was not to be part of any department or faculty, but would answer directly to the Vice-Chancellor (see Great Moment 1). There were strong representations made by Mr David A. Cameron, then a senior lecturer in dental pathology and an electron microscopist himself, that it was essential for the new chap to have a biological or clinical background. This sounded reasonable as most of the expected demand for this new technique came from fields like botany, histology, and pathology. Nevertheless, the advice from the then national leaders in electron microscopy at the Chemical Physics Section of the CSIRO in Melbourne emphasized the need for the electron microscopist to also have a grasp of physics, in other words, to be a biophysicist. Consistent with creating a centralized service, able, at least in principle, to meet the research needs of all departments, the University's executive took the CSIRO's advice that the leader of the new laboratory should blend skills in biology and physics.

The new position was advertised nationally and internationally in early 1957 (*Minutes of the Senate*, 1957a). We will never know how Dr Drummond learned of this opportunity. Possibly he saw an advertisement in Great Britain or perhaps he heard about it by word of mouth through his connections in the British and international electron microscopy communities. Whatever the route, he applied for the job, as did 18 other applicants. A selection committee, which probably included some or all of the Electron Microscope Committee, met to assess the 19 applications and recommended that Dr Drummond should be appointed as the electron microscopist. We can infer, from the Professorial Board's report to the Senate, that this recommendation was based on Dr Drummond's long experience in electron microscopy, his formative contributions to the British electron microscopy community, and his recognition among his peers. On the last point, one of Dr Drummond's referees said that he had "acquired certainly a European and to some extent a worldwide reputation as an electron microscopist". I suspect that the committee also considered this senior British microscopist to be a genuine biophysicist who might, with a bit of luck, satisfy the divergent expectations for this role across the University. Recall that Dr Drummond, though qualified as a physicist and experienced as a microscopist, had spent almost his entire career at the Shirley Institute in the study of textiles, fibers, bacteria and so on. The Professorial Board met 22 July 1957, ratified the selection committee's recommendation and passed it on to the Senate for the final appointment to be made (*Minutes of the Senate*, 1957b). Professor of Biochemistry Jack Still and the Vice-Chancellor passed the motion to appoint Dr Drummond during the Senate's meeting of August 1957. Needless to say, the appointment of Dr Drummond did prove to be a smart choice, for he was able to work successfully with the predominantly biologically oriented users of the EMU from many different departments and to forge the unit's identity in its early years.

After providing suitable medical reports, Dr Drummond was formally appointed as the electron microscopist in late August 1957 and, suddenly, Dr Drummond found himself inundated with tasks to do before departing Merry England. The logistics of uprooting his family to move to Australia and the need to conclude his employment in the Shirley Institute proved to be a challenge as he noted in correspondence with Mr W. Harold Maze, Assistant Principal of the University of Sydney: "I am sorry I have not yet got requests for other items of equipment off to you, but I cannot unduly neglect my present job & am also finding the business of selling up & removing across the world very time consuming" (Drummond, 1957b). And subsequently: "This letter has been written under some difficulty while winding up both my personal affairs here, & the work of my previous post . . ." (Drummond, 1957c). Dr Drummond also had to conclude his association with the British Electron Microscopy Group; in his final task as chairman of the group, he presided over the conference and annual general meeting in Bangor, Wales, during the second week of September 1957.

Finally, he had a number of tasks to do in preparation for his new job. When Mr Maze first wrote to Dr Drummond in early September, he provided a detailed update on the planning around the new microscopy service and supplied a swag of documentation such as basic floor plans for the rooms, layouts from other Australian electron microscopy laboratories, tentative lists of supplies and auxiliary equipment for confirmation or revision, other technical documents, and correspondence from the Australian vendor of the Siemens microscope (Maze, 1957). Mr Maze requested that Dr Drummond use the floor plans of the space allocated in the basement of the Chemistry School to devise a suitable layout for the new laboratory. He also asked Dr Drummond whether he wanted to install the new microscope, a Siemens Elmiskop I (see Great Moment 4), given that "the local agents will be of no help at all", or whether he would need to hire a Siemens technician for the installation. Mr Maze suggested: "If you are happy about installing the instrument yourself you might care to do one of the short courses which I understand Siemens runs, before you sail for Sydney". Indeed, Dr Drummond was keen to do the Siemens training course and install the microscope himself, though this added further to his 'to do' list because he had to communicate with the Siemens factory and then attend the training in Berlin during the last two weeks of November (Drummond, 1957c). Always thorough, Dr Drummond also organized to attend the installation of an Elmiskop I at the National Physical Laboratory in Teddington in early October and to visit a couple of other laboratories around Great Britain to see the latest in specimen preparation techniques before his departure for Sydney (Drummond, 1957a). When he returned from Berlin at the start of December, Dr Drummond wrote a final letter to Mr Maze from temporary digs in Essex – his house in Cheshire had been sold while he was in Germany – apprising him of his successful visit, the expected (now months late) date of dispatch of the Elmiskop, and his own arrival date in Sydney.

On or about 18 January 1958, Dr Drummond and his family arrived in Sydney after a month's sea voyage on the good ship *Iberia*. Once there, he began the process of acquainting himself with the Australia clime and lifestyle and of meeting the many challenges in setting up the electron microscopy service that soon became the EMU (see Great Moment 3). In the early days, Dr Drummond exerted considerable effort in educating the University community about electron microscopy – he started formal training courses for users in 1960 (see Great Moment 6), and he wrote general interest articles in The Gazette to explain the power of electron

Dr Drummond and his wife at the 1978 EMU Christmas party, several years after his retirement.

microscopy and the services provided by the EMU (Drummond, 1960, 1971). He also continued to travel to international electron microscopy conferences and visit other laboratories to keep abreast of the latest developments, to maintain his network in the electron microscopy community and to gather intelligence when considering the acquisition of major new instruments for the unit (see Great Moment 9).

Dr Drummond remained as the head of the EMU until his retirement in September 1973, and many of the milestones of his directorship are detailed in other great moments. Dr Drummond died in Sydney on 20 November 1987 (Cockayne, 1988).

17.2 *Great Moment 3*: 'The early days of the Electron Microscope Unit' by David Gordon Drummond[7,8]

When I arrived in Sydney in 1958 to take up the job of establishing an electron microscope laboratory as a central service in the University of Sydney, there was no effective electron microscopy in this city. True, there

[7] Reprinted from 50 Great Moments, pp. 22–29, courtesy of Sydney University Press.

[8] An edited reprint of: D. G. Drummond, "The Electron Microscope Unit of Sydney University from 1958 to 1973. Random recollections", Australian EM Newsletter, **10**: 11–15 (1986).

was a small instrument in the Technical College (which shortly evolved into the University of New South Wales), but this was a low-power instrument, not being effectively supported administratively, and not capable of serious work. The leading instrument on the market at that time was the Siemens Elmiskop I, and I had accepted the post at the University of Sydney with the assurance that one of these was on order. In fact, I went to Berlin for a short course on the instrument (J. V. Sanders of Melbourne was on the same course) just before I sailed for Australia on the *Iberia*. There was then more active interest in electron microscopy in Melbourne and when the *Iberia* docked there, John Farrant [a CSIRO electron microscopist] met the ship and took me and my family off to his home for the evening, where I was introduced to several other people active in the field. I had not expected to be met at Melbourne, and, from the ship's deck, had not recognized the gesticulating figure on the quayside below as John until in desperation he seized some handfuls of light-colored fertilizer from a bag nearby and with it began to spell out DRUMMOND DINNER on the roadway; by the time he got to DRUM-, contact had been established!

At Sydney University, I was nominally "responsible directly to the Vice-Chancellor" – Professor (later Sir) Stephen Roberts – but for immediate administrative purposes came under the aegis of Mr Harold Maze in his capacity as Assistant Principal, in which he had charge of buildings and services. Clearly the electron microscope laboratory had been thought of in the initial proposals as another general service, more or less in the same category as a centralized photographic service (or even an office of works) with little appreciation of the interdisciplinary academic implications. This seemed to me to be an inappropriate view and it was not long before I was able, with Mr Maze's concurrence, to change the name – which had not really come into use anyway – from the 'electron microscope service' to the 'Electron Microscope Unit' (with capital letters!) which gave us a more clear-cut identity, and a better recognition of our function. I was, however, very happy to remain under Mr Maze's administrative care, and he always gave me understanding support. In turn, I tried to keep within the budget and not to make unreasonable requests. My policy was to allow the unit to grow organically as the demand quickened, and not to try to force development in directions for which there was, as yet, no special call. So, in spite of the name change, which I established at the first opportunity, I did feel in the early days that it was our primary duty to provide a service, though an informed service with interdisciplinary links through which we were aware of, and could draw attention to, parallel developments both within our own clientele and also worldwide, and so cross-fertilize the whole field. If additional equipment or technical assistance was required to meet an obviously emerging need, I would endeavor to acquire it for the central laboratory.

Basically, the unit was an applications laboratory, and we never concerned ourselves with trying to develop the microscope itself as an instrument – which was the legitimate concern of some laboratories, especially some of those in Melbourne. The demand within the University of Sydney came almost exclusively from the medical and biologically-based departments (there being no metallurgy to speak of in the University, which, I confess, surprised me when I first arrived, and very little demand from chemistry or other non-biological departments), so that virtually all the assistants I appointed to the staff were biologists of one sort or another – with equipment requirements to match. As a physicist myself, I deliberately abstained from personal researches which would have diverted resources and detracted from the build-up of a strongly established, if biologically oriented, unit, on what might be called a 'popular' basis, directed to meeting existing demands. Maybe this did little for my personal standing in the academic hierarchy, but I had previously been away from a purely academic environment long enough to lose what seemed to me the exaggerated veneration paid to 'publication', per se, as a yardstick of success. Not of course, that I would decry publication when justified on its merits. The advent of the scanning electron microscope and the microprobe changed the picture. When I acquired the first of these instruments in the last year or two of my time, the unit's clientele widened to bring in the non-biological departments, and opportunities for personal research in the non-biological fields were considerably enhanced. By then the unit was sufficiently established and respected in the University to accommodate the new developments without question. I had allowed my biologically minded staff to have their own research projects, often in collaboration with client departments. Now it became possible to sustain this freedom in other fields, and I am happy to think that my successors could take advantage of this.

The material considerations

Before I left England, I had been sent a plan of the area that was to be made available to me, and asked to draw up a proposed layout for the new laboratory. I did this with provision for one microscope room and for ancillary equipment and services, but also with a contingency provision for making a second microscope room within the same area, should expanding demand make this necessary. I had, of course, never seen the site, and the result would have been horribly crowded, but I think the committee responsible was somewhat impressed by this approach. Fortunately, the contingency never needed to be implemented and our subsequent growth took place by expansion into adjacent areas. When I first saw the space provided, in the basement of the Bank Building, it was a rather dingy cellar. Natural light came only from two windows at the far end of a long room, and the walls were of exposed brickwork, quite unsuitable for ensuring the 'clean' conditions of working that I considered necessary for EM preparative procedures. And meanwhile I had nowhere to use as an office while workmen prepared the room in accordance with my plans – somewhat modified following inspection of the site. Mr Maze, however, soon found me temporary accommodation in a small (very small!) room in an offshoot of the Bank Building that overlooked the Vice-Chancellor's quadrangle, and here I set out a desk and chair, and one or two shelves. The room had a tall, narrow window and I realized that the sunlight falling where I first placed the desk was uncomfortably strong, but I subconsciously supposed that it would soon move off, instead of which it moved farther and farther on. I had failed to allow for the fact that I was now in the southern hemisphere. Another hazard was that the Vice-Chancellor's quadrangle was shut and locked in the evenings against public access, and on one occasion I left by a side door into that quadrangle, banging the door behind me, only to find myself trapped. Thankfully, I did not have to spend the night there, as one of the late-working chemists in a lab overlooking the quadrangle eventually saw my plight and came to the rescue.

One of my first requirements was to check the area of my main laboratory for stray magnetic fields, which, according to Siemens' specifications, might affect the performance of the microscope. (If one is between the main incoming power-cable to a building and the return cable, one is sitting in a large magnetic loop with a fluctuating field.) The site I had planned had a fairly low field, but I nevertheless consulted the electrical engineers on the thickness of iron shielding required to reduce the magnetic field to negligible proportions. They came up with a figure of 12 feet! So I compromised by screening the walls, floor and ceiling with a 16-gauge galvanized sheet – at least it would be an electrostatic screen! To bring in air-conditioning, a hole of about half a square foot was cut through the screen. It was covered with expanded metal, electrically bonded to the screen, but insulated from both the inner and outer runs of ducting. This was satisfactory at first, but, in the course of a year or two, a build up of fluff and dust on the expanded metal gradually shut off the air flow, and the EM room became unbearably stuffy – inexplicably so, until we took down a section of the ducting and revealed the trouble. Dr Frank Mercer of the Botany Department was using the microscope one afternoon, before the air-conditioning trouble had been rectified, and, being overcome by the close atmosphere, fell asleep. His ear must have been close to the microscope table, immediately over the alarm bell which on rather rare occasions signaled vacuum trouble, and which chose that moment to carry out its appointed function. I have never before or since seen anyone shoot out of an EM room so fast or looking so shaken; but he was not amused.

When I had finalized my requirements as to layout, room divisions, screening, benches etc., the office of works moved in to carry out the work. This included lining all the rough brick walls and other partitions and ceiling with smooth Masonite panels, painted in light colors, and the siting of air-conditioning ducts close up to the ceiling where they were then boxed in so that no dust-trapping horizontal surfaces were exposed above the benches.

But perhaps the biggest headache was the supply of cooling water to the microscope. At first I proposed simply to use tap-water and run it to waste, as I had done for my EM installation in Manchester, but apparently this caused panic at the Water Board who came and solemnly inspected the equipment and the requirements and gave, one felt rather grudgingly, permission to run one liter per minute to waste. The Sydney water supply had had problems up till then, but the Warragamba Dam was just coming on line and there was no longer such an urgent need for water economy. However, the quality of the water left a good deal to be desired. I think the sediment stirred up during the initial filling of the dam had not yet settled. We had gauze filters in the line to the microscope but these became clogged very quickly and one could find snail shells as big as three or four millimeters across among the debris. There was also a need to cool the water below its normal mains temperature to cope with the heat output of the microscope (mainly arising from the lens windings) and we tried various devices for this – including, at one stage, a beer-cooler such as is used in pub equipment. But it became clear that we were going to have to use a recirculating water system with an adequately big cooler. I then made the mistake of putting distilled water into the circulating tank and this proved to be far too active and caused corrosion in the lens bodies, which was a considerable worry.[9] Finally I reverted to tap water (but one

[9] Associate Professor Cedric Shorey recalls that the water corroded right through the cooling tubes in the objective lens, allowing water to enter the evacuated microscope column. He

recirculating tank-full only; not a continuous muddy flow!) with an anti-corrosive agent added. All this took time to sort out and I was getting impatient. At lunch in the staff club one day, I was sitting opposite Mr Maze and burst out, "I'm fed up with all these water problems. I came out here to do electron microscopy, not plumbing!" and from the quick glance that he exchanged with (I think it was) Professor O'Neill, the authorities had been beginning to have similar thoughts about the rate of progress. However, soon afterwards, though nearly a year after my arrival in Sydney, we got things right and began to operate as electron microscopists.

I preferred at first that the actual handling of the microscope should be by my trained staff only, with the client sitting alongside to indicate what fields were to be photographed, as I could not afford to have the equipment put out of commission by inexperienced operation when we had only one microscope and no back-up facility. There were of course exceptions to this rule, and Dr Mercer, for instance, was one of them.

As regards specimen preparation, we were initially feeling our way almost as much as the client departments, and I have to acknowledge the debt which we owed to their collaboration in many instances. After two or three years, however, I felt sufficiently confident to propose running a vacation school in electron microscopy for all comers, both from within our own University and from outside, Australia-wide. I was encouraged to do this by my experience in the summer schools in electron microscopy that Dr V. E. Cosslett had run several years previously at the Cavendish Laboratory, Cambridge, and at which he had invited me to assist. We could provide the venue and equipment and the basic course, but still welcomed help from other experts. Dr Neil Merrilees of the Department of Anatomy, University of Melbourne, especially was able to come and help considerably on the biological side. I was very conscious also that there were people out there who wanted to know about electron microscopy, but whose interests were non-biological (more especially the metallurgists), and who should be catered for, and I was very grateful for the assistance of Frank Nankivell of the Aeronautical Research Laboratory in Melbourne, who came, bringing

discovered the problem on a Monday morning just before he turned on the Elmiskop. He reported to Dr Drummond about the water in the column, but that he could not see any goldfish. Once Dr Drummond inspected the damage, they had to import a new objective lens from Germany. This was expected to put the microscope out of action for several weeks. Dr Maret Vesk, who was about halfway through her doctoral research at that stage, remembers that it actually took months because the new lens was sent by sea rather than by air.

some items of specialized equipment, and virtually ran all that aspect of the School.

Similar vacation schools followed in succeeding years (not quite every year) and we gradually became less dependent on outside assistance on the biological side, though on the metallurgical side this remained necessary, and, in the later schools, Frank Nankivell's role was taken over by Tony Malin of the University of New South Wales's School of Metallurgy, who was able to call on the facilities of that laboratory to back up the course. These vacation schools seemed to be appreciated, and were certainly enjoyed by my staff, who welcomed the change from normal routine. By the time I retired we had put through a total of about 200 participants mostly from Australia, but one or two from New Zealand, and even one from the USA. We tried to accommodate interstate and overseas people in the colleges during the course, and on the last occasion, to speed things up, decided to hire a minibus to ferry them between college and lab. My technician went to collect the bus on the afternoon of the first day and, to my mounting anxiety, did not reappear until a few minutes before the first ferry trip was due. The bus had been stolen from its home garage that day and only recovered by the police in the late afternoon!

Expansions

The original space I was allocated was immediately under the CBC Bank premises, and one day, during the initial fitting out of the lab, I was told that in fixing some of the alterations the workmen had drilled through into the bank strongroom to the general consternation upstairs. It makes a good story, but I was not an eyewitness and, as no official complaint reached me, I cannot vouch for its truth.

Our first expansion came when the firm of Andrew Thom wished to set up an Akashi Tronscope for demonstration to potential purchasers and, rather than go to the trouble and expense of installing the instrument on their own premises, approached us to accommodate it in our unit on the understanding that we could have full use of it, provided that it would be available to them for demonstration purposes whenever they wished. I welcomed this suggestion and was able to acquire the adjacent space, about 12 feet wide, parallel to our original room, as the site for its installation. In the end wall of this space, we made a removable panel – normally bolted into position – to give access for major items of equipment from the corridor outside. As the years went by, several microscopes in succession were housed in this area. The Tronscope, which was the first of these, arrived on one huge crate that, by some wizardry, the office of works contrived

The sandstone columns and façade of the Bank Building in the 1960s – home to the new EMU. (Image is courtesy of the University of Sydney Archives, G3/224.)

to manoeuvre into the corridor outside the laboratory. Here it completely blocked the passage, the total clearance being only about six inches, so that we had to place a ladder at each end to allow passersby to climb over it. I think it was on this occasion that we had a mobile crane in Science Road, easing the crate down the ramp to our basement, when Princess Alexandra, on an Australian visit, was due to pass up Science Road in procession. I had been unaware of this when the crane was booked and we just got clear in time.

The next area beyond the Akashi room was occupied by the University Stores (before they moved over to Darlinghurst), which was very convenient for us, but had its hazards. There was an old disused sink on the stores side of the intervening wall, the water-pipe from which had once been routed through the wall and through our area, but this pipe, instead of being removed entirely or completely blocked off, had simply been sawn off flush with the wall on our side, behind the masonite paneling. For several years all was well, until one day we suffered inexplicable flooding across our floor where no water pipes were supposed to be. Someone in the stores had rediscovered the old sink and found it a convenient place to wash bottles!

When the stores finally moved out, we were able to expand farther into the vacated area, which by that time was much needed for microtome rooms, and for some preparation benches and office space, as well as for the Philips EM 200, which had been diverted into our care (see Great Moment 4). By this time, the pathway through the laboratory from our main, and only, entrance door to the far end of the extensions had become long

The Badham Building, which housed part of the EMU as the unit expanded. (Image is courtesy of the University of Sydney Archives, G77/1.)

and devious, and I was conscious of the safety aspect and fire risk, so instead of making a bolted panel access from the corridor to the EM 200 room, had a wide door fitted, which could be used as a fire exit if need be. When we got the Philips EM 300 microscope, there was no further convenient space under the Bank Building and we acquired, as an annexe, a site in the neighboring Badham Building. And I relieved the pressure on space in the original laboratory by moving all the photographic processing (which was becoming 'big business') into the annexe also. Later again, when the first scanning electron microscope was to be set up, we had quite run out of convenient space, and had to start looking at inconvenient space. Moving eastwards from our main (extended) laboratory, under the Bank Building, one came to an under-croft of the Pharmacy lecture theatre. Here the area available was not too bad, but the headroom not much more than six feet. It was the only space on offer, however, and with some reluctance I accepted it, having made sure that the height was in fact sufficient (though only just) to accommodate the JEOL microscope I had in mind. Fortunately, in the event, the health authorities stepped in and forbade putting personnel into a space so cramped for headroom, and the University had to excavate some three feet of rock over the intended area. Then came the problem of supporting the floor and the seating of the lecture theatre above – a serious consideration when the load was apt to be increased by the weight of a couple of hundred students. Three brick piers had been carrying the weight and these were now left standing on bulky isolated plinths of rock, which interfered considerably with the layout of the rooms. It was decided that one of these could remain, but that the other two were to be replaced by much slimmer steel stanchions. Accordingly, the two stanchions were installed alongside the two piers to be removed and a laborer sent in to demolish the latter. But he had misunderstood his instructions and started on the wrong pier, so, after some moments of panic, a third RSJ was rushed up and installed, and we of course benefitted by getting rid of our last major obstruction.

Over the years, various other vicissitudes came our way. The air-conditioning in the ex-stores area was always a bit troublesome and I not infrequently had to phone the office of works to come to deal with its problems, so that they also were getting tired of it. But one day the conversation went something like this:
Myself: "We have more trouble with the air-conditioning in the EM Unit."
Voice (wearily): "Oh! What is it this time?"
Myself: "A cat has had kittens in the duct."
Voice: "What!"

A feral cat about the University grounds had birthed its kittens not actually in the duct, as it proved, but in the ceiling space above and we could hear them mewing. This meant taking down part of the ceiling and dealing with a wildly ferocious mother cat and her litter, and, of course, blocking off the ingress where the duct had been carried rather loosely through the outside wall. A problem that recurred from time to time was that solvent vapors (most often ether) got into the air of the laboratory, mysteriously, as we were not using ether. The Bank Building above us was occupied (apart from the CBC Bank and the Post Office) by the Chemistry Department at first and, when they moved out, by Pharmacy, and it seemed certain that the trouble was coming from there – but how? The labs upstairs were not themselves affected. It turned out that the air intake for our air-conditioning system was sited too close to a grid over a drain from the labs above, and vapors from waste solvents passing along this gully were being drawn in and recirculated. Both Chemistry and Pharmacy were cooperative in trying to forbid the disposal of solvents through this drainage system, but sooner or later someone would ignore the interdict and our troubles recurred. It was only cured when we had that particular grid cemented over, I believe quite illegally!

Supplies

When I arrived in Sydney I was, of course, completely unfamiliar with the city and the availability of the various supplies that would be needed for the laboratory. Most of the run-of-the-mill requirements could be got through the University Stores, but, during the fitting out of the laboratory, I took the opportunity to spend time, very profitably, in searching out and making contacts with the scientific instrument firms as well as suppliers of the more specialized minor items of equipment, such as tweezers, eye-lenses, etc. Grids, osmium tetroxide and the more specifically EM requirements had to be the subjects of special orders, usually from overseas. Grids, in particular, were always a problem. Virtually the only supplier, worldwide, was Smethhurst Highlight in Lancashire, England, which produced grids of very high quality but limited quantity. Mr Smethhurst operated on quite a small scale, with rigorous selection of only prefect grids, which involved a substantial rate of rejection. Many of us would have been only too glad to get hold of some of the rejects, and to put up with the occasional distorted hole or missing bar, but he refused to prejudice his reputation for quality, maintaining that he could not charge as much for the rejects as for the perfect ones, but that it cost just as much to reject a grid as to pass it. When the Americans became aware of his product they placed orders which absorbed practically the whole of his output, and delivery for the rest of us (and maybe for the Americans too) became inordinately long. He resolutely refused to expand, preferring to work on a cottage-industry scale, and was extremely reluctant to supply any 'middleman' (such as university stores) for reissue or resale. He liked to deal directly with the end-user, and a personal visit made to his premises to meet him, during one of my trips back to England, paid dividends. We got on well together and he enjoyed a joke so that later on, when we were getting desperate for grids, which had been on order for a long time, I sent him a card that showed a row of little birds sitting on a tilted balance beam, with the legend: "Get well soon. We can hardly wait for you to come back on the beam!" That produced a few hundred grids by return and saved the situation.

Projects

As I have said, nearly all our clients were concerned with biological projects and I do not attempt here to list all, or indeed any, of these, but among the exceptions were the CSIRO 'rainmakers'. They flew prepared grids on stratospheric flights to try to collect and study the nuclei on which raindrops could form. Regular flights were made, and the work had been going on for least a year and showing small and fragile looking particles when Mount Agung in Indonesia erupted and threw into the upper atmosphere a cloud of rock dust, which slowly drifted south until about two months later it reached the collection area (we had magnificent sunsets in Sydney at that time). Our grids then began to show quite a different kind

Some of the EMU's staff in 1964 (from left): Cedric Shorey, Tony Webber, Gordon Drummond, Kyra Korniloff and Susan Maxwell.

of particle – solidly opaque and massive by electron microscope standards, which obviously had come from the volcano. I think this must be the only time a volcanic eruption has been studied in the electron microscope, but it put paid to that particular project for the time being.

Staff

I enjoyed my years with the EM Unit, and felt that, by and large, we had a 'happy ship'. I have not mentioned staff by name, but this is not because I do not appreciate what many of them had done to make the unit what it is. Some have left to achieve positions of considerable responsibility elsewhere, but it was pleasant to find that at least four or five of the people I had appointed were still working happily in the unit 12 years after I retired.

References

References to Sections 1–16

Anderson, T. F. (1953). *Paris, 1950*. (p. 567).

Barrett, C. S. (1943). *Metals Technology, 10*. Tech. Publ. No. 1637, p. 5.

Cuckow, F. W. (1947). *Nature (London), 159*, 639.

Drummond, D. G. (1950). *Journal of the Royal Microscopical Society, 70*(3), 1–141.

Fernández-Morán, H. (1956). *The Journal of Biophysical and Biochemical Cytology, 2*(4, Suppl. 29).

Gettner, M., & Hillier, J. (1950). *Journal of Applied Physics, 21*, 68.

Latta, H., & Hartmann, J. F. (1950). *Proceedings of the Society for Experimental Biology and Medicine, 74*, 436.

Martin, L. C., Whelpton, R. V., & Parnum, D. H. (1937). *Journal of Scientific Instruments, 14*, 14–24.

Müller, H. O. (1942). *Kolloid-Zeitschrift, 99*, 6.

Newman, S. B., Borysko, E., & Swerdlov, M. (1949). *Journal of Research of the National Bureau of Standards (U.S.), 43*, 183.
Nutting, J., & Cosslett, V. E. (1950). *Institute of Metals (London), Monograph and Report Series, 8*, 65.
Palade, G. E. (1952). *The Journal of Experimental Medicine, 95*, 285.
Porter, K. R., & Blum, J. (1953). *The Anatomical Record, 117*, 685 (See also p. 698).
Williams, R. C., & Wyckoff, R. W. G. (1944). *Journal of Applied Physics, 15*, 712.

References to Section 17

Minutes of the Senate. University of Sydney, (4 March 1957).
Minutes of the Senate. University of Sydney, (6 August 1957).
Challice, C. E. (1954). Summarized proceedings of a conference on electron microscopy – London, November 1953. *British Journal of Applied Physics, 5*(5), 165–170.
Challice, C. E. (1957). Summarized proceedings of a conference on the electron microscopy of fibres – Leeds, January 1956. *British Journal of Applied Physics, 8*(1), 1–26.
Cockayne, D. J. H. (1988). Obituary. *Australian EM Newsletter, 17*, 16.
Cosslett, V. E. (1947). Summarized proceedings of conference on the electron microscope, Oxford 1946. *Journal of Scientific Instruments, 24*(5), 113–119.
Cosslett, V. E., Nutting, J., & Reed, R. (1953). Summarized proceedings of a conference on electron microscopy – Bristol, September 1952. *British Journal of Applied Physics, 4*(1), 1–5.
Cumberbirch, R. J. E., Drummond, D. G., Ford, J. E., & Simmens, S. C. (1959). Model viscose rayon filaments. *Journal of the Textile Institute, 50*, T262–T283.
Drummond, D. G. (1932). Infra-red absorption of quartz. *Nature, 130*, 928–929.
Drummond, D. G. (1934a). Infra-red spectra of silica. *Nature, 134*, 739.
Drummond, D. G. (1934b). Corrections to the refractive indices of quartz in the infra-red. *Nature, 134*, 937.
Drummond, D. G. (1936a). The 2.73μ absorption band in fused silica. *Nature, 138*, 248–249.
Drummond, D. G. (1936b). The infra-red absorption spectra of quartz and fused silica from 1 to 7.5 μ. I – experimental method. *Proceedings of the Royal Society of London. Series A, Mathematical and Physical Sciences, 153 [A879]*, 318–327.
Drummond, D. G. (1936c). The infra-red absorption spectra of quartz and fused silica from 1 to 7.5 μ. II – experimental results. *Proceedings of the Royal Society of London. Series A, Mathematical and Physical Sciences, 153 [A879]*, 328–339.
Drummond, D. G. (1940). Anisotropy of cellulose sheet. *Nature, 145*, 67.
Drummond, D. G. (1950). The practice of electron microscopy. *Journal of the Royal Microscopical Society, 70*(3), 1–141.
Drummond, D. G. (1953). Summarized proceedings of a conference on the optical and electron-microscopical properties of textile fibres – Manchester, October 1952. *British Journal of Applied Physics, 4*(4), 119–124.
Drummond, D. G. (1957a). *Letter to the Assistant Principal.* 15 September.
Drummond, D. G. (1957b). *Letter to the Assistant Principal.* 17 November.
Drummond, D. G. (1957c). *Letter to the Assistant Principal.* 1 December.
Drummond, D. G. (1960). *The Electron Microscope Unit. The Gazette, University of Sydney, 1*(19), 276–278.
Drummond, D. G. (1971). *The Electron Microscope Unit. The Gazette, University of Sydney, 3*(2), 21–23.
Drummond, D. G. (1985). Personal reminiscences of early days in electron microscopy. *Advances in Electronics and Electron Physics, Supplement 16*, 81–101.
Drummond, D. G., & Hamlin, M. (1952). Airborne bacteria in cotton mills, I: Survey of counts of viable bacteria. *British Journal of Industrial Medicine, 9*(4), 309–311.

Drummond, D. G., & Liebmann, G. (1952). Summarized proceedings of a conference on electron microscopy – St Andrews, June 1951. *British Journal of Applied Physics*, *3*(1), 25–29.

Drummond, D. G., & Warwicker, J. O. (1959). An electron-microscope examination of the effect of boiling water on some anthraquinone vat dyes. *Journal of the Textile Institute*, *50*, T487–T493.

Kay, D. (1961). *Techniques for electron microscopy*. Oxford: Blackwell Scientific Publications.

Maze, W. H. (1957). *Letter to D G Drummond*. 4 September.

Monteith, J. L. (1986). Howard Latimer Penman. 10 April 1909–13 October 1984. *Biographical Memoirs of Fellows of the Royal Society*, *32*, 378–404.

Mulvey, T. (1985). The industrial development of the electron microscope by the Metropolitan Vickers Electrical Company and AEI Limited. *Advances in Electronics and Electron Physics, Supplement 16*, 417–442.

Reed, R. (1985). Some recollections of electron microscopy in Britain from 1943 to 1948. *Advances in Electronics and Electron Physics, Supplement 16*, 483–500.

Sayer, J., Brown, A. F., & Todd, J. B. (1948). Report of the electron microscopy conference – London, March 1947. *Journal of Scientific Instruments*, *25*(1), 23–27.

Megavolt electron microscopy

Gaston Dupouy[a] **and Peter Hawkes (Afterword)**[b,*]
[a]Formerly Laboratoire d'Optique Electronique du Centre National de la Recherche Scientifique, Toulouse, France
[b]CEMES-CNRS, Toulouse, France
*Corresponding author. e-mail address: hawkes@cemes.fr

Contents

Advances in Imaging and Electron Physics, Volume 220
ISSN 1076-5670
https://doi.org/10.1016/bs.aiep.2021.08.008

1. Introduction

In the physical universe, man holds a very special place, halfway between the atomic world and the world of the planets. That is the reason why we have been instinctively attracted by both these utmost poles of our knowledge.

The notion of the infinitely great probably occurred to the first men who, on a beautiful night, gazed at the starry sky; history brings us various evidence showing that the most remote civilizations have proved to be quite inquisitive about the structure of living or inanimate matter and of its ultimate elements.

Scientists have always endeavored, and are still endeavoring by every means at their disposal, to progress in the knowledge of the infinitely small. In this race toward the vision of the infinitely small, the classical electron microscope has allowed a considerable distance to be covered. Thanks to that instrument, man could pass from the contemplation of the "macroscopic" world to the vision of the "microscopic" world, and thus were revealed to scientists the constitution of the tissues of living material, the texture of alloys, the fine structure of bacteria, and so on.

It has been pointed out, with some reason, that it was almost providential that pathogenic bacteria were within the limits of vision of our microscopes: if this had not been so, bacteriology would hardly exist as a science, and the stage of microbial life would not have been revealed. Since the time of van Leeuwenhoek, many attempts have been made to develop the field of our observations in this direction.

However, until about 1930 it seemed we had come to a standstill. The limits of the light microscope had effectively been predicted by Ernst Abbe in 1873, yet he hoped that some progress could be accomplished later and had written

> *Perhaps in the future the human spirit may succeed in using some processes and some forces which will allow man, through utterly different ways, to overcome the limits which now seem to us impassable. That is my own opinion. But I believe that the instruments that will assist our senses some day, with more efficiency than the present microscopes, to explore the ultimate elements of the physical world, will have nothing in common with them but their name.*

Fortunately, the discovery of the electron microscope now allows us, through absolutely new techniques, to increase considerably the acuteness of our prospecting means and to investigate nature with more powerful instruments that reveal the presence, the shape, and the structure of corpuscles of minute dimensions, invisible with a light microscope: a new stage toward the vision of the infinitely small is indeed opening before us.

We must stress, however, that the electron microscope does indeed have nothing in common with the light microscope but its name. In the latter a light beam is used; in the former, to obtain images we appeal to electrons.

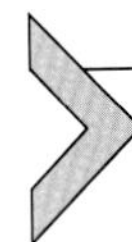

2. Early developments in electron microscopy

2.1 The birth of the electron microscope

In a recent work, which has been brilliantly translated by Thomas Mulvey, Ernst Ruska (1980) describes the efforts he made with his co-workers to conceive and to design the electron microscope. He thus makes an important contribution to the history of science. We shall often refer to this book.

Quite curiously, the construction of the electron microscope is the result and the extrapolation of research on the cathodic oscillograph. The point was, in particular, to determine the most convenient diameter of the spot which was used in the oscillograph and to endeavor to concentrate the beam in such a way that the electron density at this spot would be the best possible.

This work was carried out at the Electrotechnical Institute of Berlin Technological University. During the years 1924–1929, several young research workers, who are well known today, devoted themselves to investigations which, under various aspects, dealt with the cathodic oscillograph. These include D. Gabor, who was preparing a Ph.D.; Professor A. Matthias; Max Knoll; and Bodo von Borries, who became Ruska's friend.

The concentration of the beam by a magnetic field was adopted as the most convenient method, but many hesitations remained for a while about the best way of producing and making the best use of this field.

On April 7, 1931, Ruska, quite astonished and happy, obtained the first images of a thin metal grid with electrons. The still modest magnification was $M = 16$.

Drawing his inspiration from a theoretical work by H. Busch (1927), about which I will write later, Ruska improved the magnetic lenses of his instrument. His perseverance was at last rewarded: in December, 1933, he obtained images of cotton fibers, on a fluorescent screen, with a magnification $M = 8000$, and of aluminum foils with $M = 12{,}000$. This was obtained with the first two-stage electron microscope: its magnification was larger than that of the light microscope.

The work of Manfred von Ardenne (1941, 1942) must be mentioned: he gave some results about a "new electron super-microscope installation" and others about a "universal microscope for operation in bright field, dark field, and stereo-imaging."

2.2 The role of wave mechanics

Nevertheless, the interest in electron microscopy, which is currently undergoing prodigious development, clearly appeared only in the light of wave mechanics, for there remained a critical stage to go through: it had to be shown that the resolving power of the electron microscope could surpass that of the light microscope.

At the beginning, Knoll and Ruska considered this to be highly probable and, curiously enough, they thought that such a gain depended on the very small dimension of the "electron" corpuscle. They thought that this optimum resolution should be smaller than the distance between two adjacent molecules or atoms in liquids or solids. When the correctness of L. de Broglie's work (1924) was demonstrated, and his famous formula giving the wavelength of the wave associated with an electron in movement was verified, Ruska was disappointed because, once more, the limit of the resolving power was governed by a wave phenomenon. But, applying de Broglie's formula, he was relieved to see that the wavelength of the wave associated with the electron was five orders of magnitude shorter than that of a light wave.

In 1923, Louis de Broglie was the first to clear up the mystery of the electron, through his brilliant conception of wave mechanics. Wave mechanics solves with supreme neatness the problem of wave-particle dualism, which had long bewildered physicists. According to that theory, the electron is both a wave and a particle; these two aspects may be considered today as two complementary aspects of the same physical reality. So, with every particle of moving matter, we can now associate a wave; de Broglie found a very simple formula allowing the wavelength of the wave associated with the electron to be calculated as a function of its velocity or of the potential difference by which it was accelerated. Let us consider this formula, which is now completely standard.

$$\lambda = h/mv$$

where m is the mass of the particle, v its velocity, and h is Planck's quantum of action: $h = 6.624 \times 10^{-27}$ erg sec. As far as an electron is concerned, this can be written

$$\lambda = 12.25/\sqrt{V}$$

where V is the accelerating voltage of the electron in volts and λ is expressed in angstroms. We find thus that, if $V = 50{,}000$ V, the wavelength $\lambda = 0.055$ Å.

We know that the wavelength of the green radiation emitted by a mercury-vapor lamp is $\lambda = 5461$ Å, i.e., ~5500 Å; that is to say, about 100,000 times longer than the length of the wave associated with an electron at 50,000 V. Thus, other things being equal, it should be theoretically possible to multiply by 100,000 the resolving power and the magnification of an electron microscope by "lighting" the object with waves associated with rapid electrons.

From such a simple calculation the immense possibilities and the range of applications of electron microscopy can be established.

The work of H. Busch (1927) allowed electron optics to be developed. Busch demonstrated the fundamental fact that every static electric or magnetic field having symmetry of revolution about an axis exerts a real focusing action on paraxial electron beams and constitutes an electron lens able to produce images. If a and b are the respective distances of the object and of the image from the lens, the relation between these distances and the focal length f corresponds to the well-known relation in optics

$$1/a + 1/b = 1/f$$

As a result of Busch's work, several researchers developed the construction of electron lenses; some were electric, others magnetic, and they allowed various objects to be observed.

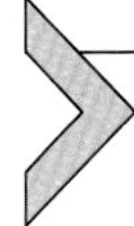

3. How I came to electron microscopy

3.1 My first research work in electron microscopy

I came to electron microscopy through thoroughly unexpected ways. Before 1937 most of my work dealt with magnetism and magneto-optics. It was carried out under the guidance of Professor A. Cotton at Paris University. In 1930 Cotton asked me to help him with the direction of the Laboratory of the Large Electromagnet of the French Academy of Sciences.

In 1925 a "Pasteur Day" was organized in France. In memory of the great scientist who, for so many French people, is the personification of

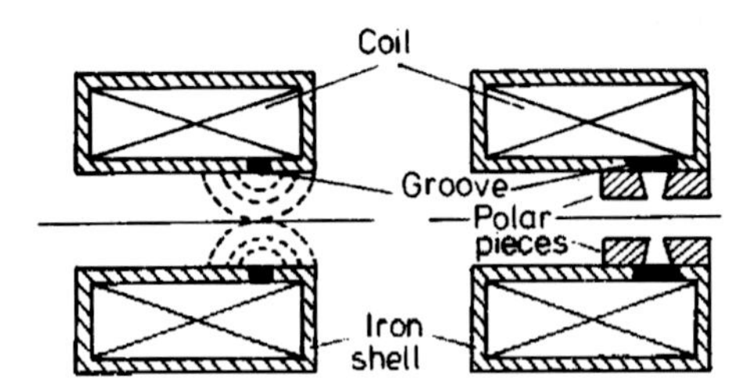

Figure 1 Diagram of the first magnetic electron lenses.

science, thousands of donors contributed generously to support our laboratories. The Academy of Sciences placed at Cotton's disposal important funds for the construction of a large electromagnet. This instrument, which was the most powerful at that time, was built in collaboration with G. Mabboux in Bellevue, halfway between Paris and Versailles, at the National Office of Scientific and Industrial Research and Inventions (Cotton, 1928; Cotton & Mabboux, 1928).

I rapidly made myself familiar with the production of intense magnetic fields. In collaboration with Cotton I measured the magnetic fields obtained by using the various polepieces with which the electromagnet is equipped, including those which have an axial canal to let a light beam pass (Cotton & Dupouy, 1930). I often thought of these measurements when, later on, I designed magnetic electron lenses.

It was around 1934 that I became aware of Ruska's work, which led me to read Busch's transactions. I was at once sure that a magnetic electron lens is nothing but a rotationally symmetric electromagnet, the polepieces of which have an axial canal to let the electron beam pass. Curiously, this idea did not emerge clearly at the beginning. Other workers started from the idea of a solenoid, but this is a thick lens, producing a wide magnetic field, generally not intense enough to obtain short enough focal lengths. Such lenses were soon abandoned.

In a second stage, in order to localize the magnetic field in a smaller portion of space, the solenoid was enclosed in a shell of iron or of steel forming a magnetic shielding (Fig. 1). Inside, there was a groove in which a ring made of a nonmagnetic substance (brass, copper, bronze, etc.) was set. The power lines of the magnetic field open out, as shown in the figure, in a limited volume. It is better to localize the magnetic field still more and, at the same time, increase its intensity. For this purpose, polepieces made of mild steel were set on both sides of the groove, inside the coil (Fig. 1).

Finally, as I mentioned above, it was realized that a magnetic electron lens is an electromagnet whose polepieces and magnet core have an axial hole through which the electrons can pass.

It follows from this that, to make magnetic electron lenses, one had to apply the well-known rules used to design electromagnets (Dupouy, 1968). We shall have an opportunity to describe such lenses in Section 4.

The research work I carried out in Bellevue led me to think I should not meet any major difficulty in designing magnetic electron lenses and the column of an electron microscope, as soon as the circumstances were propitious.

On the other hand, I made an effort to know as much as possible about the most recent results obtained in Germany by Ruska and co-workers, as well as by other German scientists. Krause (1936) succeeded in obtaining good images of diatoms and also the first micrographs of cells and of epithelial cell walls.

The field of electron microscopy spread to biology and medicine. In 1932, Marton, in Brussels, developed methods of preparing biological specimens in such a way that they were not destroyed by the electron beam and produced images with the best contrast. He stained the specimen with osmium, observed images of a cross section of the root of a bird's nest orchid, and, in 1936, obtained images of *Chromobacterium prodigiosum* with a resolution of 1 μm. So Marton contributed to introducing the use of the electron microscope in biology, thus playing a decisive part in its promotion (Marton, 1934a, 1934b, 1935, 1936, 1937).

At the end of 1933, Ruska and von Borries endeavored to arouse an increased scientific interest in electron microscopy. At the beginning they came up against some difficulty in convincing biologists and physicians. In 1936 Ernst Ruska was much supported by his brother, Helmut Ruska, who was a physician and who obtained some decisive help from Professor Richard Siebeck, the director of the first medical hospital in Berlin, whose student H. Ruska was. The Siemens Company accepted the task of designing a most elaborate instrument, with a maximum magnification $M = 30{,}000$ and a point-to-point resolving power of 130 Å. Those who, a few years before, had doubted the capabilities of the electron microscope were now convinced. All this seemed to me most interesting and I made up my mind to work in this field.

In 1937 I was appointed to the post of Professor in the Department of Sciences in Toulouse University. Naturally, I gave up all my other research activity in order to devote all my time to the construction of the electron microscope. But, because of difficulties of all kinds, which I shall describe shortly, this instrument was finished only in 1942.

First of all, the scientific equipment of the laboratory appeared to me practically nonexistent. Moreover, there was no workshop, and hence no technical assistance. Worse still, the building itself was in urgent need of repair.

A post for a mechanician was granted to me by the Centre National de la Recherche Scientifique (CNRS) and, first of all, we built a modest instrument with an electron source, a condenser, and an objective lens. But in 1939 the war broke out and everything became more and more difficult. To get materials we had to ask for "coupons." The high-voltage apparatus was likewise obtained piece after piece. However, I succeeded in building a generator allowing us to work at 50, then at 70, kV.

The main parts of the microscope column were assembled in the Laboratoire Central de l'Armement, which had been withdrawn during the war to Caussade—not very far from Toulouse—in a period when it was particularly difficult to manufacture such delicate pieces. General Engineer Nicolau, who was the director of this laboratory, helped us very effectively in such a rough starting period. He has left us today but I want to express my gratitude to his memory. And, once again, I thank General Inspector de Lombares for the precious help he gave us in numerous circumstances.

3.2 The first results

The electron microscope has been described in several publications to which the reader may refer (Dupouy, 1946a, 1946b). We reproduce here (Fig. 2) a photograph of the column. Of the investigations carried out in our laboratory at that time I will only mention our observations of diatoms, a study concerning the structural modifications of aluminum and of its alloys under the influence of various factors, and some results in bacteriology.

3.2.1 Diatoms

Historically, diatom shells have been routinely used as test objects in light microscopy to measure the resolving power of the instrument. Such shells have indeed a multitude of little holes, generally smaller than 1 μm. These nice little objects are also a first-class material in electron microscopy for the following reasons: their shells are made of silica and they can stand quite a long electron bombardment without being altered; also most of them show details in their structure which were unknown until study by

Figure 2 Column of the first electron microscope in Toulouse (1942).

electron microscopy and which are quite suited to determine the resolving power of the microscope.

Fig. 3 is a micrograph of a diatom. It shows rather large holes (about 1 μm). Each of them is covered by a sort of lacework on which many details, invisible in light microscopy, can be seen.

3.2.2 Metallographic studies

Fig. 4 is a micrograph obtained under the following conditions: a sample of commercial aluminum was annealed for a long time, and a replica of its surface was made using Mahl's method (1941, 1942). An aluminum cubic crystal, which was broken in the commercial sample, clearly appears again.

3.2.3 Bacteriology

In investigations in bacteriology, we obtained images of *Proteus vulgaris* showing a mitosis and flagella (Fig. 5). My co-workers and myself were

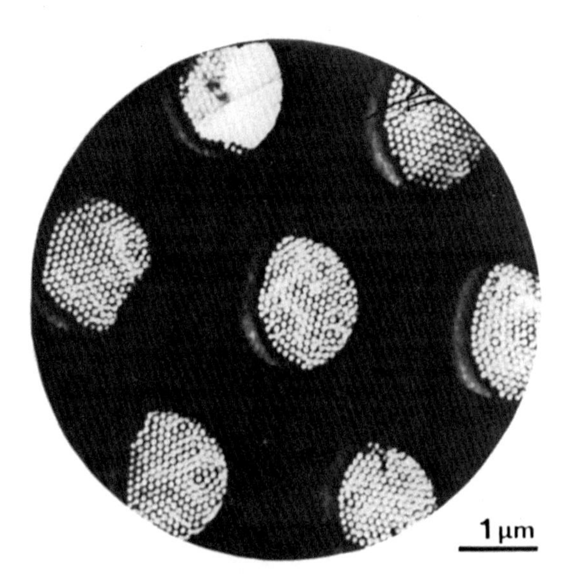

Figure 3 Image of a diatom.

Figure 4 Crystals of aluminum in an annealed specimen.

very enthusiastic about these first results and we were determined to strive harder than ever to go as far as possible.

I cannot forget those who helped me in this initial effort, from which all further results have issued, namely, the construction, during the war, of the instrument we now usually call "the ancestor." It could work at

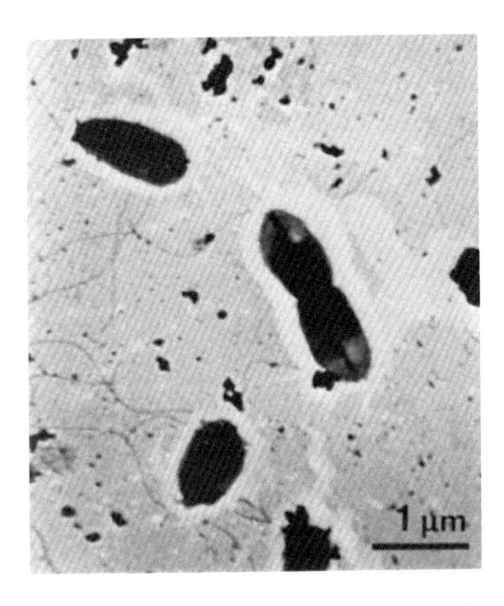

Figure 5 *Proteus vulgaris* bacterium showing a mitosis and flagella.

50,000 V. The conditions in which we had been working, overcoming the worst difficulties with beggarly means, underline the merits of the "first team." Each team member carried out all sorts of business, until all hours, very often late at night because of numerous power cuts. They may be sure that my grateful thoughts go out to them very often.

Armand de Gramont, Member of the Académie des Sciences, Chairman of the Board of Directors of the Compagnie Optique et Précision (OPL) at Levallois, was greatly interested in problems related to the world of the infinitely small, as shown by his book "Vers l'Infiniment Petit" (de Gramont, 1945). He decided at that time to design a magnetic electron microscope, which was used with great success in several laboratories. Full information about the OPL microscope is to be found in "Le Microscope Electronique" by P. Selme (1963; see also Fert & Selme, 1956).

Let me also recall that, at that time, some theoretical and experimental investigations in electron optics had been carried out in France. They were reported by their authors during meetings, with de Broglie as chairman. The transactions of these meetings have been published (de Broglie, 1946).

We must emphasize the work of Pierre Grivet, Claude Magnan, Andre Lallemand, Paul Chanson, P. Lepine, and J. J. Trillat. The designing of the electrostatic microscope started in 1943, under the guidance of Grivet (1946), in the research laboratory of the Compagnie Générale de TSF. It was carried on during the difficult years of war, until the prototype was finished. With this microscope good images were obtained and were displayed at a Congress on Surface States in 1945. Magnan (1946) and coworkers Chanson and Ertaud designed a proton electrostatic microscope in which, instead of an electron beam, a proton beam was used. The goal was to increase the resolving power of the microscope. We must also mention the work of Professor P. Lepine, of the Institut Pasteur, who was the first in France to carry out investigations in bacteriology. During a meeting which was held in Oxford in 1947 Professor Lepine gave a very remarkable lecture on bacteriophages (Lepine & Croissant, 1955). A. Lallemand, on the other hand, went on with his studies on the electron telescope at Paris Observatory. Finally, J. J. Trillat designed, in 1933, a magnetic electron microscope in collaboration with Fritz, his assistant at the Faculty of Sciences of Besançon. Its magnification was about 2000× (Trillat, 1981).

3.3 My administrative career

In September, 1945, I was elected Dean by my colleagues of the Faculty of Sciences. I was then in Great Britain with Frédéric Joliot, the Director of the CNRS, who had asked some French researchers to give a series of conferences there, to show that in some laboratories, in spite of the war, scientific research had progressed.

When I received, in London, the telegram about my election, I was at once willing to serve our faculty and afraid of the effects my new administrative task might have on the physicist's work I was devoted to.

I think I have done my best to transform the Faculty of Sciences by constructing new buildings, by promoting several institutes into National Schools of Engineering, and by obtaining rather substantial funds in order to make our equipment more up to date and develop our teaching activities.

In 1950 I was offered the direction of the Centre National de la Recherche Scientifique by the Minister of National Education. His persuasive insistence and my devotion to research encouraged me to accept this high office, for which I had not applied. I had not contemplated such a possibility. On the contrary, I was about to leave the deanship. I wanted to develop the researches which had just begun in the Electron Optics Laboratory. Our laboratory had just been promoted to the honor of becoming a "Laboratoire Propre" of the CNRS, which would give us sufficient means to develop our investigations. For more than 7 years, all my activity was devoted to the CNRS. It is not for me to judge whether my action has been good, yet I am sure that I have served French scientific research with all my heart, striving unceasingly to make it more prosperous and to make it outstanding.

In 1957 I wanted to return to research work in order to carry out a project to which I was much attached. In spite of the insistence of the Minister of National Education and of some friends, I returned to Toulouse.

4. Megavolt electron microscopy

In the meantime, a beautiful institute had been built in Toulouse. It was almost finished and, in January, 1958, we could settle in these new premises. This was the beginning of an adventure which was to lead to a new exploration of the world of the infinitely small and which led to particularly important new results.

I must, however, mention that, in various countries, several electron microscopes had been designed to be used at 200 and even 500 kV. These include the works of von Ardenne (1941), Müller and Ruska (1941), Hillier et al. (1941), van Dorsten, Oosterkamp, and Le Poole (Le Poole et al., 1947), Coupland (1956), Tadano and Sakaki (Tadano et al., 1956, 1958), and Popov (1959).

I planned to design an electron microscope working at one million volts, and more if possible. This was then a bold enterprise and the forecasts from all sides were absolutely pessimistic. Many difficulties were advanced: (1) the designing of lenses strong enough to focus high-energy electrons, the velocity of which is nearly that of light; (2) the stabilization of such high voltages; and (3) if such difficulties could be overcome and if images could be obtained, they would of course lack contrast, and everything in them would be uniformly gray.

Naturally I had, for my own part, thought a lot about all these problems before I tried this bold experiment. Many people believed my plans would lead to failure; they turned out very well indeed.

4.1 The one-million-volt electron microscope

In collaboration with Professor F. Perrier we designed the first microscope of this type. It has been working at 1 MV—and 1.2 MV—in Toulouse since 1960. It allowed us to demonstrate the advantage of working with high-voltage electrons for numerous investigations in electron microscopy and diffraction. Observing at high voltages is better from several points of view: (1) The penetration power of the electrons into the object is much greater. (2) Theoretically, the resolving power and the useful magnification

Figure 6 Column of the 1-MV electron microscope.

are increased when the voltage is raised. (3) Chromatic aberration, due to energy losses and to the scattering undergone by the electrons in the object, decreases when the voltage is very high. (4) The warming and ionization of the sample under the electron beam are also decreased, and the specimen is obviously less damaged.

Another important point deserves to be noticed. The great microscope in Toulouse allowed, for the first time, micrographs to be obtained of living bacteria. For that purpose, we enclosed the bacteria in a vacuum-sealed box—a sort of little flat for bacteria—where they are surrounded by their usual medium: air, atmospheric pressure, and a suitable relative humidity.

In December, 1960, I presented to the French Academy of Sciences an account of the first performance of the 1-MV microscope (Dupouy et al., 1960) (Fig. 6). This instrument at once awakened great interest in the world of science and became the leader of a new generation of electron microscopes: such instruments are now designed in Japan, Great Britain, and the United States. Generally speaking, we can say that these instruments working at voltages around 1 MV have opened important new research prospects.

In August–September, 1962, our American colleagues organized, in Philadelphia, the Fifth International Congress for Electron Microscopy under the chairmanship of Professor Thomas F. Anderson. I was invited to present, in the course of an opening lecture, our new instrument and some significant results we had been able to obtain.

1. Studies concerning metals and alloys: twin boundaries and grain boundaries, dislocations and stacking faults in crystals that had been altered by various stresses, extinction contours, and precipitates in aluminum–copper and aluminum–silver alloys. We could observe 3-μm-thick aluminum foils; the results thus obtained should bear a much closer relationship to those for bulk material (Dupouy & Perrier, 1962).
2. Our microscope was also well adapted for diffraction and microdiffraction experiments. A micrograph was projected showing diffraction rings of a 1000-Å-thick gold foil at 1.2 MV. More than 50 rings can be observed (Dupouy et al., 1962).
3. For the first time we could obtain images of living bacteria in conditions in which cells may remain alive after having been observed under the electron beam.

At the end of that lecture a short film was shown that had been made in our laboratory by the Overseas Services of the British Broadcasting Corporation. It gave a detailed view of the microscope equipment and its working conditions, as well as of photographs of the first results mentioned above.

4.2 The three-million-volt electron microscope

The interest raised by high-voltage electron microscopy during the following years induced us to build in Toulouse an electron microscope with an operating voltage which reached three million volts. With the help of the CNRS, we had to erect a new building to house the microscope column as well as the high-voltage generator.

Fig. 7 is an aerial photograph of the Laboratoire d'Optique Electronique du CNRS in Toulouse: L is the main building, A the workshop and electronics department, and S the spherically shaped building which houses the 1-MV electron microscope; the new building N has been erected for the 3-MV instrument and is surrounded by new laboratories L′.

Figure 7 Aerial photograph of the "Laboratoire d'Optique Electronique du CNRS," Toulouse.

The microscope began to work at 2 MV in May, 1969, and we were able to take, from the beginning, many high-quality photographs.

In June, 1969, an international conference on "Current Developments in High-Voltage Electron Microscopy" took place in Monroeville, Pennsylvania. Our colleague R. M. Fisher, from the Fundamental Research Laboratories of the United Steel Corporation, was the successful organizer and chairman of that congress. The photographs seen in Figs. 7–9 and 21 were first presented during that meeting (*Micron*, Vol. 1, No. 3, 1969).

Both high-voltage electron microscopes have already been described in numerous publications (Dupouy, 1973, 1974; Dupouy et al., 1962; Dupouy, Perrier, Ayroles, et al., 1970; Dupouy, Perrier, Durrieu, 1970; Dupouy, Perrier, Tremollieres, 1970); we shall just recapitulate some details of their construction and of their performances.

4.2.1 The microscope column of the 3-MV instrument

The microscope column (Fig. 8) consists of six lenses: the double condenser (C_1, C_2), the objective lens (O), two intermediate lenses (L_1, L_2), and the projector (P). The latter fits on the observation chamber (C). The whole arrangement rests upon a steel plate (A). The height of the column, from

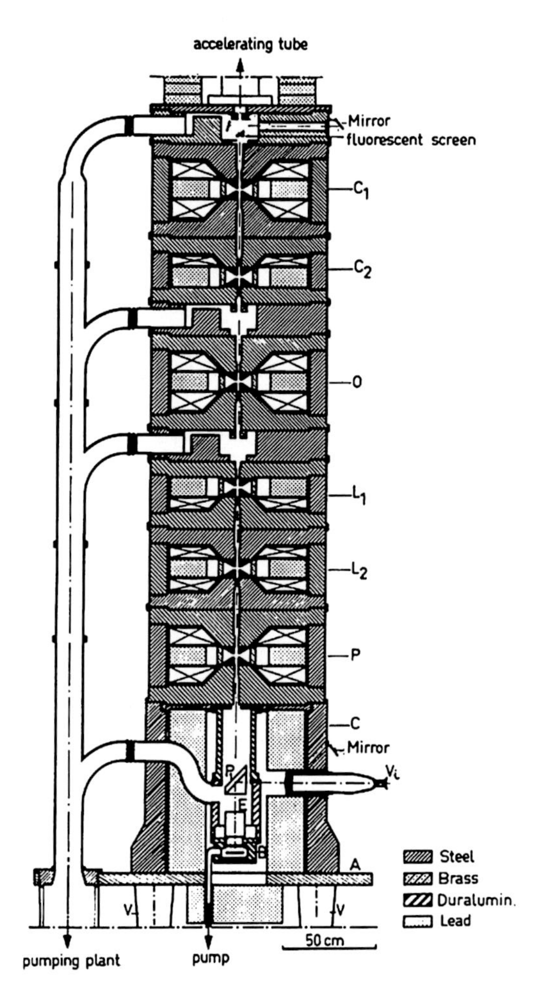

Figure 8 Cross section of the 3-MV electron microscope column.

the plate A to the top, is 3.91 m. Inside the observation chamber is the viewing screen (E) on which images are focused, as well as the photographic chamber (B). The direct magnification of the viewing screen can exceed 300,000×. The second intermediate lens (L_2) is useful for microdiffraction. The photograph in Fig. 9 shows the 3-MV electron microscope ready for work.

4.2.2 Protection against X rays

At 3 MV, X rays are very hard and penetrating. The operators must therefore be protected and a constant watch kept upon the efficiency of this

Figure 9 Photograph of the 3-MV electron microscope.

protection. This problem is much more disturbing than it is for the 1-MV microscope. Above the first condenser (C_1) we have built a lead wall consisting of 10-cm-thick bricks. Moreover, inside each lens, all the available space between the magnetizing coils has been filled with lead. The viewing axis has been laterally shifted; the image which is formed on the screen (E) is observed by means of a low-magnification (×5–10) telescope (V_i) and through a total-reflection prism (P_1). The intensity of emitted X rays at various levels in the column is measured by means of ionization counters. The dose, at the level of the operator, must never be higher than is medically acceptable. Should this dose be exceeded, an automatic device would start an alarm bell to warn the operator.

We may add that the column has been equipped with some devices which make the handling of the apparatus easier. At the upper part of the microscope a retractable Faraday cage has been installed: it is used to measure the electron beam intensity. Another Faraday cage can also be introduced under the objective lens and a third one in the observation chamber.

A tilting object holder, mounted in the gap of the objective lens, is used to observe stereoscopic images of the specimen. Inside the pumping chamber, under the objective lens, deflecting coils associated with a detector allow us to measure the exposure time on the photographic plate. The range of exposure times is from 0 to 1 sec by tenths of a second, from 1 to 10 sec by 1-sec intervals, and from 10 to 100 sec by 10-sec intervals. The precision in measuring the exposure time is $\pm 4 \times 10^{-3}$.

4.2.3 Some characteristics of the objective lens

We use a magnetic electron lens of conventional type. Its polepieces are truncated cones, made of iron–cobalt or of a special alloy known as Hyperm 0. We have at our disposal different pairs of polepieces: their gap is denoted by S, and the diameter of the axial hole by D. The values of S and D, for some polepieces, are $S = D = 12$ mm; $S = 15$ mm, $D = 12$ mm; $S = D = 15$ mm; $S = D = 20$ mm. We can thus choose the polepieces most suitable for different experiments.

There are 34,000 turns of wire, distributed on two magnetizing coils. Each coil is water cooled. Finally, it should be mentioned that this lens can also be operated as a single-field condenser-objective up to a voltage of 1680 kV (Dupouy et al., 1967).

4.3 High-voltage supplies

For our 1-MV electron microscope (Dupouy, Perrier, Ayroles, et al., 1970; Dupouy, Perrier, Durrieu, 1970; Dupouy, Perrier, Tremollieres, 1970) we had chosen an open-air generator, operating at atmospheric pressure (Fig. 10). All its components can be easily reached for modification or improvement. For a similar generator, operating at a voltage of 3.5 MV, however, security distances of about 20 m would be required between the elements at high voltage and the walls of the building. We therefore adopted a different solution. The generator and accelerating tube are enclosed in metal tanks containing an insulating gas, sulfur hexafluoride, SF_6, at a pressure of 4.8 bars. The distance between the high-voltage terminals and the grounded walls can thus be reduced to 0.8 m (Fig. 11).

Stabilization of the current in the lenses and of the high voltage

It is well known that the focal length of the objective lens depends both on the accelerating voltage V and on the magnetic induction B_z at each

Figure 10 High-voltage supply for the 1-MV electron microscope.

point along the axis, between the polepieces. We have always striven to stabilize independently the voltage V and the current intensity I in the magnetic coils, in order to keep constant the ratio B_z^2/V.

The theoretical resolving power of a high-voltage electron microscope may be increased appreciably because the wavelength of high-energy electrons is shorter, but, in order to obtain this resolving power, it is necessary to achieve a very good stabilization of V and I during the exposure time necessary to photograph the image. This exposure time is, in bright field, of the order of 1 sec or less. Many precautions had to be taken to obtain such results. They have been described in detail in our publications (Dupouy, Perrier, Ayroles, et al., 1970; Dupouy, Perrier, Durrieu, 1970; Dupouy, Perrier, Tremollieres, 1970). We have now reached the following performance levels of stabilization:

a. *Lens Current.* As far as the 1-MV microscope is concerned, the objective current does not vary by more than 1 to 2×10^{-6} in half an hour. In the

Figure 11 General view of the high-voltage (3-MV) installation. Generator and accelerating tube in metal tanks. Height: 8 m.

3-MV electron microscope the magnetic lenses are supplied by highly stabilized generators which have been designed and constructed by the electronics staff of the laboratory (Dupouy, Perrier, Tremollieres, 1970). In relative value, the stability of the current is better than 10^{-6} during 10 min. Such performances are applied to all the intensities used under the various working conditions of the microscope.

b. *High Voltage.* In the first (1-MV) microscope, the high-voltage variation is $\Delta V/V = 3 \times 10^{-6}$ during 1 min; the drift is around 1×10^{-5} in 1 hr. In the second instrument the *maximum* relative variation during 3 min is $\Delta V/V \leq 4 \times 10^{-6}$. The drift is also around 1×10^{-5} per hr.

The construction of the high-voltage (3.5-MV) generator and of the accelerating tube was undertaken by the Compagnie Générale de Systèmes et de Projets Avancés (GESPA), an affiliate of the Compagnie Générale d'Electricité. It is a most pleasant duty for me to express my gratitude to the managers of the Compagnie Générale d'Electricité, who took a very deep interest in our work and made the necessary arrangements to carry out the construction in the shortest possible time. I express my sincere thanks to the engineers and technicians of GESPA, whose ability and willingness ensured the success of this undertaking. I am most grateful to the managers of the Atelier de Fabrication of Toulouse, whose very precious help in the construction of the microscope column has been much appreciated. Our thanks are also addressed to all the staff of that factory.

A project such as I have just described is a collective achievement. I am happy to address my warmest thanks to all the staff of the laboratory; each of them has enthusiastically helped us by contributing intelligence and professional ability: I express my deepest gratitude to all of them.

To conclude this discussion we can say that these high-voltage instruments are a second generation of electron microscopes (Hirsch, 1974). Several others built in various laboratories deserve mention, in particular, a 750-kV microscope at the Cavendish Laboratory in Cambridge, designed and built by Cosslett (1967). Researchers in Japan and the United States have also been active in this field. Their design of these instruments has reached such a degree of perfection that they fulfill the hopes of investigators. They have obtained beautiful results, which account for the rapid development of HVEM.

4.4 HVEM development worldwide

There are presently about 60 high-voltage (500-kV and above) electron microscopes in operation in the world. Such instruments have been marketed by several companies. Detailed information about the investigations carried out with these microscopes in various countries is to be found in various documents. V. E. Cosslett (1968) has given a complete account of the development of HVEM in Britain. We must also mention the opening lecture of P. B. Hirsch, "High-Voltage Electron Microscopy in the U.K.," at the International Conference in Oxford (1974).

H. Hashimoto, Chairman of the Organizing Committee of the Fifth International Conference on High-Voltage Electron Microscopy (Kyoto, 1977), has described the progress accomplished in electron microscope design in Japan since 1950. R. Uyeda has shown the importance of the work that has been carried out in microdiffraction and HVEM.

A six-million-volt electron diffraction instrument has been constructed by JEOL. N. Anazawa, M. Morimura, S. Kato, T. Honda, and S. Gonda have planned a combination of a pulsed electron gun and a linear accelerator, instead of a Cockroft-Walton-type accelerator.

S. Horiuchi, Y. Matsui, Y. Bando, Y. Sekikawa, and K. Sakaguchi (1977) have constructed a high-voltage electron microscope specially designed for the observation of crystal-structure images at a resolution level of 2 Å. This microscope can work at accelerating voltages between 200 and 1250 kV.

E. Sugata, K. Fukai, H. Fujita, and K. Ura presented at the International Congress that took place in Grenoble in 1970 an electron microscope which can be operated at 3 MV. They have described in detail the column and the high-voltage accelerator of this instrument, constructed by Hitachi. This microscope is operating at Osaka University. With this instrument, Fujita and his co-workers have solved many important problems concerning materials science.

In the United States, HVEM applications are being studied in many universities, in particular the important laboratory of the University of Colorado at Boulder. Keith R. Porter has obtained many new results in the biological sciences. There is also an important range of equipment, including a 1.2-MV electron microscope, in Dr. Parsons' laboratory at Albany, New York. Another high-voltage electron microscope is in operation at the University of Wisconsin at Madison.

R. M. Fisher, of the U.S. Steel Corporation, Fundamental Research Laboratory, Monroeville, Pennsylvania, has created a modern HVEM installation. His work on the metallurgical applications of HVEM has contributed to the understanding of many problems in metallurgy.

In the Department of Biophysics of the University of Chicago, H. Fernández-Morán is working on HVEM at liquid-helium temperatures (1971). He has published many new results concerning cell membranes and macromolecular structures of biological systems.

E. Zeitler and A. V. Crewe have presented in several publications (1976, 1977) a new type of microscope, "The Chicago One-Million-Volt Scanning Microscope." In a communication at the Eighth International Congress on Electron Microscopy (Canberra, 1974) Crewe (1974) wrote "The high-resolution STEM is now a reality." Scanning electron microscopes are, in many situations, superior to transmission microscopes. This type of microscopy is also undergoing rapid development.

J. B. Le Poole and H. B. Zeedijk have designed and constructed a 1-MV compact electron microscope, which was first described in Grenoble (Le Poole et al., 1970), then in Kyoto (1977). The aim of the design was to keep the size of the complete microscope small enough to fit into an ordinary laboratory room. The electron beam emerging from the accelerating tube passes through a 180° double-focusing magnet. The users claim to reach a 10-Å resolution.

A magnetic energy filter is used, in Toulouse, for the 1.2-MV electron microscope, by G. Zanchi, J.-Ph. Perez, and J. Sevely (Zanchi et al., 1975). For thick biological specimens, a filtering device has been adapted to the microscope in order to eliminate inelastic scattering and to obtain better images.

To end this section we should also mention work on superconducting lenses. In France, Laberrigue et al. (1974) employ this type of coil at liquid-helium temperature. They have built an electron microscope working at 400 kV, but the optical column is capable of operating at up to 1 MV. In Toulouse, under the direction of J. Trinquier, electron lenses with superconducting coils have been designed. Their characteristics have been determined by the "vibrating probe coil" method. A 300-kV electron microscope with a superconducting objective lens has been operated in the laboratory. The first micrographs obtained are very good (Trinquier & Balladore, 1968; Trinquier et al., 1972). Such a lens could be used as an objective lens on the 1-MV microscope. These investigations have been continued, and some interesting results have been obtained, especially as far as the resolving power is concerned.

In Germany, I. Dietrich and co-workers have developed a shielding superconducting lens system, tested at 400 kV, and they conclude that the shielding lens may be applied, with only slight geometric variations, from 300 kV up to the range of a few megavolts (Dietrich et al., 1974). In 1975, I. Dietrich made a proposal for a high-voltage electron microscope with a superconducting microwave linear accelerator and superconducting lenses (Dietrich et al., 1974). This is a new development which may lead to interesting results. Recently G. Lefranc, K. H. Müller, and I. Dietrich (1981) have built a superconducting lens system that can be operated in the fixed-beam and scanning modes.

T. Mulvey and C. D. Newman have designed new experimental lenses for high-voltage electron microscopes. A substantial reduction in lens volume can be achieved. Good micrographs have been obtained. These researchers conclude that "the performance of single-pole lenses at 1 MV can be comparable with those of the superconducting lenses. It therefore seems feasible to design electron-optical columns for 1 MV with dimensions comparable to those of present 100-kV instruments" (Mulvey & Newman, 1974).

5. Applications of HVEM

In this section we give a review of the results obtained in some domains of research which particularly display the advantages of HVEM.

5.1 Penetrating power

One important advantage of using high-energy electrons is that much thicker specimens than usual can be examined. In Toulouse, observations made on metals and alloys show that good images can still be obtained when the thickness of the specimen is as large as 6 or even 9 μm (Dupouy, 1973). To give a comparison, the maximum thickness observed in ordinary microscopes is of the order of 0.1 μm. Thicker specimens are more representative of bulk material. R. Uyeda (1968) and V. E. Cosslett (1969, 1974a, 1974b) have provided much interesting information on this subject. The increase in penetration can be explained as follows: (1) The mean free path Λ_t, of an electron in a given substance increases with its energy: for instance, in the case of carbon and an aperture angle $\alpha = 5 \times 10^{-3}$ rad, Λ_t is 18 times greater at 3 MV than at 100 kV. (2) Another point must be mentioned, although it is only relevant in the case of crystalline specimens. For a crystal, the absorption coefficient for electrons decreases as the voltage is increased; alternatively we may say that the object becomes more transparent as the energy of the incident electrons is raised.

5.2 Diffraction contrast

Images in dark field of crystalline substances (metals, alloys, etc.) show very good contrast. We have studied, in collaboration with F. Perrier, R. Ayroles, and A. Mazel, the variation of diffraction contrast with respect to the accelerating voltage (Dupouy, Perrier, Ayroles, et al., 1970; Dupouy, Perrier, Durrieu, 1970; Dupouy, Perrier, Tremollieres, 1970). With a constant effective aperture, *contrast increases with voltage* V. When V increases from 300 kV to 2 MV, the gain is equal to 2 if the thickness $t = 100$ nm, while it reaches 5 if $t = 500$ nm. This is a new and significant advantage of operating at high voltages.

5.3 Chromatic aberration

We recall also that, when passing through an object, electrons may lose variable amounts of energy, depending on the nature and thickness of the specimen and on the energy of the incident beam. This results in chromatic aberration, by which the quality of the image is reduced. Chromatic aberration decreases in a spectacular way when the energy of incident electrons is increased; as far as aluminum is concerned, it is about 65 times less important at 3 MV than it is at 100 kV.

5.4 Selected areas

When the specimen under examination possesses a crystalline structure, an electron microscope provides us not only with the enlarged image of the object, but also with the microdiffraction pattern corresponding to a selected area of the specimen.

We have shown (Dupouy et al., 1963) that in HVEM the disturbing effect of spherical aberration becomes very much smaller and the accuracy in selected-area diffraction is increased.

5.5 Metallurgy and materials science

The advantages of HVEM have been exploited to examine objects a few micrometers thick. This is particularly valuable in the study of imperfections in the structures of metals and alloys, dislocations and stacking faults, for example. Observations of this kind have already made possible considerable progress in our knowledge of a large number of topics connected with metal physics and have provided solutions to various metallurgical problems.

At the beginning of these studies the mechanism of the contrast in the image was analyzed in terms of the kinematic or the dynamical theory (Fukuhara et al., 1962; Hashimoto et al., 1960; Hirsch, 1962; Kainuma & Yoshioka, 1962; Kamiya, 1963; Kohra & Watanabe, 1961; Uyeda, 1962; Yoshioka, 1957). It was shown that the diffraction contrast is sensitive to crystal thickness, movements of atoms caused by external stresses applied to the crystal, and changes of orientation of the lattice planes, as caused by bending a crystal, for example. Within a metal or alloy, therefore, it is possible to observe the following features: (1) grain junctions and grain boundaries—the latter will, in general, be indicated by fringes of equal thickness; (2) dislocations and stacking faults in crystals that have been made imperfect by the action of various kinds of constraint; (3) extinction contours; and (4) precipitates occurring in specimens after suitable thermal treatment.

5.5.1 Some examples

We present a few micrographs of various metals and alloys that have been investigated at 1 MV or above. Fig. 12 shows a stainless-steel sample with

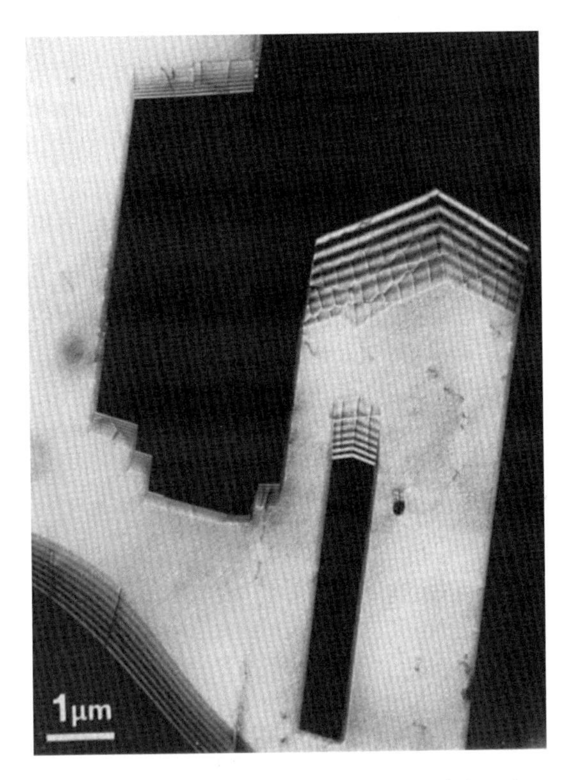

Figure 12 Stainless-steel sample: grain boundary fringes, dislocations, twin boundary (dark field); $V = 1$ MV.

grain boundary fringes and dislocations (dark field); $V = 1$ MV. Fig. 13 also shows a stainless-steel grain with equal-thickness fringes (dark field); $V = 1$ MV. Fig. 14 shows, in dark field, the classical aspects of equal-thickness fringes on the edge of stainless-steel grains; $V = 1$ MV. Fig. 15 shows stainless steel; one can see twin boundary dislocations and the diffraction pattern from the (overprinted) selected area; $V = 1$ MV. Fig. 16a and b displays screw and ring dislocations in a sample of an Al–Cu alloy; $V = 1$ MV. Fig. 17 is the image of a beautiful dislocation pileup in a sample of stainless steel. This pileup is very long and we have taken only one part of it. About 140 dislocation lines can be counted; $V = 2$ MV. Fig. 18 shows numerous dislocations in stainless steel. The following image (Fig. 19) shows precipitates in an Al–Ag alloy 4 μm thick which has undergone heat treatment. Fig. 20 is also a micrograph of precipitates in an Al–Ag alloy observed at 3 MV and showing many fine details. Fig. 21 is the first image that was obtained at 2 MV. It shows precipitates in an Al–Cu

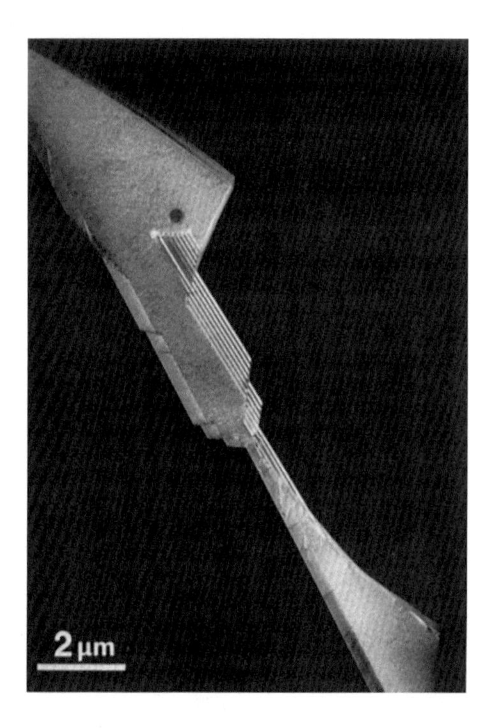

Figure 13 Stainless-steel grain; equal-thickness fringes (dark field); $V = 1$ MV.

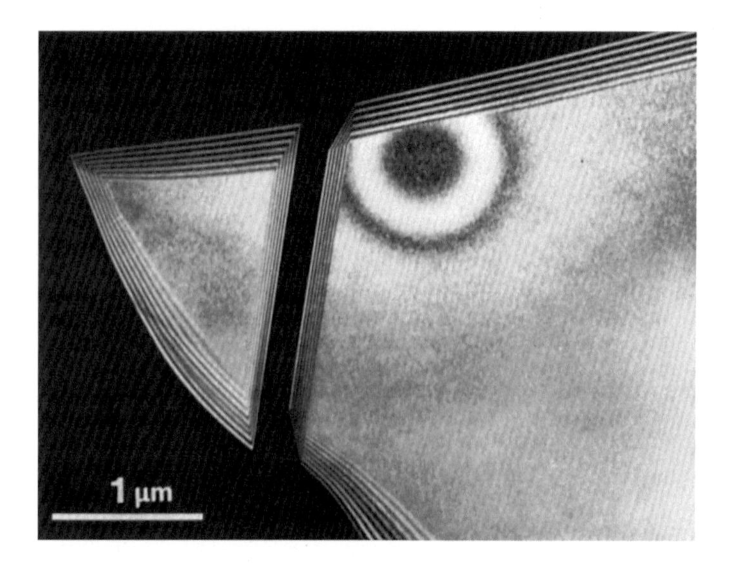

Figure 14 Stainless-steel grains; equal-thickness fringes (dark field); $V = 1$ MV.

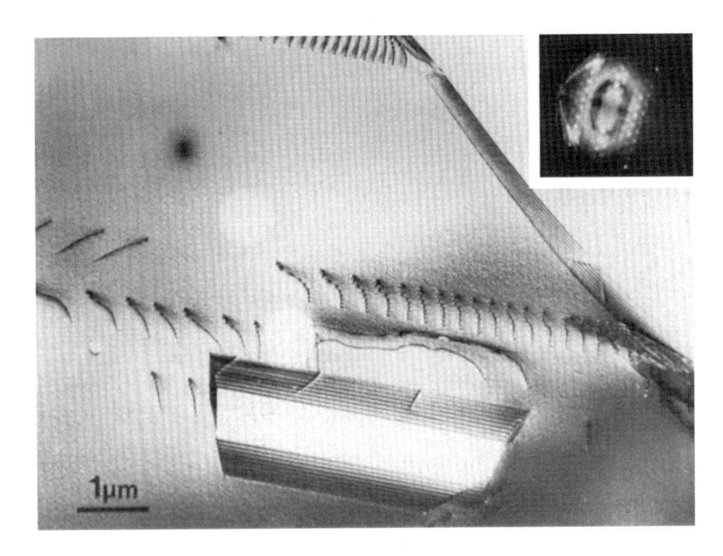

Figure 15 Stainless steel: twin boundary dislocations; diffraction pattern from the overprinted selected area; $V = 1$ MV.

alloy. I presented it at the Monroeville Conference (*Micron*, Vol. 1, No. 3, 1969).

P. Coulomb and F. Reynaud (1970) have observed interesting images of antiphase boundaries in Ni_3Mn, at 1 MV (Fig. 22a and b). The sample has been annealed for 200 hr at 500 °C. Fig. 22a is the dark-field (300) image, with orientation of the foil (110). The antiphase domains are observed in the (300) extinction contours. Fig. 22b is the corresponding bright-field image; the antiphase boundaries cannot be observed.

P. B. Hirsch (1968, 1974, 1977) has particularly contributed, showing how the HVEM succeeded in solving problems that had no prior solution. He gave a remarkable presentation of these questions in one of the general lectures at the Fifth International Conference on HVEM (Kyoto 1977), entitled "Applications of HVEM to Materials Science."

H. Fujita et al. (1977) have studied with the 3-MV electron microscope in Osaka University various applications of HVEM to metal physics. Interesting observations are given concerning work-hardening processes in metals; the deformation of tungsten, iron, and aluminum single crystals; dislocations in copper and in Cu–Al alloys; and dislocation motions. They also describe devices adapted to their microscope for *in situ* studies of different specimens.

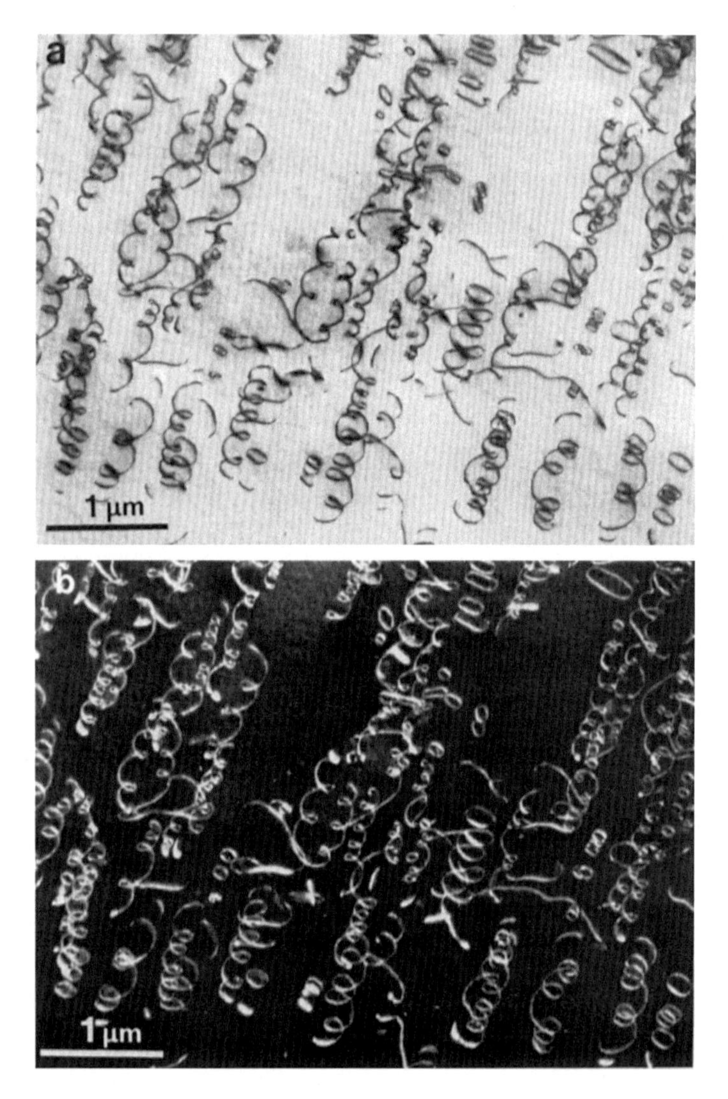

Figure 16 Screw and helical dislocations in an Al–Cu alloy: (a) bright field, (b) dark field; $V = 1$ MV.

R. M. Fisher (1977) and faculty members of several universities in the United States have studied, in collaboration, different applications of HVEM, including the plasticity of cementite, the fracture of pearlite, hydrogen attack of carbon steel, metal dusting, and gasification of graphite. Very good images show the different aspects of the interesting results they have obtained.

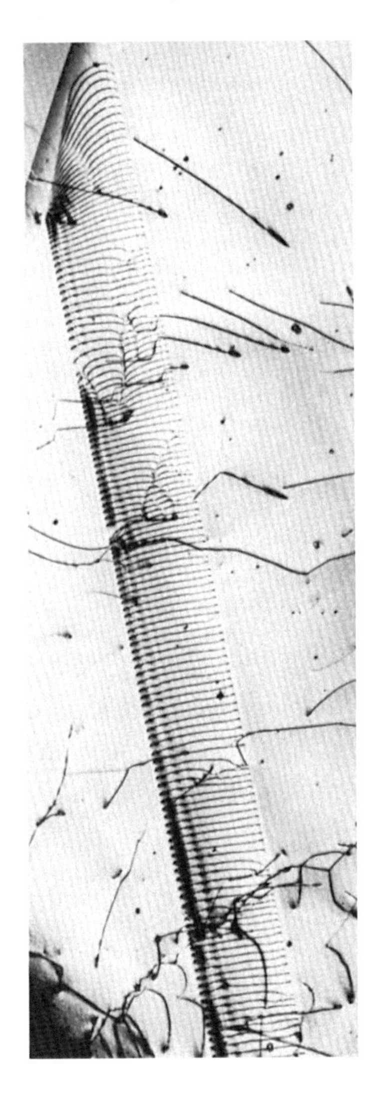

Figure 17 A beautiful dislocation pileup in a sample of stainless steel; $V = 2$ MV.

G. Thomas (1977) has successfully carried out numerous HVEM investigations in ceramics research. New results are to be found in his publications concerning phase transition in ceramic ferrites and the study of such defects as stacking faults in nickel ferrite, dissociated dislocations and stacking fault energy in $LiFe_5O_8$, refractory ceramics, and magnetic ceramics. Thomas emphasizes the significant impact of HVEM on ceramics research.

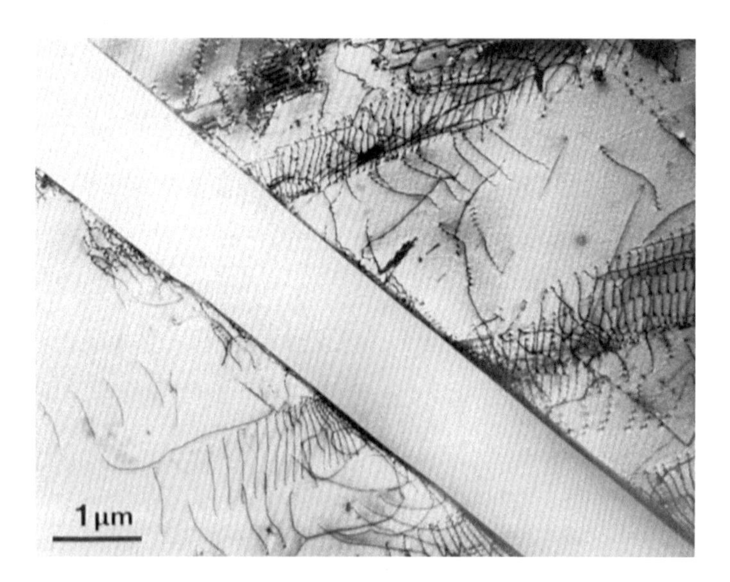

Figure 18 Numerous other dislocations and a twin boundary in stainless steel; $V = 1$ MV.

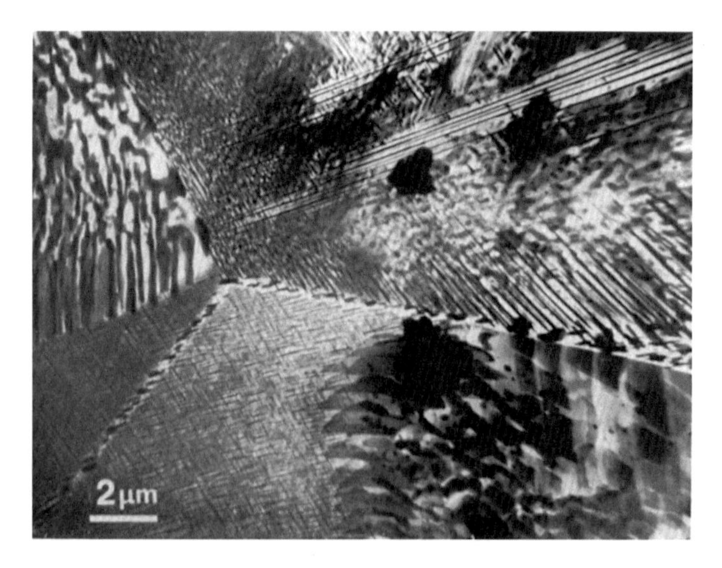

Figure 19 Precipitates in an Al–Ag alloy, after heat treatment; thickness $t = 4$ μm; $V = 1$ MV.

5.5.2 Observation of metals and alloys during ion thinning

We have developed a method which may be especially useful for studying various problems in metal physics. This technique enables us to thin down

Figure 20 Other precipitates in an Al–Ag alloy; $V = 3$ MV.

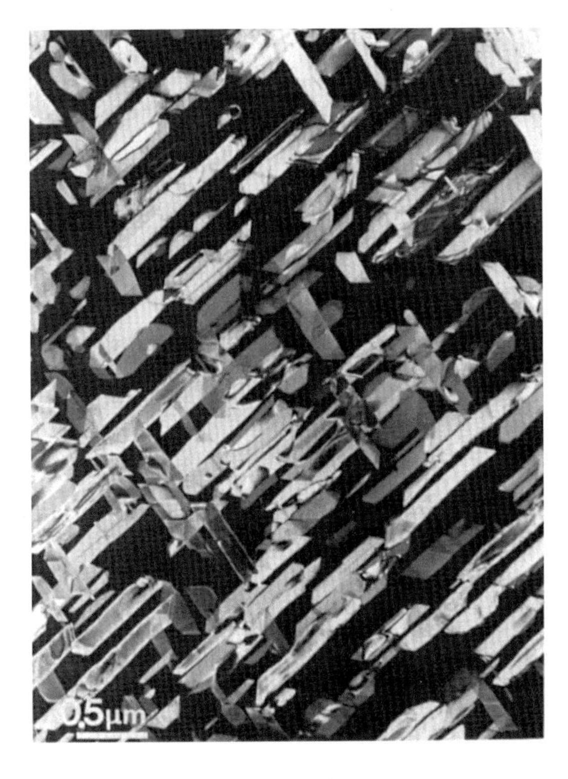

Figure 21 First image obtained at 2 MV: precipitates in an Al–Cu alloy.

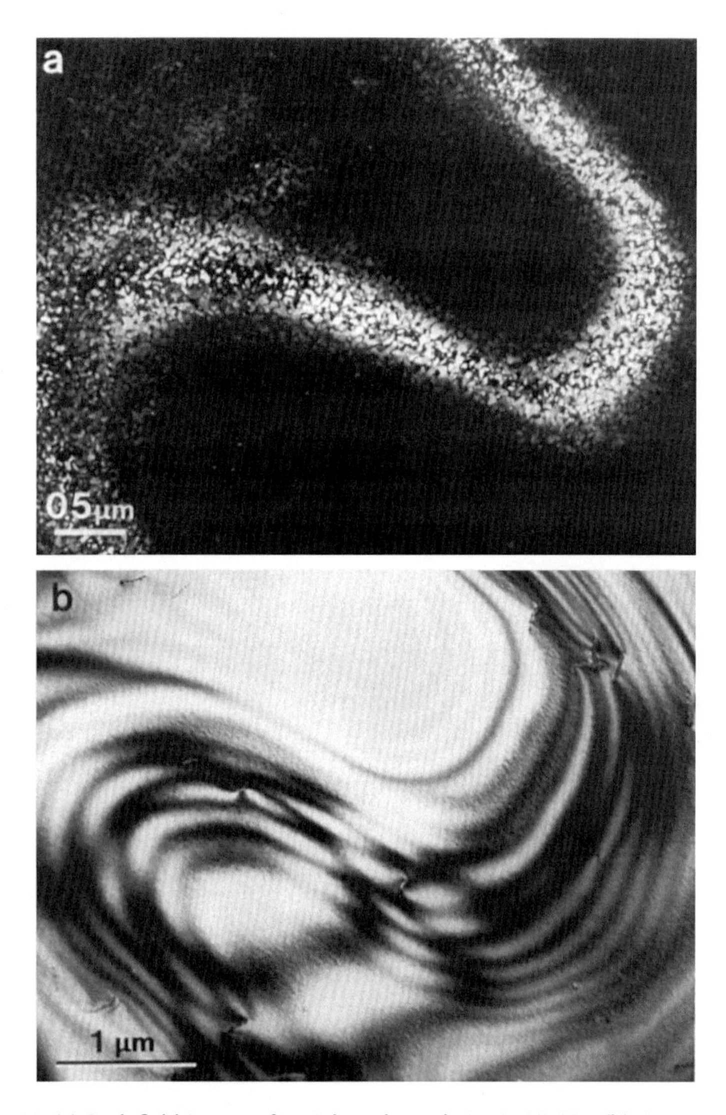

Figure 22 (a) Dark-field image of antiphase boundaries in Ni_3Mn. (b) Corresponding bright-field image; the antiphase boundaries cannot be observed.

a specimen and to follow, during thinning, the evolution of its structure (grains, twin boundaries, precipitates, etc.) and of its defects (dislocations, stacking faults) as a function of its thickness. The local thinning of the specimen is carried out in the microscope and the chosen region constantly remains in the field of observation. We use ion thinning (Castaing & Jouffrey, 1962). The experimental device is to be seen in Fig. 23. Both

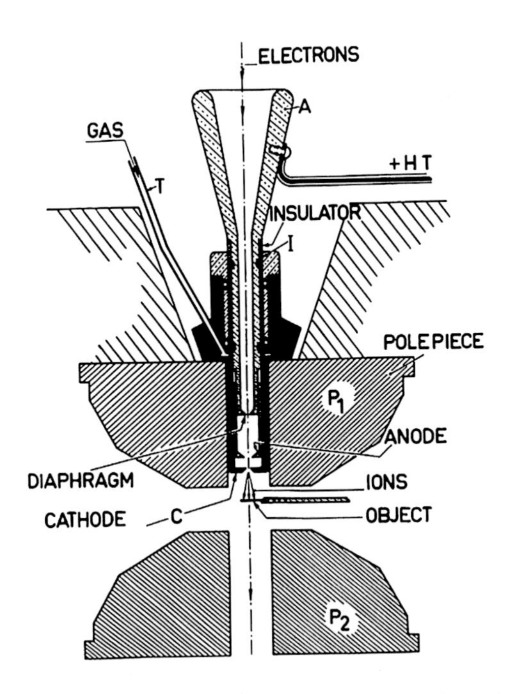

Figure 23 Ion-gun device for thinning the specimens during their observation.

polepieces of the objective (P_1, P_2) have a hole 12 mm in diameter; they are also separated by 12 mm. The ion gun is introduced in P_1. The anode (A) is brought to a positive voltage with respect to the cathode (C); A and C are insulated by a quartz cylinder (I). The gas we use is argon. It enters through an adjustable leak by a very thin copper tube (T). The ions emerging from the gun strike the specimen situated in the gap under polepiece P_1. Numerous specimens of stainless steel and alloys have been thinned by this procedure.

Fig. 24 shows three micrographs of a stainless-steel specimen which has undergone heat treatment at 1160 °C for 5 min, followed by water tempering. Many dislocations are to be seen; their length decreases as the bombardment goes on. One should notice the decrease of the density of dislocations in Fig. 24c compared to Fig. 24a. Many extinction contours appear, for small thicknesses, in Fig. 24c.

In Fig. 25 we see micrographs of a Swedish D × 9 stainless-steel specimen that has been given mechanical treatment to the stress limit. Such a treatment has promoted the development of dislocations and stacking faults.

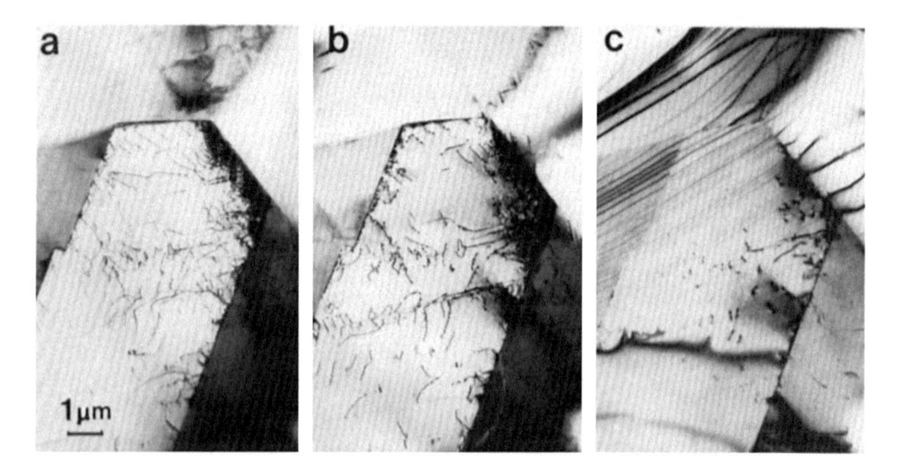

Figure 24 Stainless steel: (a) after 5 min of ion bombardment, $t = 0.55$ µm; (b) after 8 min of bombardment, $t = 0.31$ µm; (c) after 10 min of bombardment, $t = 0.1$ µm.

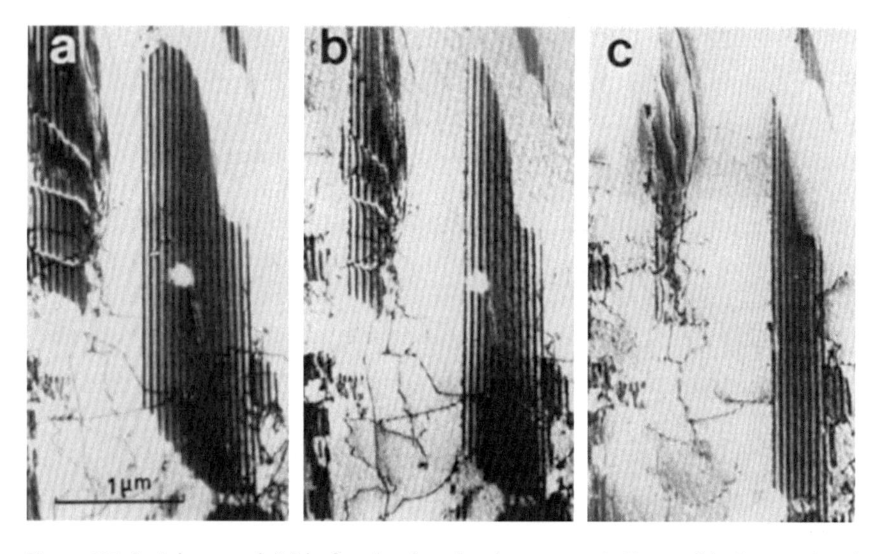

Figure 25 Stainless steel: (a) before ion bombardment, $t = 0.62$ µm; (b) after 5 min of bombardment, $t = 0.44$ µm; (c) after 9 min of bombardment, $t = 0.30$ µm.

As far as stacking faults are concerned, it is to be noticed that fringes grow less and less numerous as the thickness decreases.

It seems that such an experimental method can be most useful in metallography and solid-state physics. This work has been done by O. Vingsbo and A. Lasserre (1966). We shall add a few remarks: Ion attack is carried out through the upper part of the specimen, and so we can distinguish, on the image, points situated in the upper part from points situated in the lower part. The electric field between the anode and the cathode is equivalent to a weakly convergent electrostatic electron lens, but, with high-energy electrons, the quality of the images is not altered at all by the ion gun being on. In some cases, when the specimens are very rapidly altered, it is possible to film the phenomena that occur during this evolution. The first film made in the laboratory was presented by L. Durrieu at the Sixth International Congress on Electron Microscopy (Kyoto, 1966).

5.5.3 Observation of magnetic domains

Transmission electron microscopy is a good technique for studying magnetic domains in thin foils of ferromagnetic materials.

Almost all conventional electron microscopes are equipped with magnetic lenses. The specimen is then immersed in a region in which the magnetic field of the objective is more or less troublesome when a ferromagnetic material is being observed (Dupouy, Perrier, Seguela, et al., 1968).

For our high-voltage microscope we have made the device shown in Fig. 26. The polepieces (P_1, P_2) have been so designed that the magnetic field on the objective axis, in the object plane, is less than 1 G (0.1 mT). Holes, 4 mm in diameter and 5 mm high, have been made in the opposite face of the polepieces. In the upper polepiece P_1 the hole widens to a diameter of 30 mm. Thus two groups of coils can be used, field coils (B, B′) and compensating coils (B_1, B'_1), which are useful for studying the motion of the walls. The diaphragm (D) is placed exactly in the back focal plane of the objective lens. Moreover, this diaphragm is provided with a tungsten wire, 7 μm in diameter, fitted along a radius. This wire is used to mask the secondary spots with a very good selectivity. Only the beam giving rise to one of the central spots of the diffraction pattern is allowed to pass. The bright-field image of the corresponding domain is then obtained. The same operation, successively repeated for the various domains, enables one to observe each of them individually. The diaphragm remains practically centered. In the course of the various operations it is only shifted by a few micrometers. Thus the quality of the image is not altered. A more elaborate device has now been constructed: the contrast diaphragm always remains centered. One or several wires, moving independently, mask the spots that must be eliminated.

We show some photographs of domains in specimens of iron −4% silicon in Fig. 27; in Fig. 27a we see the defocused image, usually obtained

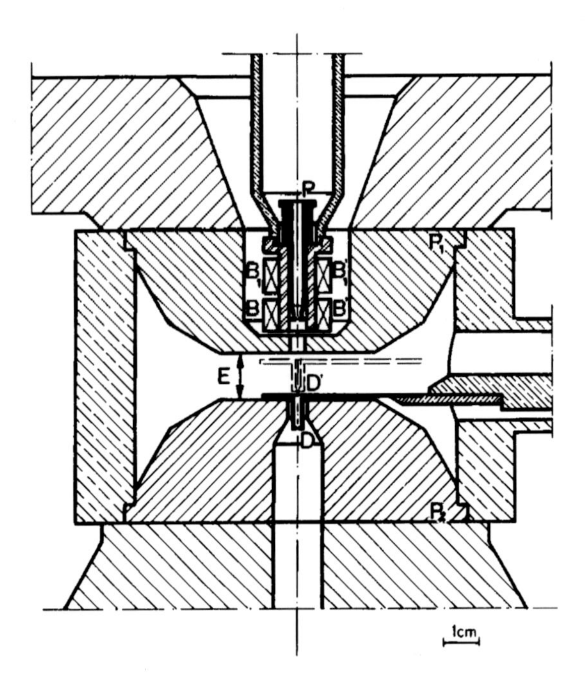

Figure 26 Device used to observe magnetic domains.

by Lorentz microscopy. The subdivision of the direct beam is to be seen in the inset, which is an enlarged image of the spots of order 0 corresponding to domains 1–3 of the overprinted area. If we choose to observe the type 3 domain, the contrast wire is placed as shown in Fig. 27b. We then obtain an image in which only the domain corresponding to spot 3 is perfectly contrasted on the focused image.

We have operated in the same way for the photographs of Fig. 28a and b, taken in another region, in which the magnetic structure is more intricate. These show the very good selectivity of our device. By using this technique we get both of the results given by the electron microscope: the focused image of both selected domains as well as the diffraction pattern of the specimen. We work with magnifications ranging around 10,000×, but it will be possible to operate with much higher magnifications (Dupouy, Perrier, Seguela, et al., 1968). A. Seguela (1971) has continued this work with the improved device. Fig. 29 is a very interesting micrograph showing magnetic structures and their evolution under the influence of a magnetic field. One can see the broadening of a domain spreading over three grains. This result is obtained by displacing the walls.

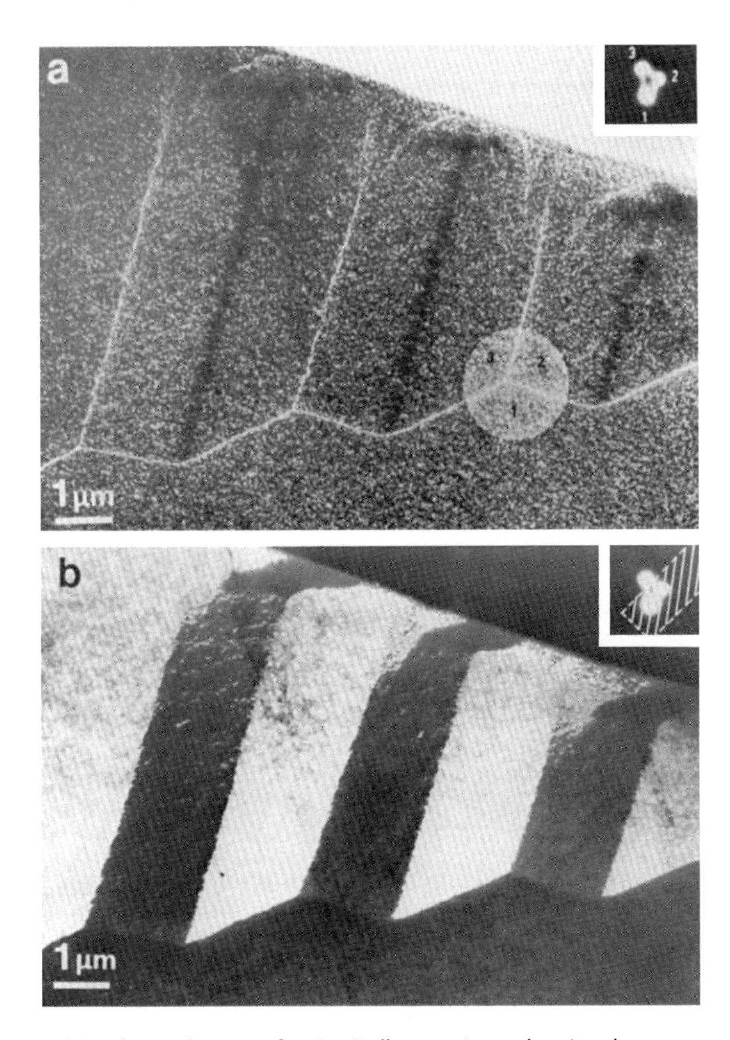

Figure 27 (a) Defocused image of an Fe–Si alloy specimen showing three magnetic domains. (b) The domain corresponding to spot 3.

R. P. Ferrier and I. B. Puchalska (1968) have studied, with the high-voltage microscope of the Cavendish laboratory, magnetic domain structures in Permalloy films of different thicknesses. This work has revealed strong stripe domains which appear clearly in Lorentz micrographs.

C. G. Harrison and K. D. Leaver (1973) have observed, by Lorentz microscopy, stripe domains in (001) nickel, which occur in a variety of structures due to the large number of easy directions present. C. G. Harrison, K. D. Leaver, and P. R. Swann (1972) have constructed a versatile

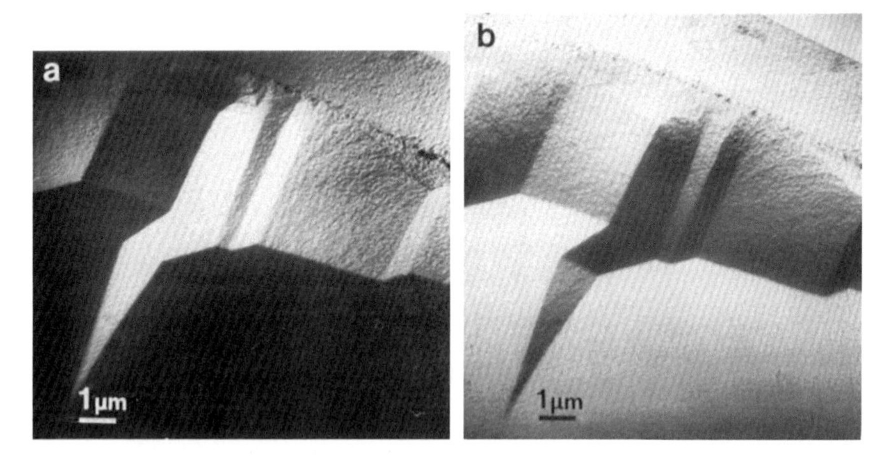

Figure 28 Another region of the specimen in Fig. 27 showing magnetic domains with various aspects.

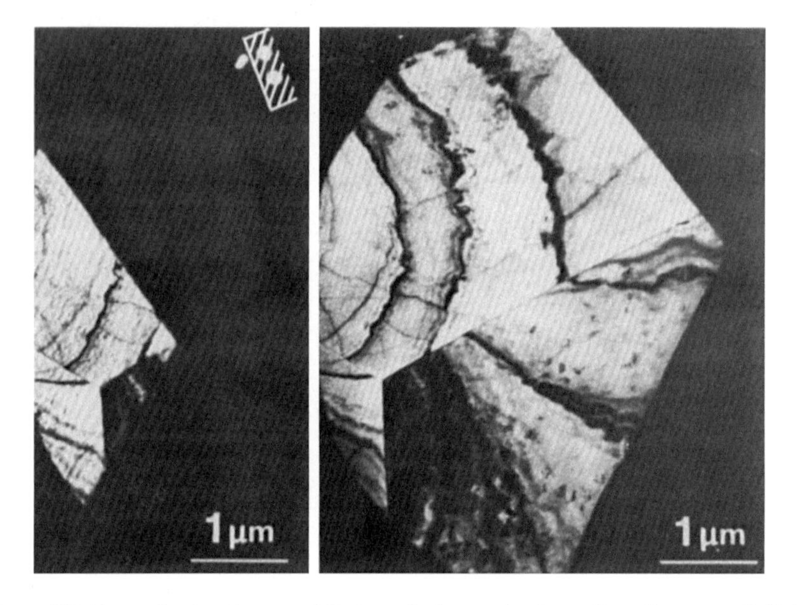

Figure 29 Magnetic structures and their evolution under the action of a magnetic field.

specimen chamber for *in situ* experiments on magnetic materials. They obtained micrographs showing, for example, a magnetite grain in natural specular hematite and domain walls in a nickel platelet. R. H. Wade (1962) has studied the effects of transverse magnetization in Permalloy films. He obtained micrographs showing that a transverse field produces magnetization rotation and wall movement and that a domain structure, antiparallel to the initial one, is obtained on demagnetization.

We have mentioned only a few examples of the investigations that can be carried out in the field of HVEM applications in metallurgy. Many very interesting studies have been reported at all the congresses organized in recent years. As R. M. Fisher pointed out (1977), "Specimen techniques and HVEM image interpretation are now sufficiently well developed for applications to many important metallurgical problems."

5.6 Biological applications

We mentioned earlier the essential role played by L. Marton in the development of electron microscopy in biological research. Marton, who first settled in Brussels, afterwards came to the United States, where he continued his activities. Among those who carried out early research in biology, we must also cite Ralph W. G. Wyckoff: he was indeed one of the American pioneers in this field. His first publications will be read with great interest as well as his book, "The World of the Electron Microscope," in which he points out the results obtained in such various domains as the study of molecules, cells, bacteria, bacteriophages, viruses, and so on.

5.6.1 Radiation damage

It is interesting to study the damage done to a specimen during its irradiation by the electron beam, especially when the specimen is an organic or biological material. The first experiments were carried out by Kobayashi and Sakoku (1965). They measured, in organic polymers, the "dose" of electrons for which, at a given operating voltage, the diffraction pattern of a crystalline specimen vanishes. The important result of their experiments is knowledge of the decrease in radiation damage when the electron energy is increased. Similar experiments have been performed by Glaeser and Thomas (1968) with valine and glycine.

In collaboration with F. Perrier, we thought that it would be useful to extend these experiments up to 3 MV. They were carried out with thymine, which is one of the components of DNA. To measure the density of the electron beam on the specimen, we used a Faraday cage that had been especially designed to work at high voltage (Dupouy, 1974).

We operated at 0.5, 1,2, and 3 MV. At each voltage we irradiated the specimen under well-defined conditions, and we arranged the exposure time so as to get a photograph every fifth second. With the diffraction pattern from thymine, we could determine the lifetime at the end of which the crystalline structure vanished. We had thus all the data necessary to calculate the dose, in coulombs per square centimeter, corresponding to the fading of the pattern. The result was that the critical dose was about four times as great at 3 MV as it was at 100 kV. This in itself constitutes a great advantage of HVEM for the observation of organic material and biological specimens.

P. Favard and N. Carasso (1972) and P. Favard et al. (1971) have shown that some samples observed successively at 2, 1, and 0.5 MV, then at 2 MV again, did not reveal any visible alteration in their structures.

Today biologists are more and more interested in HVEM, which allows them to extend their investigations in life science. Due to the important gains in penetrating power and to the reduction of chromatic aberration, thicker specimens can be studied (Audrey M. Glauert, 1974). P. Favard, N. Carasso, and L. Ovtracht have photographed thick specimens (up to 10 μm) at 2.5 MV.

Unfortunately, in the case of biological specimens, the image contrast becomes poor at very high voltage. Such a difficulty can be overcome in two ways: First, the different regions of the object can be selectively contrasted by using various impregnating methods (Carrasso et al., 1973, 1974). Second, we developed, a few years ago, with F. Perrier and P. Verdier (Dupouy et al., 1966), the very simple *contrast–stop method*, which considerably improves the contrast in images of biological specimens.

5.6.2 Contrast–stop method

The principle of this method is as follows: the electron beam coming from the condenser of the electron microscope (Fig. 30) strikes the objects at an illumination angle α_i, which has a very small aperture (some 10^{-4} rad when we work at 1 MV and above).

The electrons emerging from the object can be divided into three groups:

1. Those that pass through the specimen without interacting with it. The number of these electrons depends on the specimen thickness and density. As far as an amorphous carbon film of thickness $t = 500$ Å is concerned, it may reach 80% at $V = 1$ MV. Such electrons convey no information about the object. They just reduce the contrast. We shall call them the *direct beam*.
2. The second group includes the electrons that have been scattered by interacting with the specimen, through angles ranging from 0 to α_0

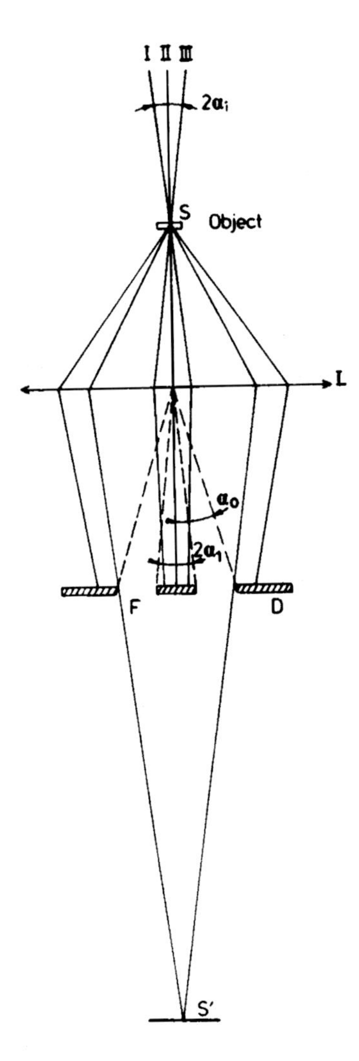

Figure 30 Principle of contrast–stop method.

(α_0 being half the objective-lens angular aperture). We call them the *scattered beam.*

3. The last group contains the electrons that have been scattered through angles $\alpha > \alpha_0$. They are stopped by the diaphragm (D). Thus electrons from groups (1) and (2) are the only ones that contribute to the formation of the image.

In the contrast–stop method we use an annular diaphragm obtained by setting in the focal plane of the objective lens a small disk (F), opaque to

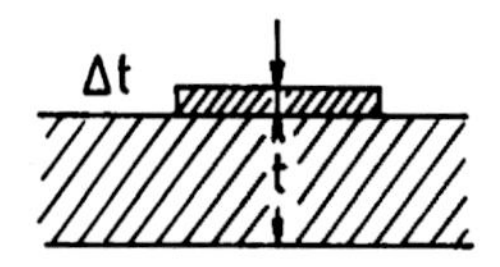

Figure 31 Schema showing a film of graphite of thickness t with a local extra thickness Δt.

the electrons and centered in the diaphragm (D). Its diameter (6 μm in our experiments) has been chosen in such a way that *all* the electrons from group (1) are stopped, as well as a small part of the scattered electrons. We call $2\alpha_1$, the angle under which the disk (F) is seen from the optical center of the objective lens.

In our experiments at 1 MV, $\alpha_0 = 5 \times 10^{-3}$ rad, and α_1 may vary from 4×10^{-4} to 2×10^{-3} rad. We have carefully studied the contrast of the image in the case of a graphite foil of constant thickness $t = 200$ Å (Fig. 31) presenting a local extra thickness $\Delta t = 100$ Å (Dupouy et al., 1966). We shall just recall a few results:

1. Contrast C is reversed when we pass from bright field to dark field (its sign is changed). Moreover, its absolute value increases in a spectacular way; at 1 MV it is, according to the individual case, multiplied by a coefficient which can reach 30. As an example, bright field $C = +0.013$, dark field $C = -0.38$. This method also gives good results at the low voltages that are used in usual commercial microscopes. At 75 kV, for the same values of α_0 and α_1 and a graphite foil, the thickness of which is now $t = 50$ Å, with an extra thickness $\Delta t = 30$ Å, the contrast can be multiplied by 6.
2. The contrast increases when the angle α_1 reaches around $\alpha_1 = 10^{-3}$ rad. Above such values the gain is weak.
3. The exposure time is naturally longer than in usual conditions. At 1 MV it may be multiplied by 4, at the utmost.

With this method the image contrast in dark field is so improved that biological specimens can be photographed, at 1 or 3 MV, without being fixed, stained, or metal shadowed.

Fig. 32 shows the bacterium *Escherichia coli* K-10. We see a flagellum which can fix MS2 bacterial viruses, arranged side by side as a single or double row along the flagellum or over the body of the bacterium ($V =$ 1 MV).

Fig. 33 shows a 36-hr culture of *Pseudomonas acidovorans* 130 of human origin which had been isolated in the Central Laboratory of Microbiology

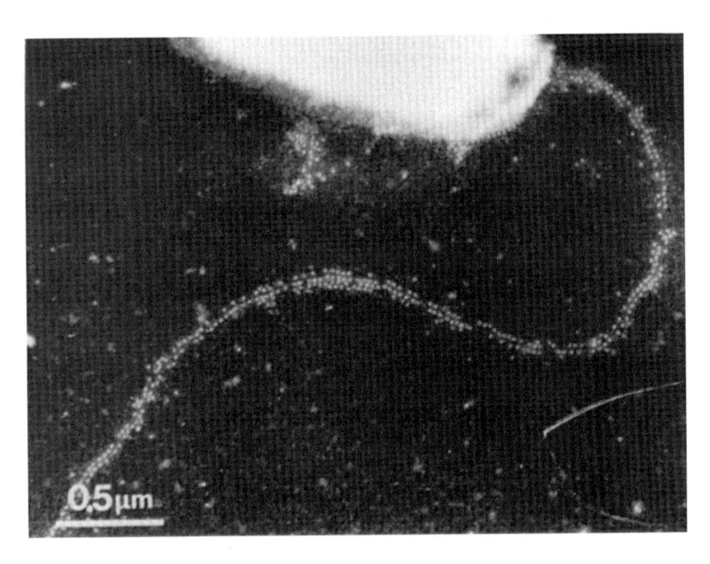

Figure 32 Bacterium *Escherichia coli* K-10: MS2 viruses are fixed on a bacterial filament; $V = 1$ MV.

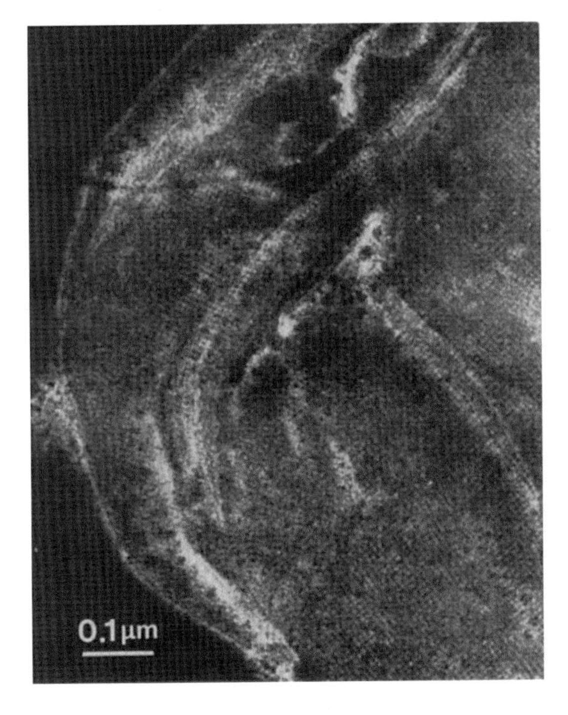

Figure 33 *Pseudomonas acidovorans* 130, showing a fine periodic structure; $V = 3$ MV.

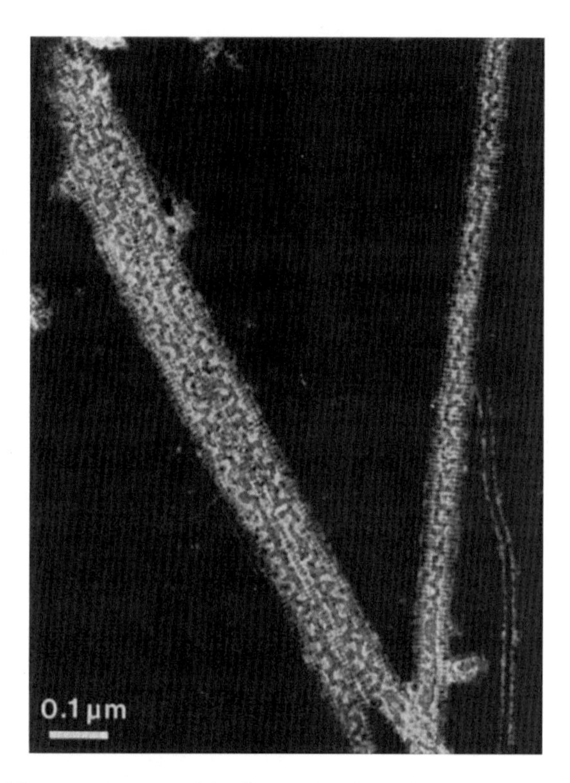

Figure 34 A ribbon coming out of the bacterium shown in Fig. 33: very fine details can be distinguished; $V = 3$ MV.

of Purpan Hospital in Toulouse (Mrs. L. Enjalbert). It has been slightly stained with phosphotungstic acid, the concentration of which was only $\frac{1}{2000}$ (Lapchine & Enjalbert, 1971). Carefully looking at the bacterium we can see a fine periodic structure, a regular pattern of squares with a spacing of about 50 Å. This bacterium has a peculiar characteristic: long ribbon-shaped elements break off the bacterium. One of them can be seen in Fig. 34. (Since these first micrographs were taken it has been possible to obtain images of this bacterium without any staining.)

As a second example we present (Fig. 35) another bacterium, *Pseudomonas acidovorans* P-20. It can clearly be seen that the bacterial wall has a very sharp periodical structure, with hexagonal symmetry. This micrograph was taken at 3 MV (Dupouy et al., 1969).

We want to emphasize that the contrast–stop method can be used at all voltages. Fig. 36 has been obtained at 100 kV, i.e., with a conventional electron microscope. This micrograph shows filaments of DNA, extracted

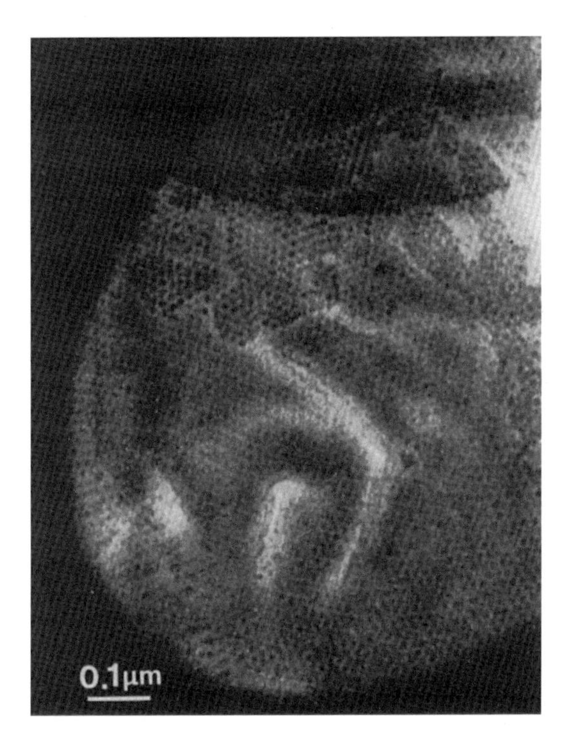

Figure 35 *Pseudomonas acidovorans* P-20: the bacterial wall shows a very sharp periodical structure with a hexagonal symmetry; $V = 3$ MV.

from T4 bacteriophages by Kleinschmidt's method. One of the T4 phages, lying near a partially unwound strand of DNA, seems to have an empty capsid. The other phage, with a bright head, seems to be extruding a DNA filament from the caudal tip.

W. H. Massover has observed in our laboratory, at 2.8 MV, dark-field images of a frog skeletal muscle thin section by the contrast–stop method. The specimens were prepared exactly as is routinely done for observation at 100 kV. Fig. 37a shows the excellent contrast of the image and the good definition of all the minute components of the muscle cell, such as thick and thin filaments, details in Z lines, glycogen granules, membraneous elements, etc. These results indicate a new scope of application of HVEM in biological ultrastructure research.

We must also mention another result of Massover et al. (1973); with co-workers from our laboratory they have been able to observe ferritin macromolecules deposited on an ultrathin film of carbon. The dark-field images obtained at 3 MV reveal that some ferritin regions contain one or

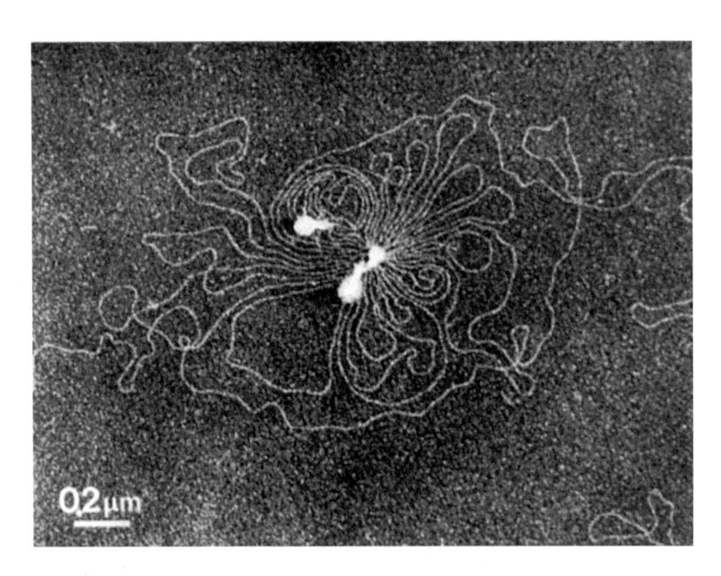

Figure 36 Filaments of DNA extracted from T4 bacteriophage by Kleinschmidt's method. The specimen has been photographed without being fixed, stained, or metal shadowed.

more granules of varying size, shape, and image intensity. Ferritin cores do have real crystalline substructure. This was confirmed in a later study. Finally, we recall a study concerning the effective resolution in biological thick sections. Massover (1974) gives experimental results with HVEM (1–3 MV).

Audrey M. Glauert (1974, 1979; Glauert & Mayo, 1973) has also studied the progress HVEM has brought in biological investigations. She has published very interesting papers on this matter. At the beginning of 50- to 100-kV electron microscopy, biologists had been led to study biological specimens in terms of thin films. A. M. Glauert has emphasized the main two advantages of HVEM: the possibility of obtaining both high-resolution images of thin specimens with reduced damage and sharp images of thick specimens. Thus she very properly wrote "This restoration of the third dimension in the appreciation of biological organization has been the major contribution of the HVEM so far."

A stereopair of micrographs of a stained section of a chondrocyte in an embryonic chick limb bone is shown in Fig. 37b. Its overall architecture can hence be studied (Glauert, 1974).

We have already mentioned the work of P. Favard, N. Carasso, and coworkers on thick biological sections. Taking into account the important

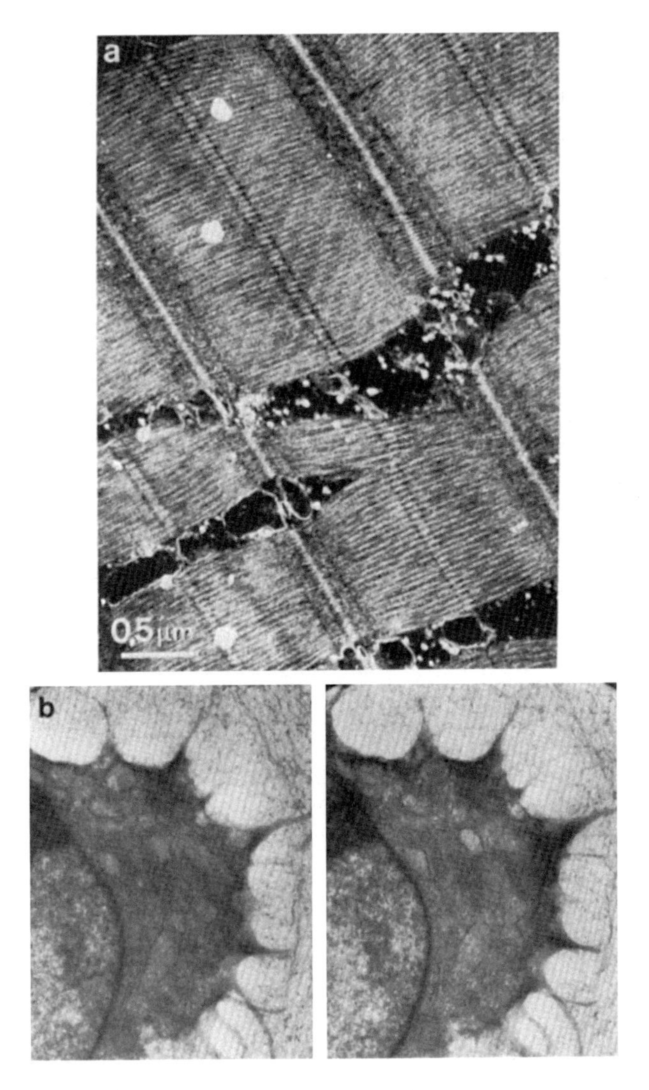

Figure 37 (a) Dark-field image of frog skeletal muscle section. Note the clear definition of fine details in the filaments, Z lines, and glycogen granules; $V = 2.8$ MV. (b) Embryonic chick limb bone chondrocyte section. Block stained with uranyl acetate, section stained with alcoholic PTA; the section is 0.5 μm thick; stereo tilt angle, 16°; $V = 1$ MV (reproduced from Audrey M. Glauert et al., with permission of the authors).

results they have obtained, we must look at them again and give some more details. These investigations have been carried out at Toulouse at voltages of 1–3 MV. The maximum thickness that could be observed was approximately 10 μm. The spatial organization and dynamical evolution

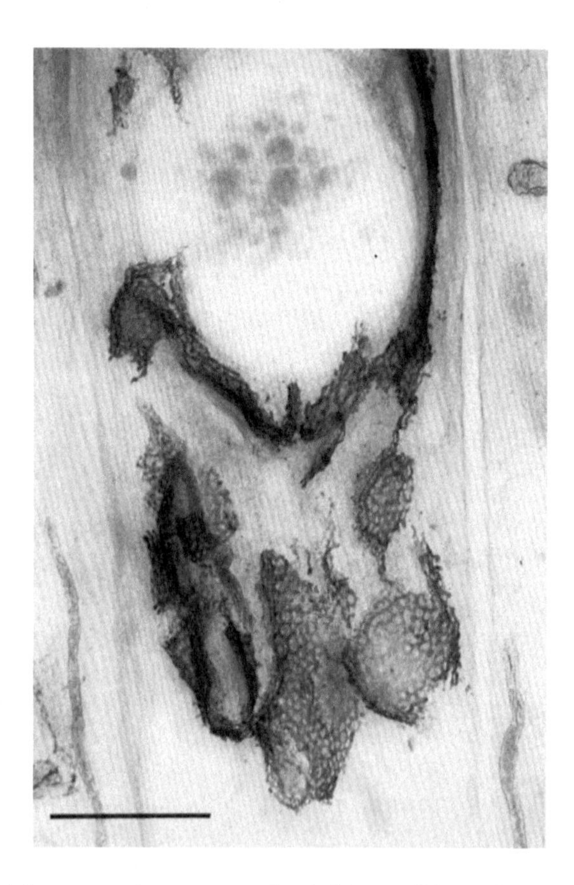

Figure 38 Golgi apparatus in a mucous gland of the snail; bar = 2 μm.

of cell compartments have been investigated in Araldite thick sections at 2.5 MV (Carrasso et al., 1973, 1974). The compartments were selectively contrasted, either by osmium impregnation or by Gomori's lead method.

The resolving power depends on the thickness and accelerating voltage. At 2.8 MV, sections several micrometers thick can be investigated, with a sufficient resolution for understanding the spatial organization. As we have said, stereoscopic viewing allows a better interpretation but with a convenient depth of field. With the 3-MV electron microscope one can get a depth of field of 4 μm with a resolution of 2.0 nm, and 12 μm at a resolving power of 6.0 nm.

Fig. 38 shows the Golgi apparatus in a mucous gland of the snail. Osmium impregnation reveals the fenestrated structure of the saccules and the tubules interconnecting the dictyosomes; $V = 2.5$ MV. Fig. 39a is a 4-μm-

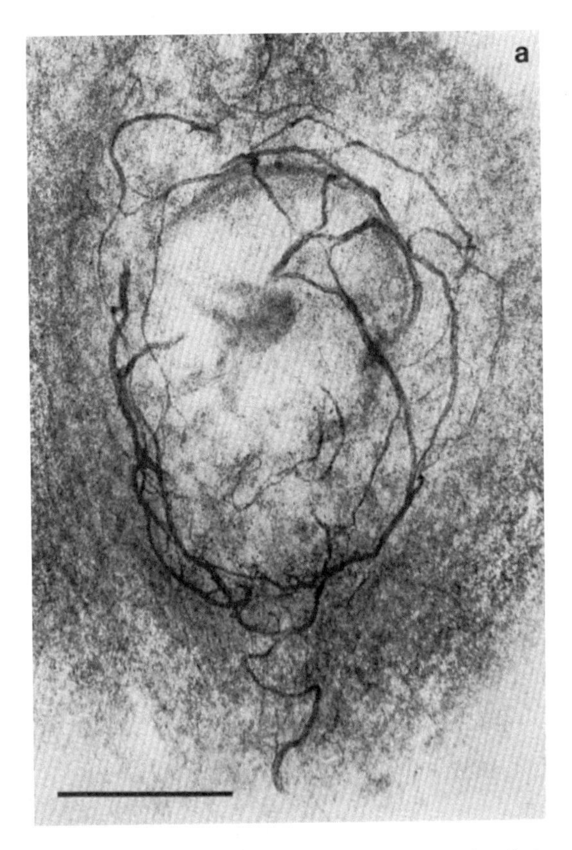

Figure 39 (a) A 4-μm-thick section of a small neuron of the leech impregnated with silver; bar = 4 μm. (b) Micrograph of part of a BSC cell showing numerous microtubules (reproduced from K. R. Porter et al., with permission of the authors).

thick section of a small neuron of the leech, impregnated with silver. The neurofibrils form a perinuclear network.

In the United States the laboratory of Keith R. Porter at Boulder, Colorado, is devoted entirely to biological research. The Boulder group, working with a MV electron microscope, has published important papers (Peachey et al., 1974) describing the internal skeletal structure of the cell. Today we can realize that the cell has an elaborate and dynamic network of skeletal elements (Porter et al., 1983). Fig. 39b depicts the fine structure of a part of a BSC cell. It is possible to view the smaller elements of the cytoskeleton, as well as a mitochondrion (M).

We mention also that Professor Porter and his colleagues have obtained stereo-images of the human diploid cell line (Volosevick & Porter, 1976).

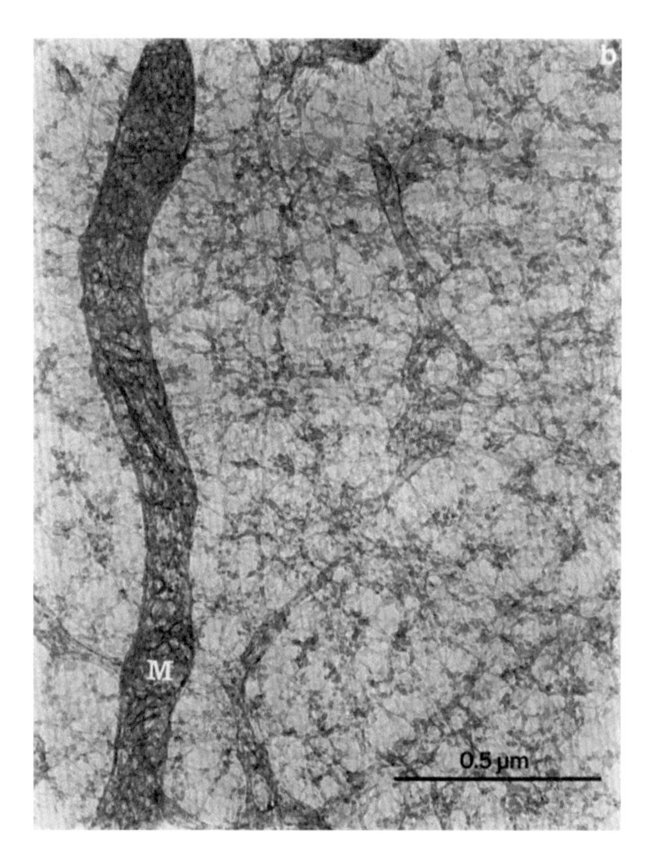

Figure 39 (*continued*)

Another advantage of the contrast–stop method deserves to be mentioned. By more or less shifting the contrast–stop it is possible, on the image of an object having local extra thickness—a sort of relief—to make this relief visible (Dupouy, Perrier, Verdier, 1968). The contour of the relief is outlined, on one side by a bright fringe, on the other side by a dark one. We give an example of this in Fig. 40a and b, showing the large-scale image of a portion of the diatom *Aulacodiscus*. Fig. 40a was obtained, in dark field, with the contrast–stop centered in the diaphragm (D). In Fig. 40b the same portion of the diatom is seen in dark field, but the contrast–stop, which had been set in an intermediate position, is no longer centered on the diaphragm. The contours of the parts of the diatom clearly appear in relief.

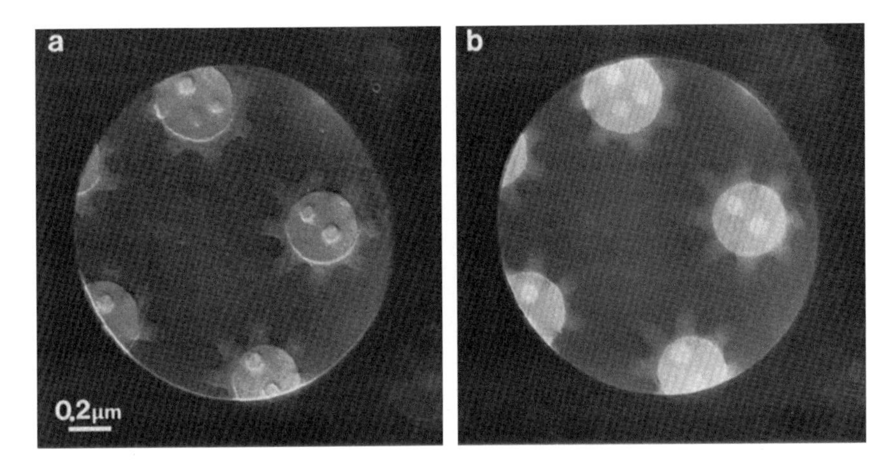

Figure 40 Dark-field images, at high magnification, of the diatom *Aulacodiscus*: (a) image obtained when the beam stop (F, Fig. 30) is not centered in the diaphragm. One can see the parts in relief; (b) usual image by contrast–stop method.

A very interesting study of dark-field imaging has been done by A. M. Johnson and D. F. Parsons (1969). We invite the reader to refer to this document. A. K. Kleinschmidt (1970), of the New York University School of Medicine, has done interesting work on bright-field and dark-field electron microscopy of biomolecules which deserves to be mentioned. We must also point out an important paper of J. Dubochet et al. (1970) on high-resolution dark-field electron microscopy.

5.6.3 The observation of living bacteria

By working at 1 MV with the Toulouse microscope we were able, for the first time, to photograph living bacteria (Dupouy et al., 1960). To achieve this, we placed the bacteria in a leak-proof container—a sort of microchamber for bacteria—where they could exist under their normal living conditions: in air, at atmospheric pressure, and with a suitable level of humidity. *Corynebacterium diphtheriae*, *Bacillus subtilis*, and staphylococci have been observed and photographed. After irradiation by the electron beam, the bacteria were placed again in a suitable culture medium; they remained alive and were capable of reproducing. L. Lapchine and L. Enjalbert (1974) have studied the survival of bacteriophage T4 after irradiation at 1 MV at variable doses. They were able to determine the maximum dose for which the infectious particles of the virus could still be observed.

Their investigations yielded encouraging results and it seems possible to hope that viable particles will be visible at high magnification. Such work is now being carried out in several laboratories; it will, of course, have to be completed and extended.

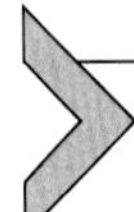

6. Resolution and contrast in megavolt electron microscopy

6.1 Resolution

In electron microscopy, the ultimate resolution is obtained for thin specimens, and depends on diffraction and spherical aberration. The combined effects of these constraints, which correspond to the minimum value d_{m} of resolving power, are obtained for an optimum aperture

$$\alpha_{\mathrm{M}} = 1.41\lambda^{1/4} C_{\mathrm{s}}^{-1/4} \tag{1}$$

and the minimum distance d_{m} which can be resolved between two points is then

$$d_m = 0.61\lambda/\alpha_{\mathrm{M}} = 0.43 C_{\mathrm{s}}^{1/4}\lambda^{3/4} \tag{2}$$

C_{s} is the spherical aberration constant of the objective lens and λ the wavelength of the wave associated with the electron. The product $R = C_{\mathrm{s}}^{1/4}\lambda^{3/4}$ is known as the resolution parameter.

Various workers (Black & Linfoot, 1957; Cosslett, 1951, 1968, 1974a, 1974b, 1982; Haine, 1961; Haine & Mulvey, 1954; Ruska, 1966) have made a detailed analysis of this problem. Other expressions for α_{M} and d_{m} are to be found in the literature, depending upon the hypothesis made in combining the effects of spherical aberration and diffraction. Nevertheless, all the methods lead to expressions of the form:

$$d_{\mathrm{m}} = AC_{\mathrm{s}}^{1/4}\lambda^{3/4}$$
$$\alpha_{\mathrm{M}} = B\lambda^{1/4} C_{\mathrm{s}}^{-1/4}$$

Henceforward, we shall take $A = 0.43$ and $B = 1.41$.

The value of C_{s} depends, to a certain extent, on the choice of the shape of the polepieces and on the value of the gap we use. This choice is conditioned by the nature of the work to be carried out. If, in the investigations, we are led to take images by tilting the specimen at various angles with respect to the electron beam, the gap should be 20 mm or so. If we intend to obtain the ultimate resolving power of the microscope, as is possible for very thin objects only, a smaller gap is convenient, that is, $s = 10$ mm and even less. Thus, at a given voltage, the value of C_s can be chosen, to a certain extent. In the same manner, for a given gap, C_s varies with the excitation of the objective magnetizing coils. It can be found, in some cases, that C_s goes through a minimum value according to the number of ampere turns.

With the 3-MV electron microscope, taking into account the preceding remarks, for an objective lens, the characteristics of which have already been given (Hyperm 0; $S = D = 12$ mm), the following minimum values of C_s can be obtained: $V = 1$ MV, $C_s = 2.3$ mm; $-V = 3$ MV; $C_s = 5.4$ mm.

In the following table, we give the values of α_M and d_m for voltages of 0.1, 1, and 3 MV. For $V = 0.1$ MV we have taken the value of C_s corresponding to a good conventional microscope, $C_s = 1.6$ mm.

V (MV)	α_M (rad)	d_m (Å)
0.1	9.8×10^{-3}	2.3
1	6.2×10^{-3}	0.85
3	4.02×10^{-3}	0.54

Such results permit us to emphasize again the advantage of working at high voltage in electron microscopy to reach very high resolving powers. When one passes from 0.1 to 1 MV, the theoretical gain in resolving power is $2.3/0.85 = 2.7$. When working at 3 MV, it becomes $2.3/0.54 = 4.3$.

6.2 Contrast

In order to be able to observe effectively two points separated by the microscope, it is not sufficient that their minimum distance be d_m; it is also necessary that the contrast C between the images of these two points be good enough, and it is well known that, in the usual conditions of observation, the contrast of the image decreases appreciably as the voltage is raised. But we have shown that it is possible, in dark field, to increase the contrast considerably (see Section 5.6.2). Thus the difficulty we have just mentioned can be overcome by very simple means.

6.3 Resolution in dark field

We are now faced with a new problem: will not the resolving power of the microscope be lessened by the presence of the contrast–stop?

This question has already been approached in classical optics, when studying the mirror microscope. The calculation has also been made in the case of the electron microscope. At a voltage of 1 MV and in conditions where the contrast practically reaches its optimum value, not only is there no loss but, on the contrary, there is a slight gain in resolving power. When α_1 is increased the diffraction rings have a slightly greater intensity, but this can be neglected in our experiments.

At the end of a study of high-resolution dark-field electron microscopy, J. M. Cowley (1973) writes "The dark field image may show detail on a finer scale and so have a better 'resolution' in some cases." Let us add two more remarks.

a. *Defocusing.* We have always worked, in dark field, with focused images. The contrast is so good that it would not be improved by defocusing. We agree, indeed, with several workers, particularly Johnson and Parsons (1969), that phase contrast obtained through defocusing leads to images which are difficult to understand as far as amorphous biological specimens are concerned. In biological investigations methods that increase contrast without defocusing are better.

b. *Coherence.* Some workers, especially Cosslett (1982), were right when they pointed out that the difference in intensity between two images also depends on whether the illumination of the object is coherent or not. In our experiments at 1 MV the degree of coherence at a point M is higher than 0.5 for all points situated inside a circle of center M and diameter 6 Å. The coherence width is greater than the resolution δ.

6.4 Atomic resolution

It has been shown recently by several workers that a resolving power allowing us to photograph atoms can be reached. Several studies have been carried out (1) with conventional 100-kV or very high-voltage (3-MV) microscopes and (2) by scanning transmission electron microscopy with A. V. Crewe's microscope (STEM).

Let us also mention the experiments of R. M. Henkelman and F. P. Ottensmeyer (1971), who have photographed, in dark field, palladium, platinum, and osmium atoms. They have worked at 80 kV.

Tanaka et al. (1975) have, for their own part, obtained images of atoms of mercury bound to DNA filaments (100 kV).

D. Dorignac and B. Jouffrey (1975, 1976) have photographed, at low voltage, atoms of strontium, the lightest atoms that have so far been observed. They have also made, at 3 MV, images of barium and uranium atoms.

A. W. Crewe and co-workers (1974) have shown several dark-field images of isolated atoms taken with the microscope they have designed.

According to Uyeda and Tanaka (1978) atom images taken by the conventional transmission electron microscope are less distinct than those obtained by STEM and they explain the reasons for this difference.

Such results show the limits to which modern electron microscopes can penetrate into the world of the infinitely small.

7. Future prospects

In the preceding pages we have shown how HVEM has been developed, since we designed our first 1-MV microscope in 1960, in domains as different as solid-state physics and biology. In both cases a great many important new results have been obtained.

In biology new prospects have been opened by using stereo-imaging methods which reveal the three-dimensional structure of thick sections. Keith R. Porter mentions, for instance, that "a promising but still embryonic field of research concerns differences in the structure of the microtrabecular lattice in normal cells and in cancer cells. The lattice of cancer cells is denser than that of normal cells and shows signs of a lack of normal organization, particularly in the distribution of polysomes." Biologists and physicians will certainly be willing to develop and carry out research in this field, the interest in which need not be pointed out.

In solid-state physics *in situ* techniques have been developed and offer new experimental possibilities for examining specimens in a controlled environment.

We must now, for some types of research, modify the position that has hitherto been adopted by numerous microscopists. U. Valdré (1975) clearly defined the reorientation that is required when he wrote "It is time that electron microscopes were no longer designed exclusively for the highest possible resolution. Designers should also take into account the new trends of applications which involve not just 'specimen observation,' but 'experiments on specimens' inside the electron microscope." According to this point of view, electron microscopes should effectively incorporate a microlaboratory. For this purpose various devices have already been designed for HV electron microscopes: environmental cells, specimen chambers, and special stages. They are used, according to the different cases, for chemical reactions, heating, cooling, straining, irradiating, magnetization, etc. With all these devices researchers will no doubt go further than before.

A high-voltage electron microscopy facility, as it is designed and built at present, has two disadvantages: it is cumbersome and expensive. But I do think that improvements will be accomplished; we must rely upon the imagination and work of physicists and engineers: they will find new techniques and new designs for the microscope column as well as for the high-voltage supply and accelerating tube. The cost of such equipment will then decrease.

In spite of all the progress that has been accomplished, I think that we are still at the beginning of a great scientific adventure. But I am sure that, due to our endeavors and to the work of all researchers, we shall reach undreamt of horizons and travel over a fruitful new stage of the long hard road leading from the knowledge of matter to the knowledge of truth.

Acknowledgments

I wish to express my thanks to Dr. P. W. Hawkes, who has been kind enough to help me with the presentation of this article.

8. Afterword by Peter Hawkes

Abundant information, both formal and more personal, is available about Gaston Dupouy The best source is the booklet that records La Remise de l'Epée d'Académicien to Dupouy on 10 May 1952. The sword itself is decorated with symbols that recall Dupouy's life: a female figure symbolizing scientific thought; a Romanesque capital (Toulouse); pine cones, for Dupouy, like Castaing, loved the Landes. An agate reminds us of the subject of Dupouy's doctoral dissertation, the magnetic properties of certain crystals, and spirals evoke the trajectories of electrons in magnetic lenses. Diatoms are also present for these were among the specimens studied with Dupouy's first microscope. 'La vie et l'œuvre de Gaston Dupouy' by Pierre Grivet is also highly recommended (La Vie des Science, *Comptes Rendus*, série générale, **3**(6), 1986, 597–606)

Dupouy was born on 7 August, 1900 in Marmande, a small town in south-western France. His father, a carpenter, died young and Gaston and his five siblings were brought up by their mother, who kept the home together by working as a seamstress. Thanks to the support of one of his schoolmasters, Dupouy was able to continue his studies in Paris and in due course, joined Aimé Cotton at a laboratory specially created for magnetic field studies. Here he made his first acquaintance with the early publications on the electron microscope. In 1937, he was appointed Professor of Physics in the Faculty of Sciences in Toulouse and that same year, he applied to the Ministre de l'Education Nationale for funding to build an electron microscope with magnetic lenses. His letter is reproduced in the obituary of Dupouy by his successor in the Laboratoire d'Optique Electronique, Bernard Jouffrey (*J. Microsc. Spectrosc. Electron.* **12**, 1987, 508–509) and in Jouffrey's chapter in *The Growth of Electron Microscopy* (*Adv, Imaging & Electron Phys.* **96**, 1996, 101–130). That issue of the *Journal* was dedicated to Dupouy and contains recollections by R. Uyeda as well as contributions related to Dupouy's interests. Dupouy describes subsequent developments in his article here but he does not mention that it was he who was chosen to reply to the Pope when the participants at a European Congress on Electron Microscopy in Rome in 1968 were received at Castelgandolfo. Nor are we told that his cardiologist forbade him to deliver the opening speech at the International Congress on Electron Microscopy in Grenoble in 1970. He of course paid no attention to this prohibition and all went well (doctors, nurses and stretcher-bearers were all concealed in the wings). He does, however, mention his pleasure at becoming a "mainteneur" of the Académie des Jeux Floraux, where poetry replaces science, in his speech in response to the presentation of a medal marking the 25th anniversary of his electron to the Académie des Sciences.

He died in Toulouse on 22 October 1985.

References

Black, G., & Linfoot, E. H. (1957). *Proceedings of the Royal Society of London. Series A, 239*, 522.

Busch, H. (1927). *Archiv für Elektrotechnik (Berlin), 18*, 583.

Carrasso, N., Delaunay, M. C., Favard, P., & Lechaire, L. P. (1973). *Journal de Microscopie (Paris), 16*(3), 257.

Carrasso, N., Favard, P., Mentré, P., & Poux, N. (1974). *Oxford, 1974*. (p. 414).

Castaing, R., & Jouffrey, B. (1962). *Journal de Microscopie (Paris), 1*, 201.

Cosslett, V. E. (1951). *Practical electron microscopy*. London: Butterworth.

Cosslett, V. E. (1967). *Science Progress (Oxford), 55*, 15.

Cosslett, V. E. (1968). *Rome, 1968*. (p. 19). 59.

Cosslett, V. E. (1969). *Quarterly Reviews of Biophysics, 2*(2), 95.

Cosslett, V. E. (1974a). *Proceedings of the Royal Society of London. Series A, 338*, 1.

Cosslett, V. E. (1974b). *Journal of Microscopy (Oxford), 100*(3), 233.

Cosslett, V. E. (1982). *Journal of Microscopy (Oxford), 128*(1), 23.

Cotton, A. (1928). *Comptes Rendus Hebdomadaires des Séances de L'Académie des Sciences, 127*, 77.

Cotton, A., & Dupouy, G. (1930). *Comptes Rendus Hebdomadaires des Séances de L'Académie des Sciences, 190*, 544.

Cotton, A., & Mabboux, G. (1928). *Recherches et Inventions, 9*, 421.

Coulomb, P., & Reynaud, F. (1970). *Journal de Microscopie (Paris)*, *9*(8), 993.
Coupland, J. H. (1956). *London, 1956*. (p. 159).
Cowley, J. M. (1973). *Acta Crystallographica. Section A*, (5), 529.
Crewe, A. V. (1974). *Canberra, 1974* Vol. 1. (p. 20).
Crewe, A. V., Langmore, J., Isaacson, M., & Retsky, M. (1974). *Canberra, 1974* Vol. 1. (p. 260).
de Broglie, L. (1923). *Comptes Rendus Hebdomadaires des Séances de L'Académie des Sciences*.
de Broglie, L. (1924) (Thèse Paris). Paris: Masson.
de Broglie, L. (Ed.). (1946). *L'Optique Electronique*. Paris: Edns. Rev. Opt.
de Gramont, A. (1945). *Vers l'infmiment petit*. Gallimard: Collection l'Avenir de la Science.
Dietrich, I., Fox, F., Weyl, R., & Zerbst, H. (1974). *Oxford, 1974*. (p. 103).
Dorignac, D., & Jouffrey, B. (1975). *Toulouse, 1975*. (p. 143).
Dorignac, D., & Jouffrey, B. (1976). *Comptes Rendus Hebdomadaires des Séances de L'Académie des Sciences*, *82*, 479.
Dubochet, J., Ducommun, M., & Kellenberger, E. (1970). *Grenoble, 1970* Vol. 1. (p. 601).
Dupouy, G. (1946a). In L. de Broglie (Ed.), *L'Optique Electronique* (p. 162). Paris: Edns. Rev. Opt.
Dupouy, G. (1946b). *Journal de Physique et Le Radium*, 7, 320.
Dupouy, G. (1968). *Advances in Optical and Electron Microscopy*, *2*, 167.
Dupouy, G. (1973). *Journal of Microscopy (Oxford)*, *97*(1/2), 3.
Dupouy, G. (1974). *Oxford, 1974*. (p. 441).
Dupouy, G., & Perrier, F. (1962). *Journal de Microscopie (Paris)*, *1*(3/4), 167.
Dupouy, G., Perrier, F., & Durrieu, L. (1960). *Comptes Rendus Hebdomadaires des Séances de L'Académie des Sciences*, *251*, 2836.
Dupouy, G., Perrier, F., & Durrieu, L. (1962). *Comptes Rendus Hebdomadaires des Séances de L'Académie des Sciences*, *254*, 3786.
Dupouy, G., Perrier, F., Uyeda, R., Ayroles, R., & Bousquet, A. (1963). *Comptes Rendus Hebdomadaires des Séances de L'Académie des Sciences*, *257*, 1511.
Dupouy, G., Perrier, F., & Verdier, P. (1966). *Journal de Microscopie (Paris)*, *5*(6), 655.
Dupouy, G., Perrier, F., Trinquier, J., & Murillo, R. (1967). *Comptes Rendus Hebdomadaires des Séances de L'Académie des Sciences*, *265*, 1221.
Dupouy, G., Perrier, F., Seguela, A., & Segalen, R. (1968). *Comptes Rendus Hebdomadaires des Séances de L'Académie des Sciences*, *266*, 1064.
Dupouy, G., Perrier, F., & Verdier, P. (1968). *Rome, 1968*. (p. 155).
Dupouy, G., Perrier, F., Enjalbert, L., Lapchine, L., & Verdier, P. (1969). *Comptes Rendus Hebdomadaires des Séances de L'Académie des Sciences. Série B*, *268*, 1321.
Dupouy, G., Perrier, F., Ayroles, R., & Mazel, A. (1970). *Comptes Rendus Hebdomadaires des Séances de L'Académie des Sciences. Série B*, *271*, 465.
Dupouy, G., Perrier, F., & Durrieu, L. (1970). *Journal de Microscopie (Paris)*, *9*(5), 575.
Dupouy, G., Perrier, F., & Tremollieres, C. (1970). *Comptes Rendus Hebdomadaires des Séances de L'Académie des Sciences*, *270*, 41.
Favard, P., & Carasso, N. (1972). *Journal of Microscopy (Oxford)*, *97*(1/2), 59.
Favard, P., Carasso, N., & Ovtracht, L. (1971). *Journal de Microscopie (Paris)*, *12*, 301.
Fernández-Morán, H. (1971). *Grenoble, 1971*. (p. 91).
Ferrier, R., & Puchalska, I. B. (1968). *Rome, 1968*. (p. 351).
Fert, C., & Selme, P. (1956). *Bulletin de Microscopie Appliquée*, *6*, 157.
Fisher, R. M. (1977). *Kyoto, 1977*. (p. 597).
Fujita, H., Tabata, T., & Aoki, T. (1977). *Kyoto, 1977*. (p. 439). See also pp. 141, 151, 281, 391, 407, 451, 519.
Fukuhara, A., Kohra, K., & Watanabe, H. (1962). *Journal of the Physical Society of Japan*, *17*(Suppl. BII), 195.
Glaeser, R., & Thomas, G. (1968). *Biophysical Journal*, *9*, 1073.

Glauert, A. M. (1974). *The Journal of Cell Biology, 63*(3), 717.
Glauert, A. M. (1979). *Journal of Microscopy (Oxford), 117*(1), 93.
Glauert, A. M., & Mayo, C. R. (1973). *Journal of Microscopy (Oxford), 97*, 83.
Grivet, P. (1946). In L. de Broglie (Ed.), *L'Optique Electronique* (p. 129). Paris: Edns. Rev. Opt.
Haine, M. E. (1961). *The electron microscope*. London: Spon.
Haine, M. E., & Mulvey, T. (1954). *Journal of Scientific Instruments, 31*, 236.
Harrison, C. G., & Leaver, K. D. (1973). *Journal of Microscopy (Oxford), 97*(1/2), 139.
Harrison, C. G., Leaver, K. D., & Swann, P. R. (1972). *Manchester, 1972*. (p. 334).
Hashimoto, H., Howie, A., & Whelan, M. J. (1960). *Philosophical Magazine, 8*(5), 967.
Henkelman, R. M., & Ottensmeyer, F. P. (1971). *Proceedings of the National Academy of Sciences of the United States of America, 68*(12), 3000.
Hillier, J., Vance, A. W., & Zworykin, V. K. (1941). *Journal of Applied Physics, 12*, 738.
Hirsch, P. B. (1962). *Journal of the Physical Society of Japan, 17*(Suppl. BII), 143.
Hirsch, P. B. (1968). *Rome, 1968*. (p. 49).
Hirsch, P. B. (1974). *Oxford, 1974*. (p. 1).
Hirsch, P. B. (1977). *Kyoto, 1977*. (p. 21).
Horiuchi, S., Matsui, Y., Bando, Y., Sekikawa, Y., & Sakaguchi, Y. (1977). *Kyoto, 1977*. (p. 91).
Johnson, H., & Parsons, D. F. (1969). *Journal of Microscopy (Oxford), 90*(3), 199.
Kainuma, Y., & Yoshioka, H. (1962). *Journal of the Physical Society of Japan, 17*(Suppl. BII), 134.
Kamiya, Y. (1963). *Japanese Journal of Applied Physics, 2*, 386.
Kleinschmidt, A. K. (1970). *Berichte der Bunsengesellschaft für Physikalische Chemie, 74*, 1190.
Kobayashi, K., & Sakoku, K. (1965). *Laboratory Investigation, 14*, 1097.
Kohra, K., & Watanabe, H. (1961). *Journal of the Physical Society, 16*, 580.
Krause, F. (1936). *Zeitschrift für Physik, 102*, 417.
Laberrigue, A., Genotel, D., Girard, M., Severin, C., Balossier, G., & Homo, J. C. (1974). *Oxford, 1974*. (p. 108).
Lapchine, L., & Enjalbert, L. (1971). *Comptes Rendus Hebdomadaires des Séances de L'Académie des Sciences, 272*, 3092.
Lapchine, L., & Enjalbert, L. (1974). *Comptes Rendus Hebdomadaires des Séances de L'Académie des Sciences, 279*, 607.
Le Poole, J. B., & Zeedijk, H. B. (1977). *Kyoto, 1977*. (p. 95).
Le Poole, J. B., Osterkamp, W. J., & van Dorsten, A. C. (1947). *Philips Technical Review, 9*, 1.
Le Poole, J. B., Bok, A., & Rus, P. J. (1970). *Grenoble, 1970*. (p. 113).
Lefranc, G., Müller, K. H., & Dietrich, I. (1981). *Ultramicroscopy, 6*, 81.
Lepine, P., & Croissant, O. (1955). *Bulletin de Microscopie Appliquée, 5*, 22.
Magnan, C. (1946). In L. de Broglie (Ed.), *L'Optique Electronique* (p. 18). Paris: Edns. Rev. Opt.
Mahl, H. (1941). *Zeitschrift für Metallkunde, 33*, 68.
Mahl, H. (1942). *Naturwissenschaften, 207*.
Marton, L. (1934a). *Bulletin de la Classe des Sciences. Académie Royale de Belgique [5], 25*, 439.
Marton, L. (1934b). *Nature (London), 133*, 911.
Marton, L. (1935). *Revue d'Optique Théorique et Instrumentale, 14*, 129.
Marton, L. (1936). *Revue de Microbiologie Appliquée, 2*, 117.
Marton, L. (1937). *Bulletin de la Classe des Sciences. Académie Royale de Belgique [5], 23*, 672.
Massover, W. H. (1974). *Oxford, 1974*. (p. 163).
Massover, W. H., Lacaze, J. C., & Durrieu, L. (1973). *Journal of Ultrastructure Research, 43*, 460.
Müller, H., & Ruska, E. (1941). *Kolloid-Zeitschrift, 95*, 21.
Mulvey, T., & Newman, C. D. (1974). *Oxford, 1974*. (p. 98).

Peachey, L. D., Fotino, M., & Porter, K. R. (1974). *Oxford, 1974.* (p. 405).
Popov, N. M. (1959). *Izvestiya Akademii Nauk SSSR*, *23*, 436. 494.
Porter, K. R., & Volosewick, J. J. (1977). *Kyoto, 1977.* (p. 15).
Porter, K. R., Beckerle, M., & McNiven, M. (1983). *Modern Cell Biology*, *2*, 259–302.
Ruska, E. (1966). *Advances in Optical and Electron Microscopy*, *1*, 116.
Ruska, E. (1980). *The early development of electron lenses and electron microscopy*. Stuttgart: Hirzel Verlag.
Seguela, A. (1971). *Grenoble, 1971.* (p. 609).
Selme, P. (1963). *"Le Microscope Electronique." Collection "Que sais-je?"*. Paris: Presses Universitaires Françaises.
Tadano, B., Sakaki, Y., Maruse, S., & Morito, N. J. (1956). *Journal of Electron Microscopy*, *4*, 5.
Tadano, B., Katagiri, S., Ichige, K., Sakaki, Y., & Maruse, S. (1958). *Berlin, 1958* Vol. 1 (p. 166).
Tanaka, M., Higashi-Fujime, S., & Uyeda, R. (1975). *Ultramicroscopy*, *1*(7), 14.
Thomas, G. (1977). *Kyoto, 1977.* (p. 627).
Trillat, J. J. (1981). In *Fifty years of electron diffraction* (p. 77). Dordrecht, Netherlands: Reidel Publ.
Trinquier, J., & Balladore, J. L. (1968). *Rome, 1968* Vol. 1. (p. 91).
Trinquier, J., Balladore, J. L., & Martinez, J. P. (1972). *Manchester, 1972.* (p. 218).
Uyeda, R. (1962). *Journal of the Physical Society of Japan*, *17*(Suppl. BII), 153.
Uyeda, R. (1968). *Rome, 1968.* (p. 55).
Uyeda, R., & Tanaka, N. (1978). *Institute of Physics Conference Series*, *41*, 57.
Valdré, U. (1975). *Toulouse, 1975.* (p. 17).
Vingsbo, O., & Lasserre, A. (1966). *Journal de Microscopie (Paris)*, *5*, 527.
Volosevick, J. J., & Porter, K. R. (1976). *The American Journal of Anatomy*, *147*, 303.
von Ardenne, M. (1941). *Zeitschrift für Physik*, *117*, 657.
von Ardenne, M. (1942). *Zeitschrift für Physikalische Chemie*, *3*, 51.
Wade, R. H. (1962). *Philadelphia, 1962* Vol. II (p. 7).
Yoshioka, H. (1957). *Journal of the Physical Society of Japan*, *12*, 618.
Zanchi, G., Perez, J.-Ph., & Sevely, J. (1975). *Toulouse, 1975.* (p. 55).
Zeitler, E., & Crewe, A. V. (1976). *Jerusalem, 1976.* (p. 322).
Zeitler, E., & Crewe, A. V. (1977). *Kyoto, 1977.* (p. 99).

Cryo-electron microscopy and ultramicrotomy: reminiscences and reflections

Humberto Fernández-Morán[a] and Peter Hawkes (Afterword)[b,*]

[a]Formerly Cryo-Electron Microscope Laboratories, Research Institutes, University of Chicago, Chicago, IL, United States
[b]CEMES-CNRS, Toulouse, France
*Corresponding author. e-mail address: hawkes@cemes.fr

Contents

Advances in Imaging and Electron Physics, Volume 220
ISSN 1076-5670
https://doi.org/10.1016/bs.aiep.2021.08.009

1. Introduction

In its first 50 years the electron microscope has surpassed the resolution of the light microscope by three orders of magnitude (Knoll & Ruska, 1932; Ruska, 1980; von Borries & Ruska, 1941), making it possible to see directly structures of molecular and atomic dimensions. Despite the initial pessimistic outlook "that living matter would burn to a cinder under the electron beam," the first pioneers, Max Knoll, Ernst Ruska (1932, 1934), and L. Marton (1934, 1968), continued their work. Successive advances in preparation techniques have since revealed whole new domains of unsuspected structural detail, most of which can be correlated with the results of X-ray diffraction and biochemical methods. Considered one of the most successful inventions of our times (Gabor, 1975), the electron microscope and its contributions to biology and medicine have ushered in a renaissance, from the first imaging of a bacterial virus (Luria & Anderson, 1942) to unraveling the submicroscopic organization of cells and tissues down to the self-replicating DNA macromolecules. Historically, it is interesting to note that the Berlin biophysicist, Helmholtz (von Helmholtz, 1873), and Ernst Abbe independently showed in 1873 that the resolving power of the light microscope is limited by the wavelength of light. Yet, the discovery of birefringence of nerves by Ehrenberg (1849) in 1849 marks the beginning of an investigation by light microscopy of the ultrastructure of a native biological system. Subsequent polarized-light and X-ray diffraction studies of the nerve myelin sheath (Schmitt, 1950; Schmitt et al., 1935) furnished quantitative details of this paracrystalline structure, consisting of lipid–protein layers about 180 Å thick, and thus of submicroscopic dimensions. In fact, the postulated layers and their highly ordered concentric arrangement in the myelin sheath were observed directly in the electron microscope (Fernández-Morán, 1950a, 1952a). Correlated biophysical and biochemical investigations of this kind have contributed significantly to validate the results of electron microscopy. Past experience has shown the unique value of this approach involving the combined application of complementary disciplines and the interaction of numerous experts working together in imaginative experiments. We may also anticipate the development of a new generation of "cryo-electron microscopes" designed to reduce specimen radiation damage and thermal vibrations at liquid-helium temperatures (Fernández-Morán, 1982).

Reminiscing is ultimately the resurrection of meaningful imagery and of fruitful concepts. Within the succinct scope of this article the author has therefore selected a series of representative micrographs to illustrate the results of correlated studies using improved preparation techniques supplemented by parallel biochemical and biophysical methods. Having shared the privilege of participating in this exploratory work with a number of distinguished colleagues during nearly four decades, salient aspects of cryo-electron microscopy and ultramicrotomy were chosen as case histories. Presented as a token of deep appreciation, they cover highlights of an international quest which, in many ways, occurred at the right time and in the right places (1946–1982).

Trained as a biologist and a physician, the author became well acquainted with the classical techniques of light microscopy under expert guidance (Professor A. Dabelow and Professor B. Romeis) while working on his dissertation (Fernández-Morán, 1944).

The first publications on electron microscopy, which appeared in Germany during 1939–1940, notably the papers by E. Ruska (1934) and by H. Ruska (1939) and the inspiring book by M. von Ardenne (1940), were studied with keen interest. Later on the classical papers by Luria and Anderson (1942) and by Hall, Jakus, and Schmitt (1945) and Hall (1955, 1956) and the books by Burton and Kohl (1942) and by Zworykin, Morton, Ramberg, Hillier, and Vance (1945), opened up a new world. It is difficult to convey the impact of seeing for the first time the electron micrographs of bacteriophages, or the fascinating beauty of the periodic structures revealed by staining or shadowcasting in trichocysts, paramyosin fibrils, or connective tissue (collagen).

However, postgraduate work in neurology and neurosurgery absorbed all the energies of the author. In 1946 Fernández-Morán transferred to Stockholm, Sweden, to continue his residency and work as a foreign research fellow under the famous neurosurgeon Herbert Olivecrona, at the Serafimerlasarettet. Professor Olivecrona was not only a great physician, but also a superb organizer endowed with prodigious energy and that rare measure of self-control typical of certain outstanding Swedes (Olivecrona, 1946). Moved by the deaths caused by malignant tumors in spite of exemplary surgical treatment, Fernández-Morán, encouraged by Professor Olivecrona, turned to basic research to learn more about the organization of tumor cells.

In 1946 he visited Professor Manne Siegbahn, director of the Nobel Institute for Physics, and after a brief discussion he was invited to work in the electron microscope laboratories. Thanks to the generous hospitality and assistance granted during an 8-year period (1946–1954), Fernández-

Figure 1 Professor Manne Siegbahn and co-workers at the Nobel Institute for Physics of the Swedish Academy of Science in Stockholm, 1947 (left to right): Professors Kai Siegbahn, Nils Svartholm, Manne Siegbahn, H. Fernández-Morán, and H. Slätis.

Morán served a unique apprenticeship in electron microscopy both here and at the Karolinska Institute.

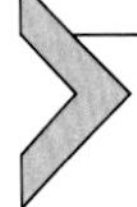

2. The Nobel Institute for Physics

2.1 Professor Manne Siegbahn

In 1937 the Royal Swedish Academy of Science created a Nobel Institute of Physics, located at Frescati in the northern suburbs of Stockholm, with Professor Manne Siegbahn as director; but in a deeper sense he was the soul of that institute, both as a great scientist and as a unique human being (see Fig. 1). Karl Manne George Siegbahn was awarded the Nobel Prize in physics in 1924 "for his discoveries and research in the field of X-ray spectroscopy." He developed special research techniques for X-ray spectroscopy, and determined with extraordinary accuracy the wavelength of X rays. His "Spectroscopy of Röntgen Rays" (1924) became the standard work on this subject. In addition to his fundamental contributions, he designed special ruling engines and invented (in 1939) an electron microscope of unusual

Figure 2 The author examining biological specimens (photographed February 9, 1947) with the electron microscope designed by Professor Manne Siegbahn (1939) and installed in special research laboratories of the Nobel Institute for Physics at Frescati, Stockholm. This instrument consistently attained resolutions of 20–50 Å when operating under optimum conditions with adequate specimen preparation techniques.

design and high resolution, equipped with a Siegbahn high-vacuum rotary molecular pump, among many other contributions. Thanks to his outstanding qualities as an organizer, the institute, devoted to researches in nuclear physics, was one of the best equipped and staffed in Europe. When I arrived in 1946 the institute was a haven of excellence and a major international center of research and training, visited by the most eminent physicists, including the Nobel Laureates in physics delivering their Nobel lecture each year. However, the personality of Professor Siegbahn remains as the lasting impression: a gentle, great man endowed with exceptional qualities of mind and character. Personally, I owe him a profound debt of gratitude. Beyond his assistance in electron microscopy, the diamond knife was developed in that unique environment where his famous ruling engines produced diamond-ruled gratings of unsurpassed quality.

2.2 The Siegbahn electron microscope

A magnetic electron microscope of unusual design was made by Professor Manne Siegbahn in 1939 (Siegbahn, 1939) and was installed in special laboratories at the Nobel Institute for Physics (see Fig. 2). The column consists of a horizontally arranged narrow tube. The precision-machined polepieces within the tube make magnetic contact with the lens coils mounted externally. The image is viewed through a thin fluorescent screen attached to a plate holder which takes up to eight photographic plates. Alignment of the horizontal column is critical but can be carried out reproducibly. A rotary molecular pump designed by Siegbahn is used to produce a high vacuum. Although it does not have a very high rate of pumping, it is relatively contamination free in the absence of vacuum leaks, since there are no oil vapors. Careful mounting and balancing of the pump eliminates disturbing vibrations. The electrical circuits are of conventional design, with adequate stability and precision focusing controls. The laboratory model operates at a regulated high voltage of 35–55 kV.

The object holders consist of rectangular metal blocks (2 cm × 2.5 mm × 2 mm) with two longitudinal openings which are covered by a fine copper net (spot welded to the holders) with openings of 100 μm in diameter. Each opening contains up to 500 screen holes. Thus, a large number of specimens can be viewed without breaking the vacuum. A major improvement was made by using very thin beryllium or aluminum (10–50 Å in thickness) supporting films, prepared according to the method developed by Professor Nils Hast (1947).

The Siegbahn electron microscope consistently attained resolutions of 20–50 Å with a useful magnification of 100,000× when operating under optimum conditions with adequate specimen preparation techniques. Over 7 years of constant operation it proved to be remarkably reliable. Moreover, since it has many features in common with an electron-optical bench, it was an ideal instrument for electron-optical experimentation and evaluating the effects of lens stability, variation of high voltage, and alignment (Fig. 6).

Elvers (1943) carried out systematic studies of plant cell ultrastructure and developed ingenious “microgrids” or “micronets” made from transverse sections of pinewood or birch, with openings down to 10 μm.

Kai Siegbahn and B. Ingelman (1944) carried out the first electron microscopic study of Dextran molecules with this instrument. The most extensive investigation on the fine structure of clay and the development of improved ultrathin metal film substrates was carried out by Professor Nils Hast (1947).

Our preliminary studies of fresh brain tumor tissue obtained from the neurosurgical clinic of Professor Herbert Olivecrona (Fernández-Morán, 1948) (Figs. 3 and 4) revealed several new components, including submicroscopic glia fibrils (Fig. 5) and cytoplasmic inclusion bodies, and furnished indications that the pathological glia cell types which are specific for each tumor differ from the normal glia cells.

ARKIV FÖR ZOOLOGI.
BAND 40 A. N:o 6.

Examination of Brain Tumor Tissue with the Electron Microscope.

By

HUMBERTO FERNÁNDEZ-MORÁN.

With 13 figures in the text.

Communicated June 4th 1947 by MANNE SIEGBAHN and HUGO THEORELL.

Examination of Brain Tumor Tissue with the Electron Microscope.

Figure 3 Report on the preliminary results obtained in the examination of brain tissue with the Siegbahn electron microscope using new preparation techniques. Communicated to the Swedish Academy of Science June 4, 1947. [From *Ark. Zool.* **40A**, No. 6 (1948).]

The main advantage derived from these preliminary studies was an evaluation of suitable preparation techniques. As depicted in the schematic outline in Fig. 4, several new methods were proposed and initiated by the author at this early stage (1947), including spreading of cell components in a protein monolayer. These proved to be useful, and were later independently developed and successfully applied in correlative studies of DNA released from bacteriophages (Kleinschmidt & Zahn, 1959) and from chloroplasts (Woodcock & Fernández-Morán, 1968).

2.3 Electron microscopy of anterior lobe cells of the rat hypophysis

An interesting study was carried out in collaboration with Dr. Rolf Luft on "Submicroscopic Cytoplasmic Granules in the Anterior Lobe Cells of the Rat Hypophysis as Revealed by Electron Microscopy" (Fernández-Morán & Luft, 1949). This study was carried out with the Siegbahn Electron Microscope using mainly the replica–adhesion method (Williams & Wyckoff, 1944) applied to cell smears of the glandular lobe cells of the rat hypophysis. It was shown that the cytoplasm of all glandular lobe cells examined consists

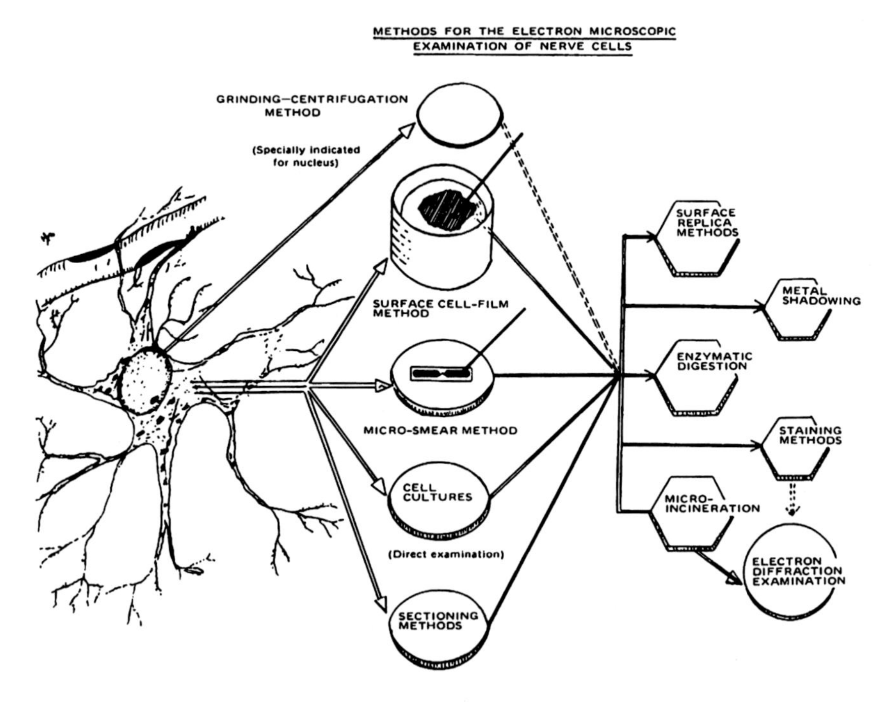

Figure 4 Schematic outline of the preparation techniques for examination of nerve cells by electron microscopy. Several methods were proposed and initiated by the author at this early stage (1947), including thin-sectioning techniques and spreading of cell components in a protein monolayer. These proved to be useful, and were later successfully developed and applied by numerous investigators in correlative studies which revealed new domains of cell ultrastructure and molecular biology.

predominantly of distinctly outlined spherical bodies of 30–300 nm in diameter, which are thus beyond the resolving power of the light microscope. These "cytospheres" (Figs. 7 and 8) contain smaller spherical granules and are embedded in a particulate submicroscopic ground substance. No fundamental differences in the three basic cell types could be observed.

3. The diamond knife

3.1 Development of the diamond knife

The preparation of ultrathin sections which were orders of magnitude beyond the lower limit of conventional microtomy was one of the most serious problems that had to be solved before cell and tissue ultrastructure could be studied by electron microscopy. The development of special

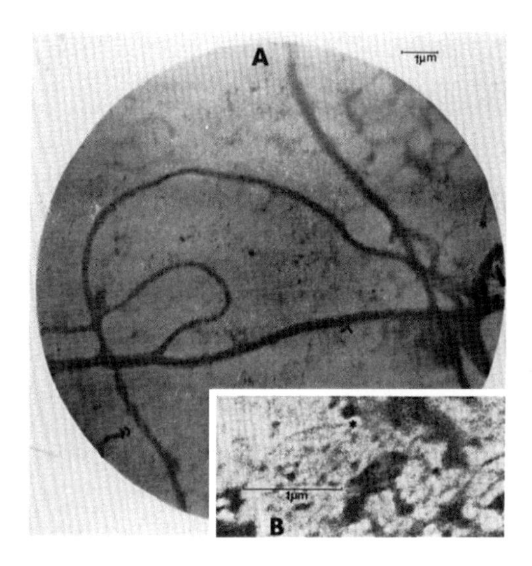

Figure 5 (A) Glia fibrils in a fresh microsmear preparation of astrocytoma deposited on 30-Å aluminum film and washed with ether–alcohol. (B) Submicroscopic glia fibrils [from 1000 down to 200 Å (∗) in diameter] in fresh, unfixed, and unstained astrocytoma preparation on ultrathin (20-Å) beryllium film. Recorded with Siegbahn electron microscope at 35 kV. [From H. Fernández-Morán, *Ark. Zool.* **40A**, No. 6 (1948).]

ultramicrotomes (Fernández-Morán, 1953b, 1956a; Hall, 1953; Newman et al., 1949; Wischnitzer, 1970) was an important contribution, but the preparation of satisfactory sections in the range of 0.01 μm required knives of exceptional characteristics. Specially sharpened razor blades were unsatisfactory because of their relatively short life, corrosion, and other limitations.

The glass knife introduced by Latta and Hartmann (1950) was a major improvement and has actually proved to be the most popular cutting edge when dealing with plastic-embedded specimens, despite their relatively short useful life and the basic limitations when hard materials have to be sectioned.

Systematic efforts were made to prepare knives from single-crystal sapphire, ruby, and other hard crystals, without success. It soon became evident that a hard crystal with a layered fine structure would be ideal, since the individual layers represent "preformed" knife edges. Thus, silicon carbide crystals could be cleaved and carefully abraded to produce sharp knives, but they were too brittle.

Diamond held a special fascination, since the diamond points used to rule with exquisite precision the diffraction gratings in Siegbahn's ruling engines exhibited such extraordinary longevity and resilience. However,

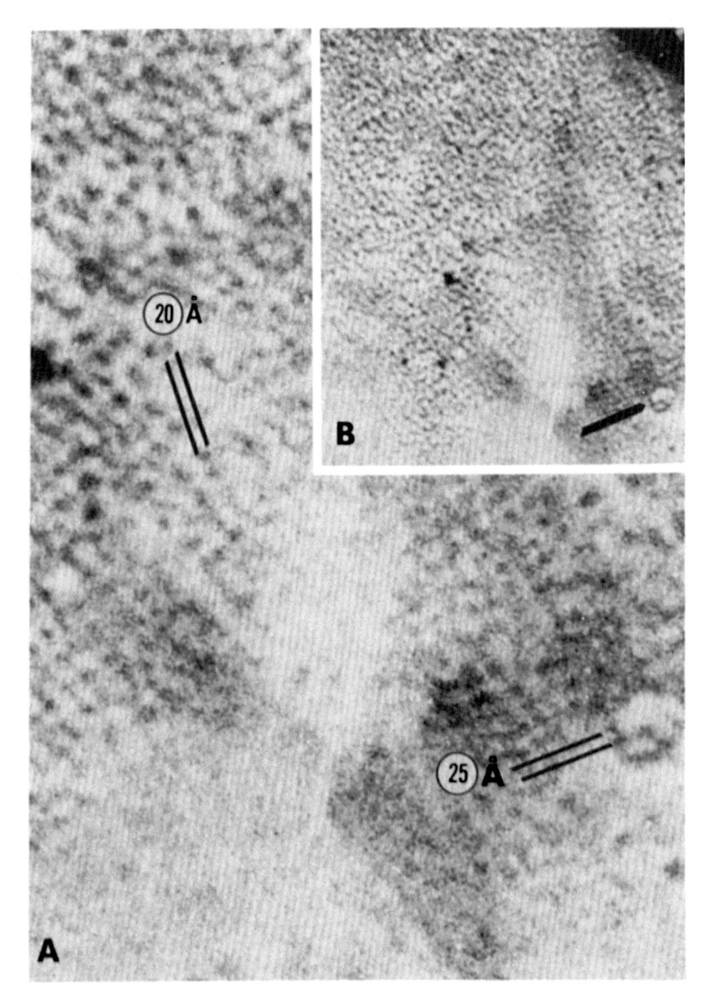

Figure 6 (A) Electron micrograph of inset (B) of gold evaporated on thin (20-Å) beryllium film, showing 20- to 40-Å point resolution. Recorded with Siegbahn electron microscope at 55 kV. (H. Fernández-Morán, 1949.)

these were natural points and not edges. Moreover, consultation with expert diamond grinders was most discouraging.

It was at this juncture that Professor Siegbahn, during one of his visits to the laboratory late at night, while we were waiting for the microscope to pump down, quietly suggested that I look into the wide variety of diamonds, particularly Brazilian boarts which exhibit a distinct layered structure parallel to the octahedron faces (see Fig. 9).

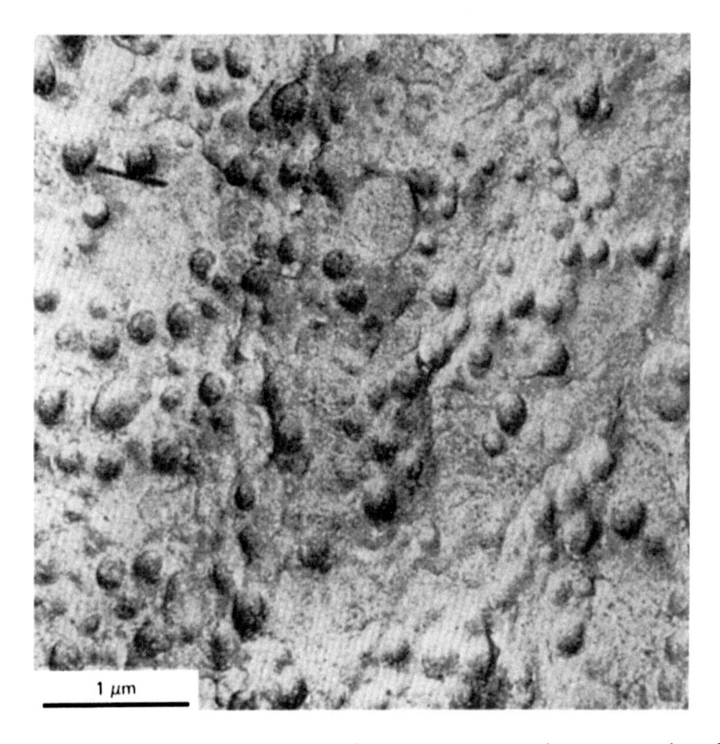

Figure 7 Electron micrograph showing submicroscopic cytoplasmic granules of a basophil cell from the rat hypophysis as revealed by surface replica techniques using very thin films of beryllium (about 10–50 Å thick). Recorded with the Siegbahn electron microscope at 35 kV. [From H. Fernández-Morán and R. Luft, *Acta Endocrinol. (Copenhagen)* **2**, 199 (1949).]

Thus began a systematic effort to determine the dimensions and orientation of the layered structures by combined application of replica techniques covering selected faces of the intact diamond to examine the lamellar surface structure by electron microscopy. This was followed by crushing the diamond into fragments thin enough for electron microscopy and diffraction. As shown in Fig. 10 all fragments exhibited a typical layered crystalline structure of diamond, stacked up in piles down to the thinnest "unit layers" about 10 Å thick. This was also in agreement with the surface replicas of the intact diamond faces. Without the resolving power of the electron microscope this submicroscopic lamellar structure would never have been detected. In turn, the existence of these "preformed" edges, which could be dissected out after cleavage along certain planes, followed by careful abrading and polishing to produce extremely sharp edges, furnished the rationale for producing a diamond knife. The unique physical and chemical

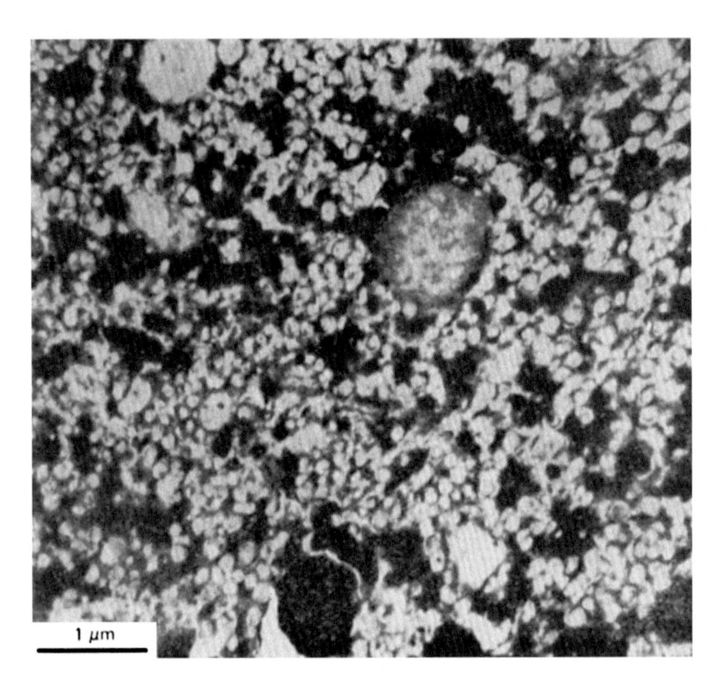

Figure 8 Cytoplasm segment in an eosinophil cell of the anterior lobe of the rat hypophysis. Two specific granules are embedded among the closely packed cytoplasmic particles, which measure only 30–300 nm and are thus beyond the resolving power of the light microscope. Metal replica–adhesion method; recorded with the Siegbahn electron microscope at 55 kV. [From H. Fernández-Morán and R. Luft, *Acta Endocrinol. (Copenhagen)* **2**, 206 (1949).]

properties of diamond favored such an ultrasharp, yet stable, cutting edge. Diamond is the hardest and most wear-resistant material known; it has a very high compressive strength; the Young's modulus of elasticity is unapproached by any other substance; its relative rupture strength is very high; it is one of the most dimensionally stable materials, with a very low thermal expansion; it has very high thermal stability; and it conducts heat faster and more reliably than any other substance. Finally, diamond is practically chemically inert except under unusual conditions (see Fig. 11).

From this point on the conviction was firm that we were on the right track. Nevertheless, it took several years and a rather substantial investment in different types of diamond before we had solved the problems of mounting the diamond, finding the optimum orientation for cleavage, developing special high-speed grinding and polishing machines, and preparing the ultrafine diamond powder by ultracentrifugation, which was essential for

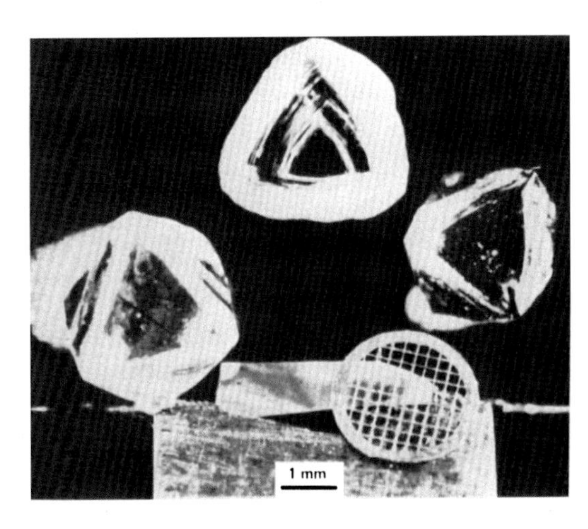

Figure 9 Diamond knife for ultrathin sectioning prepared from selected Venezuelan or Brazilian industrial diamonds by special cleavage and sharpening techniques developed by the author [*Exp. Cell Res.* **5**, 255 (1953)]. During the past three decades the diamond knife with its unique properties has found worldwide application in ultramicrotomy and opened up new fields in precision machining and microsurgery.

the final polishing operation. All the costs were defrayed by a special research fund provided by my father, Luis Fernández-Morán. The unfailing encouragement and support of Manne Siegbahn were likewise essential. Throughout, the electron microscope proved indispensable together with phase-contrast light microscopy. Fig. 12 shows an electron micrograph of a carbon replica of a diamond-knife edge exhibiting the characteristic layered structure derived from the oriented crystalline diamond lamellae. Nevertheless, when we finally started testing the first diamond knives, many of the ultrathin sections showed knife tracks, and in some cases were actually shredded. Selection of the diamond was a critical factor, since some of the knives would often shatter upon testing or in the final phases of polishing. Still, the best diamond knives were of such high quality that they approached the ideal with the following characteristics: (1) extremely sharp and uniform edge over several millimeters, with a radius of curvature of 20 to 50 Å, yielding ultrathin sections of 100 Å or less, with a minimum of compression, and forming ribbons of serial sections of practically indefinite length; (2) all types of specimens could be sectioned, including metals, crystals, glass, and, particularly, frozen sections; (3) the diamond edge was extremely durable, resilient, and stable, performing reproducibly during constant use over periods of months and years when treated with

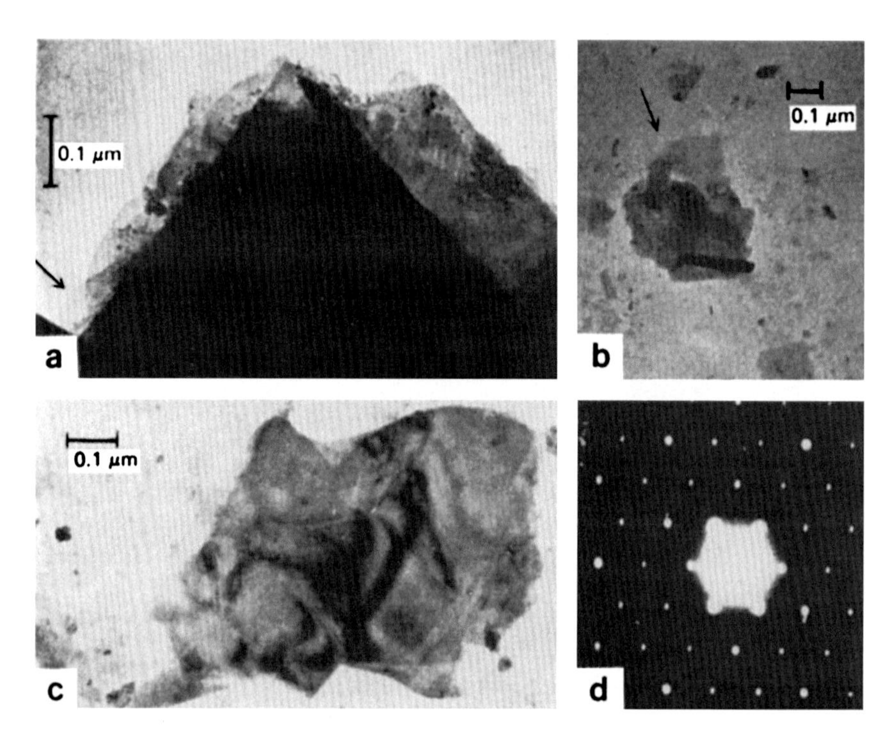

Figure 10 Correlated electron microscopy and electron diffraction (d) studies of the layered crystalline structure of diamond (a and c) featuring unit layers of ∽10 Å (b) provided the basis for producing an extremely sharp and uniform cutting edge of molecular dimensions, which is chemically inert and very durable. (H. Fernández-Morán, 1953–1958.)

care and properly used; (4) the diamond knife was chemically inert and did not contaminate the specimens. Production of the diamond knife was an edifying exercise in persistence and teamwork. It also demonstrated the indispensable role of electron microscopy in "seeing and doing" at the submicroscopic level, and marked its practical value in the field of precision diamond technology and ultraprecision machining.

At this stage, 1953, Fernández-Morán published the first paper on the diamond knife and its applications in ultramicrotomy (Fernández-Morán, 1953a). We had reached the goals set for a laboratory instrument.

3.2 The diamond knife and ultramicrotome

Now the methods had to be refined, a reliable supply for selection of suitable diamonds was required, and adequate resources were needed to achieve a satisfactory level of production of high-quality knives. An abundant sup-

Figure 11 (A) Diamond-knife cutting edge shown as very uniform bright line in dark-field micrograph. ×1000. (B) Stainless-steel knife showing typical jagged cutting edge and furrowed facet in bright-field micrograph. ×1000. (C) Diamond knife showing periodic spacing of uniform fringes indicating optical quality finish below 0.2 μm. Interference micrograph. ×600. (D) Polished razor blade cutting edge of type which was used for ultramicrotomy, showing irregular, distorted fringes in marked contrast to optical quality finish of diamond knife. Interference micrograph. ×1000.

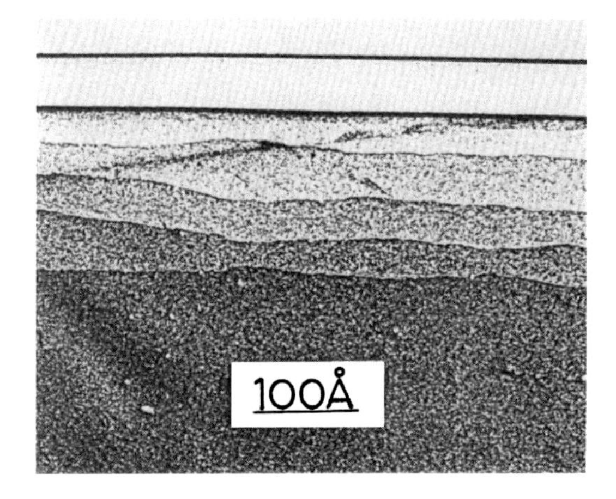

Figure 12 Electron micrograph of carbon replica of diamond-knife edge and facet showing characteristic layered structure derived from the oriented crystalline diamond lamellae. It is this combination of layered structures in diamond with its unique properties which makes it possible to "dissect out" by adequate cleavage and abrading the preformed unit layers, and thus obtain the desired ultrasharp, stable cutting edges of molecular dimensions (10–15 Å).

Figure 13 Ultramicrotome for producing serial ultrathin (20- to 100-Å) sections of plastic-embedded or frozen biological specimens, crystals, metals, and other hard materials, developed and tested by the author in 1953–1954 (Fernández-Morán, 1956a). The diamond knife is held stationary while the specimen is advanced upon it at a carefully controlled rate, operating on the principle of thermal expansion of the specimen rod. A commercial version of this instrument was built and made available by Leitz–Wetzlar (Walter, 1958).

ply of tested diamond knives would make extensive distribution possible. Having made and applied a superior cutting edge for ultramicrotomy, the diamond knife should be readily available to fellow electron microscopists in order to establish its performance and to enable sharing the experience of its use on different specimens (Fernández-Morán, 1956c).

Supplementing this project, an ultramicrotome was specially developed to fully utilize the characteristics of the diamond knife, particularly its capability for ultrathin sectioning of both soft and hard specimens. The design featured exceptional mechanical stability, compact arrangement of relatively few high-precision components, and the capability to operate without lubrication in a high vacuum or in a cryogenic environment. The Fernández-Morán Ultramicrotome (Fernández-Morán, 1956a) (Fig. 13) consists of a cylindrical rotor of specially hardened steel, machined and polished with microinch precision, which rests on two V-shaped bearings (provided with precisely aligned polished sapphire rods for optimum performance). Point bearings on both sides of the rotor prevent lateral excursion. Thus, exceptional stability and reproducibility of circular movement is guaranteed, without any lubrication and with negligible wear. The Invar specimen rod traverses the rotor through a radial hole, is firmly attached at one end, and contains an electrical heating element to control thermal expansion. The diamond knife is held stationary, firmly attached to a large trough which is rigidly clamped to a special stage. The diamond knife is moved forward until it contacts the specimen and slices off the first section. From then on the specimen is advanced toward the knife edge at a carefully controlled rate determined by the thermal expansion of the specimen rod, cutting an ultrathin section every time the precisely rotating specimen makes a single pass across the knife. The components of this prototype were built in the central workshops of the Nobel Institute for Physics, and of the Institute for Cell Research of the Karolinska Institutet, thanks to the expert help of Engineer R. Säfström. Assembly and testing of the ultramicrotome was carried out successfully by Mr. Josef Weibel, who supervised this project over the years. Professor Gunter Bahr (Bahr & Zeitler, 1965) kindly shared his vast experience in quantitative electron microscopy, and made several constructive suggestions in the early stages of this work. A commercial version of this ultramicrotome, embodying several important improvements, was built and made available by Ernst Leitz GmbH in Wetzlar (1956–1958) (Walter, 1958).

The excellent performance of diamond knife ultramicrotomy with this instrument over a period of three decades is well documented in the corresponding publications (Phillips, 1967). As shown in the illustrations, it proved to be of particular value in preparing ultrathin sections of tobacco mosaic virus (Fig. 18), of the insect compound eye with its tough chitin components (Fig. 22), and of the crystalline insect virus inclusions (Fig. 23). The basic design features of the cryoultramicrotome (Fig. 35) were derived from this instrument.

3.3 Applications of the diamond knife

Based on the methods developed by Fernández-Morán at the Nobel Institute in Stockholm, a systematic program was implemented to produce diamond knives of high quality in sufficient quantity to make extensive distribution possible. At the recently created Venezuelan Institute for Neurology and Brain Research (IVNIC) in Caracas, an abundant supply of Venezuelan alluvial diamonds (Fig. 14) and adequate research resources were available to produce and test diamond knives, many of exceptional quality. During 1955–1958 several hundred tested diamond knives were supplied, free of charge, to individual microscopists and to major scientific institutions throughout the world (Fig. 15) on the basis of an exchange of information (Fernández-Morán, 1956c). The response was most gratifying, and we received in return a wealth of information on the performance and range of applications of the diamond knife. Professor J. Francis Hartmann,

Figure 14 Phase-contrast micrograph of diamond cutting edge prepared from carefully selected Venezuelan alluvial diamonds, which furnished some of the sharpest and most durable diamond knives. Many of these naturally occurring diamonds in certain regions of Venezuela exhibit characteristic surface striations which are similar to the striations described by R. C. DeVries (1973) in framesite bort-type diamond. He considers that these oriented deformation bands are strain-hardened, as a result of plastic deformation, and have a higher abrasion resistance than even the most abrasion-resistant (111) surfaces of diamond. In view of the experimental results obtained by DeVries, F. Bundy, and R. Wentorf (1980) in work-hardening of diamond, further studies are in progress to select diamonds for improved knives.

Figure 15 Following the methods developed by the author in Sweden, production of the diamond knives in quantity was carried out at the Venezuelan Institute for Neurology and Brain Research (IVNIC) in Caracas during the years 1955–1958. Several hundred tested diamond knives were made available free of charge to major scientific institutions throughout the world, as a contribution of IVNIC to the fields of ultramicrotomy, ultrastructure research, and electron microscopy.

who together with H. Latta (Latta & Hartmann, 1950) had introduced the glass knife, wrote one of the most valuable letters, in which he discussed in detail the advantages of the diamond knife and generously stated "I con-

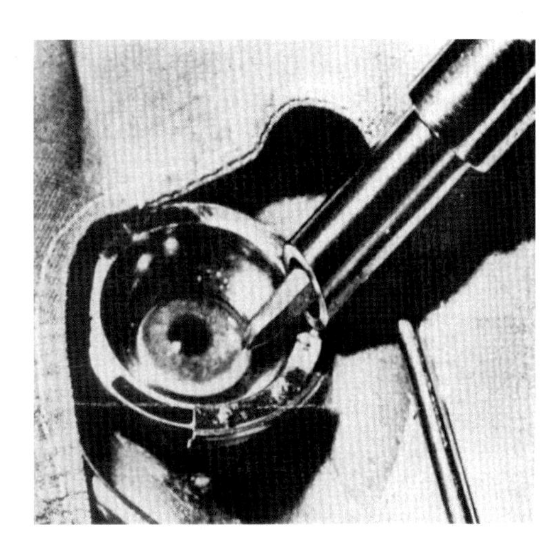

Figure 16 Vacuum fixation device developed by Davis G. Durham, M. D., applied to the eye, showing the du Pont diamond knife in position for cutting. Over 60 cataract operations have been successfully performed with the diamond knife. This is the first known use of diamond knives in human surgery. [Courtesy of Dr. Davis G. Durham, *Del. Med. J.* **38**, 202 (1966) (Durham, 1966).]

sider it an invaluable contribution to the field of electron microscopy." Dr. Keith Porter of the Rockefeller Institute also provided critical and constructive evaluation. After 1958, an important contribution was made by T. G. Lewis and associates of the du Pont company, who developed special equipment and used diamond knives as an ultraprecise machine tool to cut optical-quality finishes up to 0.2 μin. and to maintain accuracies approaching 1 μin. on ductile alloys without polishing (Lewis, 1962). The author has collaborated with the Instrument Division of du Pont in the evaluation of their diamond knife for cutting ultrathin sections during the past two decades, and wishes to acknowledge with appreciation the systematic approach supported by Daniel Friel, the improvements made by Robert Sebastian and his associates, and the significant contributions made recently by Dr. Frederick Keidel. A promising new application was pioneered by Dr. Davis G. Durham, who in 1966 introduced a new instrument for cataract incisions by use of a diamond knife (Fig. 16) manufactured by the Instrument Products Division of E. I. du Pont de Nemours & Co., Inc., Wilmington, Delaware. The 60 cataract operations successfully performed by Dr. Durham represent the first known use of diamond knives in human surgery.

The story of the diamond knife and its applications has been related in such detail because it has evolved in such an interesting and unexpected way. Developed 30 years ago merely as a device for electron microscopy, it has since received worldwide acceptance as an unsurpassed cutting tool for precision machining, for production of complex optical components, and for delicate surgical operations, and has opened up important new fields (Durham, 1966; Gargiulo, 1969; Johnson et al., 1981; Lewis, 1962; Phillips, 1967).

The John Scott Award from the city of Philadelphia was bestowed upon the author in 1967 for the invention of the diamond knife. This esteemed honor was unexpected, since the pleasure of having initiated a field and following its development, which seems to have a life of its own, is ample reward.

4. The Venezuelan Institute for Neurology and Brain Research

4.1 Foundation and research activities

Fernández-Morán returned to his native country of Venezuela in 1954, invited by the Minister of Health, the distinguished physician Dr. P. A. Gutíerrez Alfaro, and was assigned the mission of developing a regional center for research and training in neurology and brain research. The Venezuelan Institute for Neurology and Brain Research was founded in April, 1954, by decree of the President of Venezuela, as an autonomous government agency ascribed to the Ministry of Health, similar in status to the National Institutes of Health. With an initial endowment of 7 million dollars, and counting on the full support of the government and the Academy of Sciences, it was clearly understood that Fernández-Morán as chairman of an efficient executive board would concentrate on selecting a suitable site and start building the first research laboratories at the earliest opportunity. This would enable him to continue his own work, while attracting staff with the incentive of equipping their own laboratories with the most modern facilities. A site was selected with expert assistance, located in a beautiful mountain area 9 miles southwest of Caracas, 5000 feet high, with excellent climate and adequate water supply (Fig. 17). Within 7 months, the main roads and basic facilities had been built, and nearly 1 year later, between November 26 and December 2, 1955, the Laboratories of Nerve Ultrastructure (with electron microscope facilities installed and functioning), the Neurophysiology Unit, the Central Workshop (including

Figure 17 The Venezuelan Institute for Neurology and Brain Research (IVNIC), located in the mountains 12 km south of Caracas. Founded in April, 1954, it soon became an international research center for nerve ultrastructure studies by electron microscopy, X-ray diffraction, and related techniques (1955–1958). Serving as a nucleus, it has since expanded (IVIC) and has played a major role in the scientific and cultural development of Venezuela during the past decades. [Courtesy of Professor Geoffrey H. Bourne, *Nature* **176**, 1049–1051 (1955) (Bourne, 1955).]

the Diamond Knife Unit), the Library, and the Residences for staff and visitors were inaugurated (Bourne, 1955).

Research work began before Christmas, 1955, and the first paper, on the fine structure of the insect retinula, was published in *Nature* in 1956 (Fernández-Morán, 1956b). The work on production, applications, and distribution of diamond knives (Fernández-Morán, 1956a, 1956c; Fernández-Morán & Engström, 1956a, 1956b) proceeded steadily. Thanks to the devoted efforts of the administrative and engineering staff, construction work on the major institute project (which would ultimately cost more than 50 million dollars) continued independently, on schedule. One can still evoke the haunting memories of watching the ultrathin serial sections (barely 100 atoms in thickness) prepared by diamond-knife ultramicrotomy in the laboratory, and then stepping outside to see a team of tractors shaving off a barren mountain top—reflecting on when we would ever witness such contrasts in controlled manipulation across so many dimensional levels. Soon our work in high-resolution microscopy, and Gunnar Svaetichin's pioneering work in recording the spectral response curve from a single cone, began to attract visiting scientists. Professor Harold Urey and George Gamow were among the distinguished transient visitors. Professors Arne Engström, Gerhard Schramm, Helmut Ruska, Severo Ochoa, Bryan Finean, and the unforgettable Professor Erwin W. Müller (Müller, 1937), to name only a few, stayed longer and contributed much.

Here I can only refer briefly to a joint research project with the late Professor Gerhard Schramm of the Max-Planck-Institute for Virus Research which yielded interesting results on the structure of tobacco mosaic virus particles (Fernández-Morán & Schramm, 1958) and brought us indirectly in epistolary contact with the late Rosalind E. Franklin. Professor Schramm suggested that we apply the techniques of staining, embedding, and ultrathin sectioning to a study of the fine structure of tobacco mosaic virus particles by high-resolution electron microscopy. He pointed out that considerable information was already available on the structure of the tobacco mosaic virus based on X-ray diffraction studies of oriented preparations, which could be correlated with the results of biochemical studies. A concentrated preparation of tobacco mosaic virus furnished by Schramm was oriented by sucking it into thin plastic capillaries, then stained with uranyl acetate, lanthanum acetate, phosphotungstic acid, phosphomolybdic acid, and ammonium vanadate. The stained preparations were embedded and ultrathin sections prepared by diamond-knife ultramicrotomy. Transverse ultrathin sections of TMV particles from oriented preparations stained with 1% uranyl acetate show a light central area with a diameter of 25–30 Å (Fig. 18A). This central hole is surrounded by a concentric ring of dense granules ∽20 Å in diameter, with indications of a regular arrangement in a rosette-type of pattern (Fig. 18B). In longitudinal sections the central channel with a diameter of 30–40 Å could be seen, corresponding to the structures first described by H. Huxley in TMV particles dried on the specimen film and stained with 1% uranyl acetate (Fig. 18C). Although the results obtained were preliminary, the general picture of a cylindrical particle with a hollow axial core surrounded by a characteristic dense

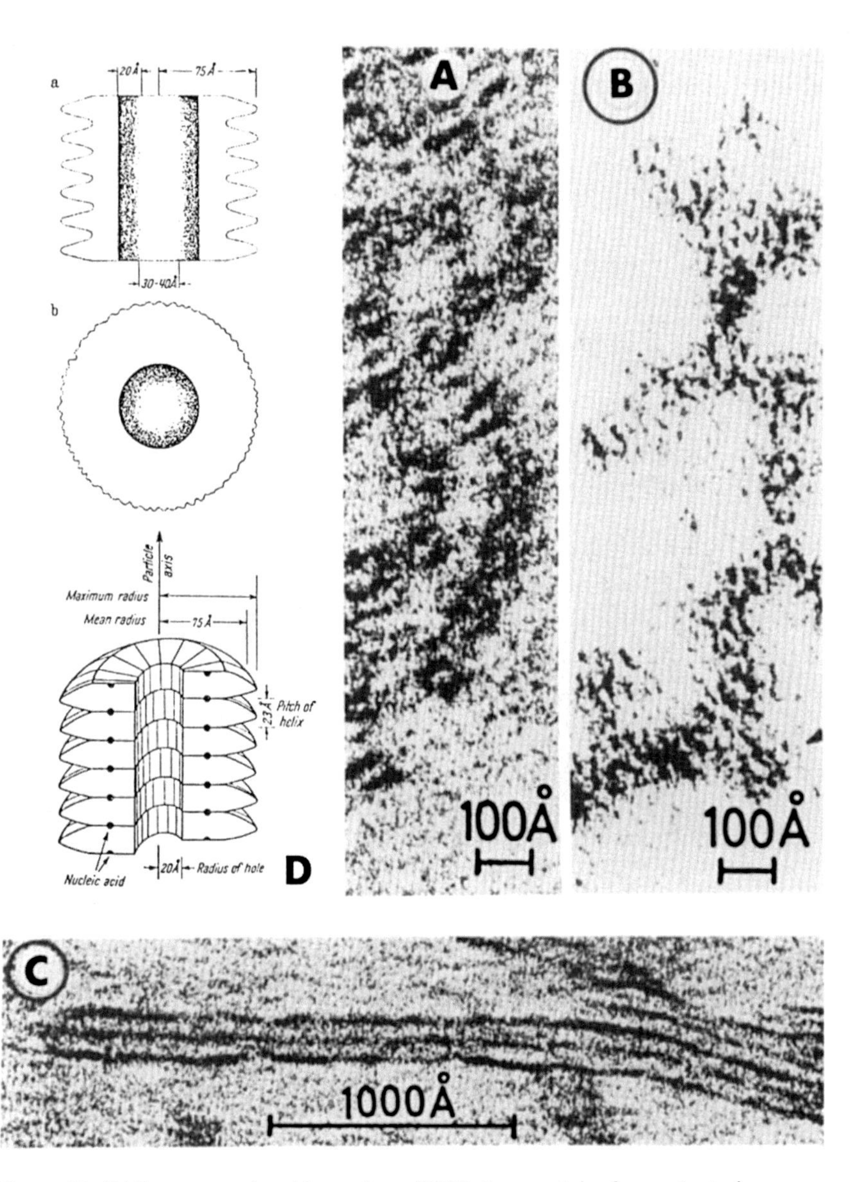

Figure 18 (A) Transverse ultrathin section of TMV virus particles from oriented preparation stained with uranyl acetate showing light central area surrounded by dense annular structure with rosette-type (B) pattern. (C) TMV particle dried on specimen film and stained showing central channel. (D) Schematic diagram of the structure of a tobacco mosaic virus particle as revealed in thin sections by electron microscopy, compared with the model (a and b) postulated by Franklin, Klug, and Holmes (1956) based on X-ray diffraction studies. [From H. Fernández-Morán and G. Schramm, 1958 (Fernández-Morán & Schramm, 1958).]

annular zone appears to be in agreement with the model postulated by R. E. Franklin, A. Klug, and K. C. Holmes (Fig. 18D) (from CIBA Foundation Symposium, 1956, Boston, Little, Brown, 1957, p. 39).

These studies demonstrated the advantages of thin-sectioning methods: (1) supplementary information on the internal structure of the TMV particles, which are relatively well preserved, can be obtained; (2) ultrathin sections make it possible to observe the fine texture of the virus particles in any desired plane, while displaying the repetitive features of internal structure which are present in regular paracrystalline arrays (Fernández-Morán & Schramm, 1958). During our long discussions with Professor Schramm I became interested in the personality and extraordinary talent of Rosalind Franklin. In 1958, when she planned to visit the United States, IVNIC wrote her a letter, inviting her to honor us by visiting IVNIC as a distinguished scientist. Part of that letter is reproduced on page 187 in Anne Sayre's brilliant and moving book on Rosalind Franklin and DNA (Sayre, 1975). Her premature death was a great loss to all who knew her, and to science. Her planned visit to South America and to Venezuela and IVNIC in particular would have been a memorable event.

4.2 International symposium on nerve ultrastructure and function

The international symposium on "The Submicroscopic Organization and Function of Nerve Cells" held at the Venezuelan Institute for Neurology and Brain Research, March 15–22,1957, under the auspices of the International Society for Cell Research, marks the acme of IVNIC. Professor G. H. Bourne's account in *Nature* (Bourne, 1957) and the publication of the proceedings as a supplement to *Experimental Cell Research* (Fernández-Morán & Brown, 1958) convey the spirit of cordial understanding, stimulating interaction, and exchange of new data which prevailed during the meeting. An international group of 46 distinguished scientists spent a week enjoying the hospitality of Caracas, and drove up to the institute every morning where they spent the rest of the day, attending the sessions and visiting the laboratories. Inspection of the institute during the symposium "showed it to be the best equipped in the world for the application of physical techniques to the study of biological problems. Probably the most intriguing new technique demonstrated was that of nuclear magnetic resonance and its application to the measurement of rapid changes in the water content of untreated intact nerve" (Bourne, 1957). Conforming to its title, the symposium centered around the fields of electron microscopy and nerve ultrastructure, electrophysiology, biochemistry, histochemistry, and spectroscopy. The papers at the symposium fell into five groups dealing with the nerve fibers, the nerve cell membrane, the neurons, the synapses, and the receptors. Nerve ultrastructure, as revealed by electron microscopy and low-angle X-ray diffraction, was an important contribution, particularly since most of the material presented was new and is still considered significant, more than two decades later. Nearly one-third of the participants were electron microscopists, including Hans Engström, Keith Porter (Porter et al., 1945), S. L. Palay, H. Ruska, E. De Robertis, and other distinguished pioneers. The classic papers on nerve fibers by Dr. H. S. Gasser of the Rockefeller Institute, by Dr. F. O. Schmitt of the Massachusetts Institute of Technology, and by Dr. A. von Muralt of The Hallerianum, Bern, were supplemented by the original contributions of Dr. J. B. Finean, who presented the results of X-ray diffraction studies of the nerve myelin sheath correlated with electron microscopy investigations carried out jointly with Fernández-Morán in the Department of Nerve Ultrastructure at IVNIC (Fernández-Morán & Brown, 1958; Fernández-Morán & Finean, 1957). Each paper contributed interesting new data on the various topics, and often led to lively discussions. Sometimes we would be witnessing the first correlation of hitherto unknown ultrastructural detail with functional data, e.g., electrophysiological recording from single neurons, or the elegant analysis of the receptor potentials from single cones using ultramicroelectrodes and techniques developed and applied by Dr. G. Svaetichin at the Neurophysiology Department of IVNIC. (See Figs. 19–23.)

The timeliness and significance of the symposium became increasingly evident as the accumulated data were discussed and the participants grew more familiar with the integration of structure and function. Moreover, we now realize the privilege of receiving the invaluable fund of knowledge and wisdom, based on a lifetime of experience, which was contributed as a legacy by such masters as Herbert Gasser, John Runnström (1928), and Yngve Zotterman, who are no longer in our midst, but ever present in our thoughts. Dr. G. H. Bourne described the results of his recent histochemical studies, pointing out the high phosphatase activity along the membranes of the pyramidal cell dendrites, and particularly in the basal cells of the olfactory epithelium. Dr. J. H. Luft described the fine structure of electric tissue of fishes, depicting a characteristic system of tubules and vesicles extending from the plasma membrane into the cytoplasm of the electroplate. According to his hypothesis, this system may account for ion transport in electric tissue.

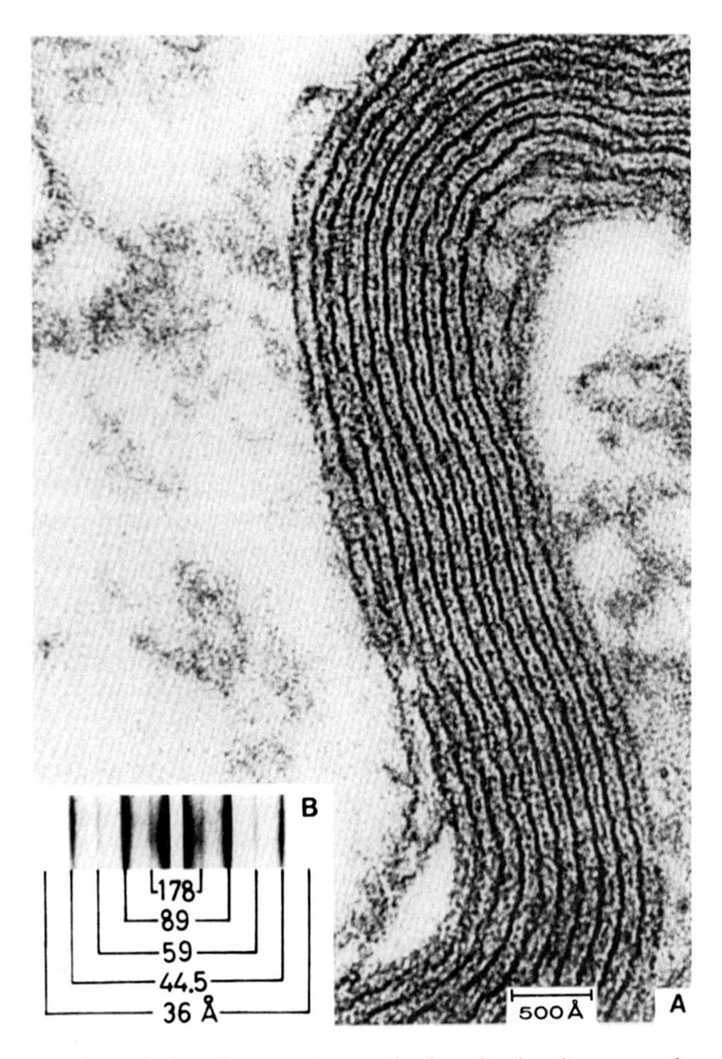

Figure 19 High-resolution electron micrograph of myelin sheath segment from transverse thin section of frog sciatic nerve showing concentric array of dense and intermediate layers in cryofixation preparation. (B) Low-angle X-ray diffraction pattern of fresh sciatic nerve featuring a fundamental period of 178 Å with typical alternation of the intensities of the even and odd orders. [From H. Fernández-Morán (Fernández-Morán, 1959a, 1961a, 1962a).]

Dr. H. Hydén, S. O. Brattgård, and J. E. Edström of the University of Göteborg, Sweden, gave an impressive account of the application of their pioneering ultramicrochemical and quantitative cytochemical methods for the determination per nerve cell of ribonucleic acids (RNAs) and the pro-

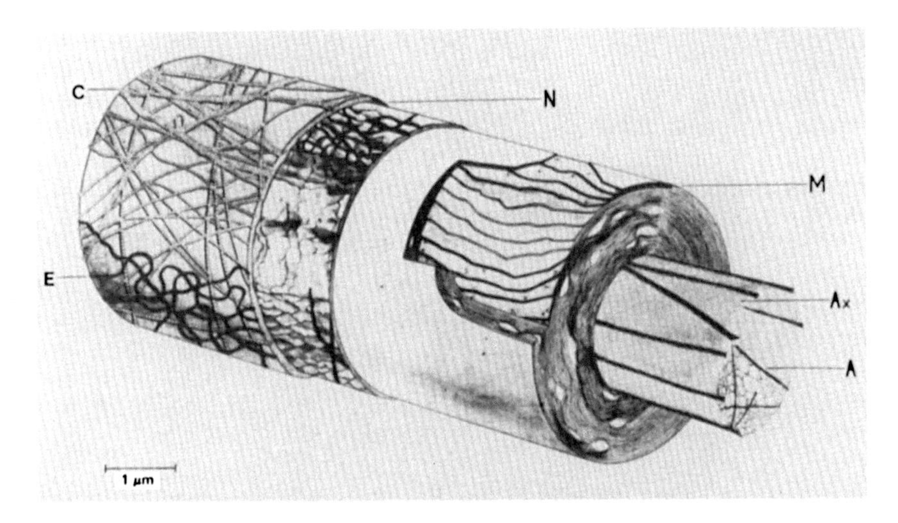

Figure 20 Schematic diagram of the internode portion of a medullated nerve fiber based on electron microscopy of ultrathin frozen sections of osmium-fixed rat sciatic nerve, correlated with the results of polarized-light and X-ray diffraction studies. The neurilemma membrane (N), the concentric laminated structure of the myelin sheath (M), and the filamentous axon structure (A) are shown. Ax, axolemma; C, collagen; E, smooth fibrils. [From H. Fernández-Morán, *Exp. Cell Res.* **1**, 309 (1950) (Fernández-Morán, 1950b).]

Figure 21 Single-unit disk isolated from the outer segment of frog retinal rod showing the granular surface structure and the marginal cord. Latex particles (2800 Å in diameter) have been added for calibration in this shadowed preparation (Fernández-Morán, 1953b).

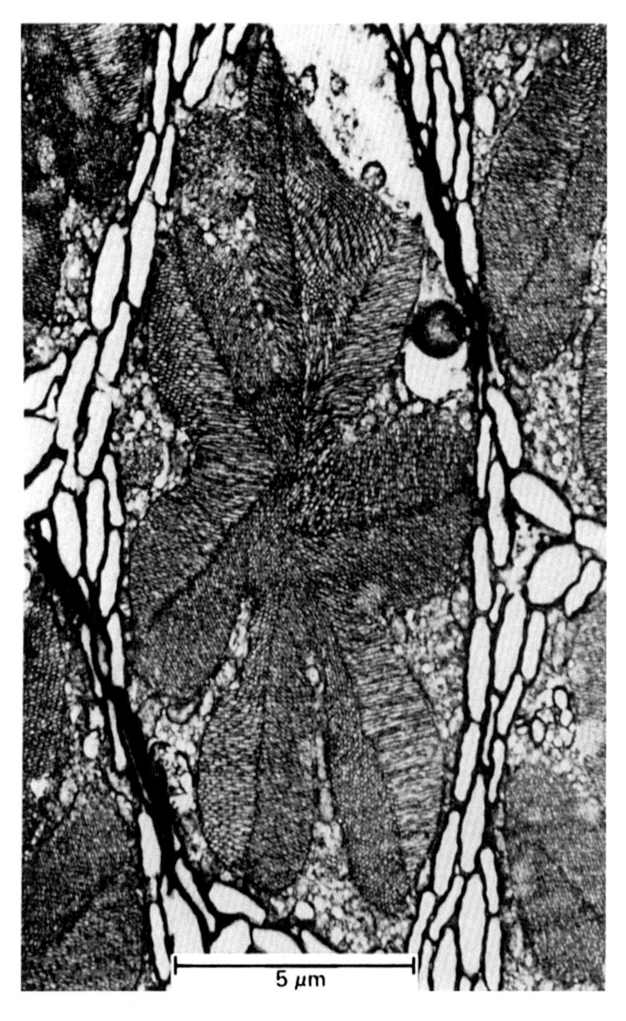

Figure 22 Oblique cross section through photoreceptors from the compound eye of the tropical moth *Erebus odora* showing the ordered fine structure of its components. Diamond-knife ultramicrotomy was essential in examination of the insect eye with its tough chitin framework (Fernández-Morán, 1958).

tein and lipid fractions to studies of nerve regeneration. The data obtained, which permitted quantitative analysis of individual neurons on the cyto-scale, were correlated to the peripheral growth of the axon. During the maturation period one of the striking features is the 100% RNA increase and the volume increase of 100%. Dr. George Wald, from the Biological Laboratories of Harvard University, gave a masterful presentation on the

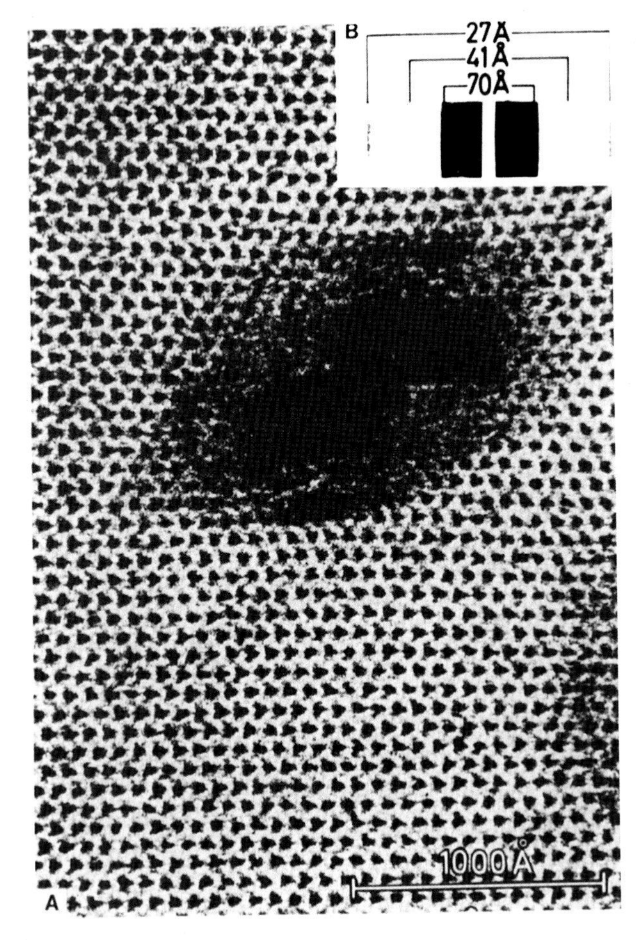

Figure 23 (A) High-resolution electron micrograph of ultrathin section prepared by diamond-knife ultramicrotomy from a crystalline insect virus inclusion showing highly ordered molecular organization of the protein matrix surrounding the virus. Correlation with the low-angle X-ray diffraction pattern (B) yielded additional information [IVNIC, 1957 (Fernández-Morán, 1964)].

"Photochemical Aspects of Visual Excitation." This paper described interesting correlations of the biochemistry of visual systems with the molecular basis of visual excitation, and it advanced numerous suggestions bearing on the structural relations between a retinal rod and a peripheral nerve cell, and between the pigment epithelium and the sheath of Schwann. Dr. Fernández-Morán described the fine structure of the light receptors in the compound eyes of insects, which is of particular interest in view of their remarkable functional capacities, including the ability of insects to orient themselves by polarized light. This paper (Fernández-Morán, 1958; Fernández-Morán & Brown, 1958) and the exhibit of electron micrographs also served to illustrate the singularly favorable conditions prevalent at IVNIC for studying the wealth of tropical insect forms with appropriate techniques, including ultrathin sectioning of the delicate yet tough insect chitin by diamond-knife ultramicrotomy. Representative specimens of compound eyes with different types of image formation, including the large compound eyes of the giant tropical moth *Erebus odora*, were examined. Recent electron microscope studies of the differentiated elements of the retinula cells called rhabdomeres by T. H. Goldsmith and D. E. Philpott, by J. J. Wolken, J. Capenos, and A. Turano, and by W. H. Miller [*J. Biophys. Biochem. Cytol.* **3**, 429, 441, 421 (1957)] were confirmed and extended. The constituent rhabdomeres in all types of rhabdoms examined are built up of numerous closely packed rod-shaped or tubular units which vary from 400 to 1200 Å in diameter. Within each rhabdom the rhabdomeres are radially arranged in a symmetrical pattern formed by matched pairs of rhabdomeres, which exhibit similar orientation of internal structure and are usually located opposite to each other (Fernández-Morán, 1956b; Fernández-Morán & Brown, 1958). The available data supported the assumption that the individual rhabdoms, which are composed of radially arranged rhabdomere pairs with molecular layers of the dichroic visual pigment similarly oriented in their periodic compartment structures, may correspond to the functional units of the analyzer for polarized light postulated in the compound eye. The remarkable ability of the insect to recognize the regional patterns of polarization of the sky as a basis for light-compass orientation would then involve a coordinated activity of these functional units at the level of the ommatidial groups (Fernández-Morán & Brown, 1958). This preliminary survey, published 25 years ago, served as a basis for systematic investigation of the submicroscopic organization of the compound eyes and of the associated neural pathways within the optic lobes.

All taking part agreed that the symposium was a considerable success: "There is no doubt that the Venezuelan Institute for Neurology and Brain Research has been scientifically launched ..." (Bourne, 1957).

Following the symposium, 1957 was a year of intense work and innocent felicity. The publications from that period attest to the wide range of activities. A comprehensive paper on "Electron Microscope and Low-Angle X-Ray Diffraction Studies on the Nerve Myelin Sheath" (Fernández-Morán & Finean, 1957) was published, reporting on a close correlation between high-resolution electron microscopy and low-angle X-ray diffraction studies of the myelin sheath after controlled experimental modifications, carried out in collaboration with Dr. J. B. Finean, who made significant contributions.

One of the most interesting and rewarding projects was initiated by the late Professor Erwin W. Müller, from the Field Emission Laboratory of the Pennsylvania State University, who in 1957 kindly accepted our invitation. Dr. Müller shared our interest in the possible application of this powerful technique to certain biological problems. He brought with him one of his low-temperature field ion microscopes, and proceeded shortly after his arrival to set up the instrument, using the Collins Cryostat for producing liquid helium installed in our Low Temperature Department to provide a liquid-helium-filled cold finger, instead of the usual liquid hydrogen, for our experiments. His extraordinary experimental skills and his virtuosity in making the pointed filaments were inspiring to watch, and the author learned how to make pointed filaments of tungsten, which were to play an important role in high-resolution electron microscopy. Fig. 24 shows the helium-ion image of a tungsten tip cooled with a liquid-helium-filled cold finger recorded by Professor Erwin Müller at IVNIC in 1957. Although these experiments were preliminary, he opened up a new world, described in his classic papers (Müller, 1960), and we shall always cherish the memories of his visit with us at IVNIC.

The New Year of 1958 was ushered in by one of those drastic phase changes, known as revolutions, not uncommon in South America. We continued our work at the institute, unaware of its magnitude. Shortly before my planned trip to attend a Biophysics Meeting in Washington, on January 13, 1958, I was called by my good friend, the late Dr. P. A. Gutíerrez Alfaro, Minister of Health, to whom I was deeply indebted for his support of IVNIC, and I was asked to accept the post of Minister of Education on a temporary basis. To this day, I consider it a privilege to have served at his side, although my tenure as Minister of Education was to last only 10 days. However, during this time I experienced the equivalent of an intensive postgraduate course in collective human behavior, trying to cope with irrational aspects of human nature. However brief, this other kind of apprenticeship, both complex and arduous, served at age 34, made a lasting impression, and conferred an added dimension of understanding and tolerance for future endeavors.

On February 14, 1958, the author turned IVNIC over to a recently appointed director and administration. Bequeathed in a spirit of objective

Figure 24 Helium-ion image of a tungsten tip (∽600 Å in radius) cooled with liquid-helium-filled cold finger (∽10 K) recorded with a field ion microscope (Müller, 1960) by Professor Erwin Müller at IVNIC in 1957. Resolution in the (111) plane (inset) is 2.74 Å.

cooperation was the nucleus of a functioning research institute, located on an ideal site of 2000 acres, only 12 km southwest of Caracas, constituting the basis for a self-contained scientific community, with its own roads, water supply (designed to serve 30,000 people), electrical power plant, and 10,000-kW substation (Fernández-Morán et al., 1957). In addition, sufficient funds were available or firmly committed to guarantee the continuation of the project. Most importantly, IVNIC, during the initial phase of less than 4 years since it was founded in 1954, had broken new ground and was the first successful demonstration in Venezuela of an institute capable of conducting basic research in an organized fashion and on a long-term basis. Scientifically and quite literally IVNIC had put Venezuela on "the map" of the international scientific world, having produced original research in the field of fundamental brain research and the neurosciences, as clearly demonstrated during the International Symposium held in 1957 (Fernández-Morán, 1957).

The institute has since expanded, and, under the name I VIC (Venezuelan Institute for Scientific Research), during the past 25 years it has provided a new generation of Venezuelan scientists with all the resources to carry out basic research and training. It has played a major role in the scientific and cultural development of Venezuela and of neighboring countries.

Two weeks later, in late February of 1958, the author and his family left Venezuela, and were cordially invited by Professor Francis O. Schmitt to visit Boston. Thanks to the gracious hospitality and the wholehearted support received from Professor Schmitt and Dr. William H. Sweet, Fernández-Morán joined the staff of the Neurosurgical Service of the Massachusetts General Hospital, where he organized the Mixter Laboratories for Electron Microscopy. He was also associated with the Biology Department of the Massachusetts Institute of Technology, where he continued his work in close cooperation with Professor Schmitt, and also with Professor Samuel C. Collins (Collins, 1960-1982; Fernández-Morán, 1960c) at the Cryogenic Engineering Laboratory. The work on high-resolution and low-temperature electron microscopy of biological systems described here was carried out in this ideal environment from 1958 to 1962.

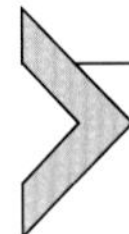

5. Low-temperature electron microscopy and ultramicrotomy

5.1 The submicroscopic organization of vertebrate nerve fibers

Based on the polarized-light analysis of the myelin sheath (Schmidt, 1937) and on long-spacing X-ray diffraction patterns featuring the first four orders of a fundamental spacing of 171 Å in amphibian nerve and 184 Å in mammalian nerve recorded by Schmitt, Bear, and Clark (1935) from undamaged whole nerve, the general features of the myelin sheath's ultrastructure could be deduced. Schmitt arrived at the following conception: "Essentially the sheath consists of lipid–protein layers about 180 Å thick wrapped concentrically about the axon" (Schmitt, 1950). This appeared to be an ideal system for correlated electron microscope studies.

Thanks to the generous hospitality granted by Professor T. Caspersson, director of the Institute for Cell Research and Genetics, Karolinska Institutet (Caspersson, 1937, 1950), the author was able to apply the new RCA EMU electron microscope (Hillier & Ramberg, 1947) and other facilities for high-resolution studies of nerve fibers. Beginning in 1949, a series of systematic studies were carried out and the following results published over a 3-year period (Fernández-Morán, 1950a, 1950b, 1952a, 1953b).

In order to avoid the severe extraction of lipids and other artifacts of usual methods requiring embedding, a technique was developed which makes it possible to *prepare ultrathin serial frozen sections of unembedded fixed or fresh nerve* (Fernández-Morán, 1950a, 1950b, 1952a, 1952b; Fernández-Morán & Dahl, 1952). In this modification of Newman, Borysko, and Swerdlow's method (Newman et al., 1949), cooling with carbon dioxide is used to operate the thermal expansion advance mechanism and at the same time to freeze the wet, unembedded specimen. Combined with supplementary methods for ultrathin sectioning, the following structures were described in the internode portion of rat and frog nerve fibers (Fernández-Morán, 1950b, 1952a):

1. (a) The myelin sheath exhibits an extremely regular concentric laminated fine structure in transverse and longitudinal thin sections, and the individual layers are approximately 80 Å thick. Unit lamellae of similar dimensions can be isolated from the fragmented sheath. Based on these observations a general concentric laminated structure of the myelin sheath is assumed (Fig. 20) which is in good agreement with the results of polarized-light and X-ray diffraction studies. However, the average thickness of 80 Å determined for the single layers by electron microscopy would correspond to only half the value of the 158-Å X-ray spacing demonstrated in dried mammalian myelinated nerve, (b) The sheath is covered with a compact granular membrane ∽200 Å thick corresponding to the neurilemma (N) with associated collagen and smooth fibrils, mainly longitudinally arranged in bundles.

2. The axon of well-preserved fresh nerve fiber sections contains filaments (100–200 Å diameter) of indefinite length, with regularly spaced axial discontinuities, and predominantly longitudinal arrangement (Fernández-Morán, 1952a).

3. (a) New types of submicroscopic nerve fibers (0.1–1 μm diameter) were found in the lateral funiculus of rat and frog spinal cord. The finest fibers consist of a single 100-Å filament within a tubular membrane, (b) Unmyelinated nerve fibers are constituted by compact bundles

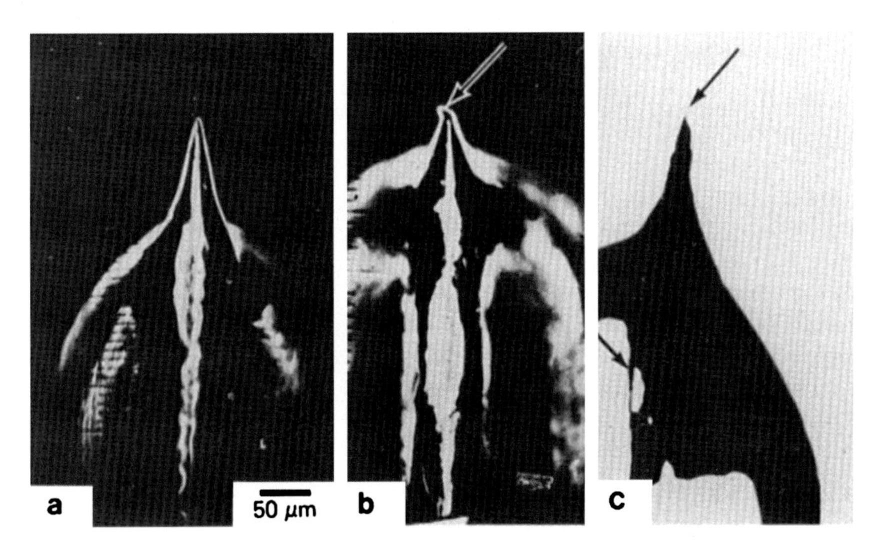

Figure 25 Pointed filaments with etched, oriented tungsten single-crystal tips showing preservation of critical cathode profile: (a) annealed tip before use, (b) filament after 40 hr, and (c) after 100 hr operation in Siemens Elmiskop I at 40–100 kV. [H. Fernández-Morán, 1960 (Fernández-Morán, 1960b).]

of 100- to 200-Å thin filaments invested with a single membrane of ∽100 Å.

4. These findings indicate that the structural elements of all types of nerve fibers are fundamentally similar, and point to a common pattern of organization in myelinated fibers of all sizes and in the so-called unmyelinated nerve fibers (Fernández-Morán, 1952a, 1953a, 1954).

Similar techniques for preparation of ultrathin frozen sections by "cryo-ultramicrotomy" were successfully applied in studies of bacteria, spermatozoa (Fernández-Morán, 1952b), pollen (Fernández-Morán & Dahl, 1952), and other labile structures.

5.2 High-resolution electron microscopy and low-dose electron diffraction

5.2.1 Low-temperature electron microscopy with microbeam illumination

A Siemens Elmiskop I operating at 40–100 kV, equipped with multiple objective apertures (Fernández-Morán, 1959b), was provided with improved pointed filaments of oriented single-crystal tungsten (Fig. 25) (Fernández-Morán, 1960b) (see also Fig. 28) and used with the double-condenser

Figure 26 Bacterial virus, a T2 bacteriophage, showing complex organization of the head (HM) and tail fiber (F) components as revealed in this negatively stained preparation by low-dose electron microscopy with microbeam illumination using pointed filament (Fernández-Morán, 1962b). CH, Collar; SC, sheath; BC, base plate.

system to provide intense microbeam illumination of high coherence. With this arrangement enhanced contrast and improved resolution (∼8–10 Å) were obtained in suitably thin sections or negative-staining embedding preparations (Fig. 26) (Fernández-Morán, 1961b, 1962a). Specimen damage by irradiation can be considerably reduced by suitable combination of low-intensity electron optics and improved specimen cooling devices. With the new pointed filaments and the double-condenser system fitted with apertures of 50–10 μm it is also possible to obtain microbeams

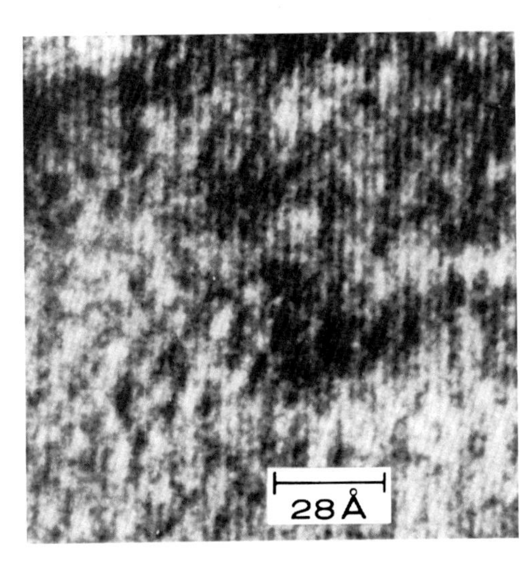

Figure 27 High-resolution electron micrograph of thin NaCl crystal showing lattice spacing of (200) planes. Spacing resolved is 2.815 Å by direct (axial) low-dose illumination method using special tantalum pointed filaments of the Fernández-Morán type (Fernández-Morán, 1960b, 1961b).

of 0.5–0.1 μm diameter and of extremely low intensity. By focusing on adjacent areas and then shifting the low-intensity microbeam (Fernández-Morán, 1958, 1959b, 1961b), useful micrographs were recorded on high-speed emulsions [i.e., Tri-X sensitized with gold thiocyanate solutions] from extremely labile components (Fig. 27). A Leisegang liquid-nitrogen stage was used to cool the thermally insulated specimen support (−130 °C) provided with special shielding devices to prevent specimen contamination (Fig. 26) (Fernández-Morán, 1962b). Several of these techniques were used in the study of multienzyme complexes [e.g., pyruvate dehydrogenase complex, studied in collaboration with Dr. L. Reed et al. (Fernández-Morán, Reed, et al., 1964b)]. Low-dose microbeam illumination was also essential in the study of mitochondrial membranes in collaboration with Dr. David Green et al. (Fernández-Morán, Oda, et al., 1964a).

5.2.2 Low-temperature and low-dose electron diffraction

The described techniques proved to be particularly useful for electron diffraction studies of organic crystals. Useful electron diffraction patterns with a complete two-dimensional reciprocal lattice net have been recorded with a limiting resolution of 1.8 Å (Fernández-Morán, 1961b).

Figure 28 Single-crystal diamond-point cathode developed by Fernández-Morán (Fernández-Morán, 1967, 1973, 1980) with a tip radius of approximately 100–500 Å after coating with tungsten, rhenium, or zirconium. These practically "permanent" point cathodes can be periodically "recoated" to provide stable, coherent microbeam illumination under optimum conditions (Fernández-Morán, 1967, 1980).

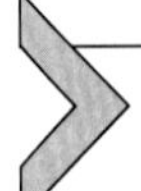

6. Cryo-electron microscopy

6.1 Study of biological systems at liquid-helium temperatures

Low-temperature preparation techniques provide one of the most promising approaches in electron microscopy of biological specimens since rapid freezing suspends all physiological activity, immobilizing and preserving labile tissue constituents. Liquid helium II, the low-temperature form, appears to represent an ideal refrigerant for rapid or ultrarapid freezing of biological systems under certain conditions because it exhibits the unique properties of heat superconductivity and superfluidity at temperatures 100 °C lower than those obtained with the standard liquid-nitrogen coolant mixtures (Fernández-Morán, 1960c). The most abundant stable isotope ($^{4}_{2}He$) of this noble gas condenses at 4.2 K into a colorless low-density liquid, helium I, which boils and evaporates rapidly because of its extremely small heat of vaporization. At temperatures below 2.19 K, called the lambda point, it undergoes a second-order transformation, and the low-temperature liquid phase, designated liquid helium II, exhibits the unique

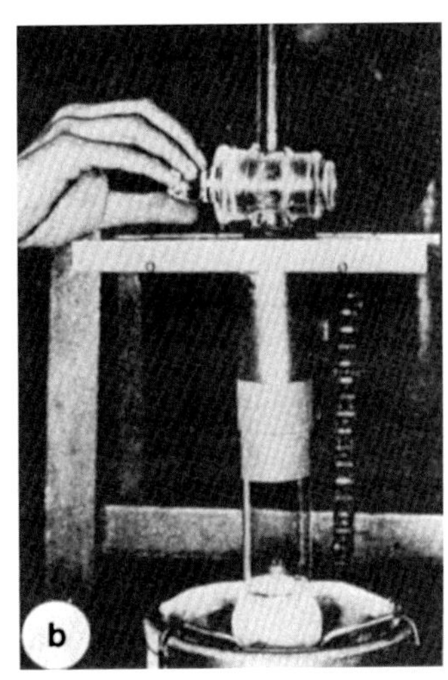

Figure 29 (a) Liquid helium II in shielded Dewar vessel at 1 K showing characteristic quiescent appearance of the superfluid helium filling the inner Dewar vessel, (b) Specimen stage for introducing minute biological specimens in helium II. [H. Fernández-Morán, 1960 (Fernández-Morán, 1960c, 1961a).]

properties of thermal superconductivity, superfluidity, formation of a thin film, and other strange physical properties.

The spectacular nature of this transformation can be seen every time liquid helium II is prepared. When liquid helium I, contained in a shielded Dewar, is evaporated under reduced pressure, it boils violently. As the temperature falls below the lambda point, bubbling suddenly ceases and a limpid column of quiescent liquid helium II fills the inner Dewar vessel (Fig. 29). This bulk liquid will conduct heat about 10,000 times better than will copper. Moreover, liquid helium II flows so rapidly through the finest capillary channels that it appears to have almost zero viscosity.

Thanks to the generous hospitality and guidance provided by Professor Samuel C. Collins and his associates at the Cryogenic Engineering Laboratory of the Massachusetts Institute of Technology, the author was able to develop special preparation techniques and become acquainted with cryogenic experimental procedures. Preliminary studies of the immersion process of microdroplets of cell components were also carried out by means of high-speed photography in collaboration with Professor Harold E. Edgerton of the Department of Electrical Engineering (Fig. 30), indicating very rapid cooling rates. During a 4-year period (1958–1962), improved low-temperature preparation techniques for electron microscopy of biolog-

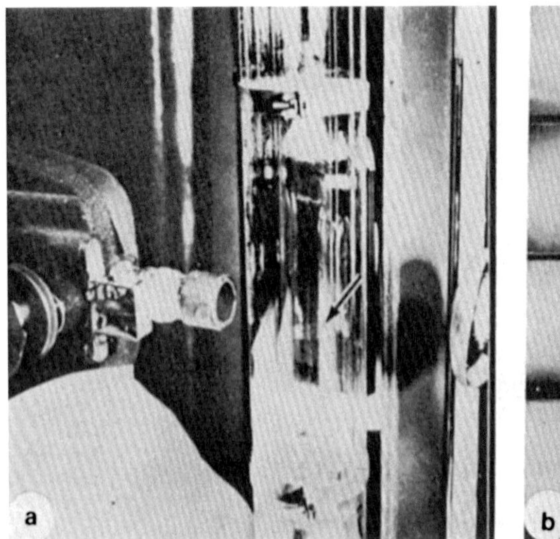

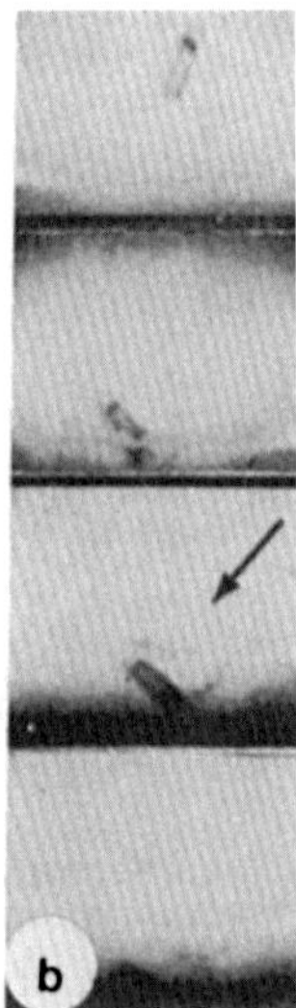

Figure 30 (a) Liquid helium II (arrow) in shielded Dewar with attached photographic devices. The characteristic quiescent appearance results because of the exceedingly high heat conductivity of helium II: a temperature difference between top and bottom of the liquid column sufficiently great to allow the formation of vapor bubbles cannot be produced, (b) Serial high-speed photographs showing immersion of minute specimen in liquid helium II without producing boiling (Fernández-Morán, 1960c, 1961a). (Courtesy of Professor Harold E. Edgerton, Department of Electrical Engineering, MIT, 1959–1960.)

ical tissues were devised and applied by Fernández-Morán (1959b, 1960a, 1961a), which yield better morphological and histochemical preservation of myelin (Fig. 19), lamellar systems, and other cell components. These "cryofixation" techniques (Fernández-Morán, 1960c, 1961b, 1962a, 1962b, 1964; cf. Valentine & Horne, 1962 and Leduc & Bernhard, 1962) are based on rapid freezing of fresh or glycerinated tissues with liquid helium II at 1–2° above absolute zero, followed by freeze-substitution and embedding in plastics under conditions which minimize ice crystal formation, artificial osmotic gradients, and extraction artifacts. The survival of living organisms, including bacteria, spermatozoa, sensitive cells, and tissues, which are first treated protectively with glycerol then frozen with liquid nitrogen or liquid helium, has since been well established. By ultrarapid cooling with liquid helium II it may be possible to preserve the native cellular organization, including the essential water component, the transient intermediates with unpaired electron spin which participate in metabolic electron transfer, and other free radicals.

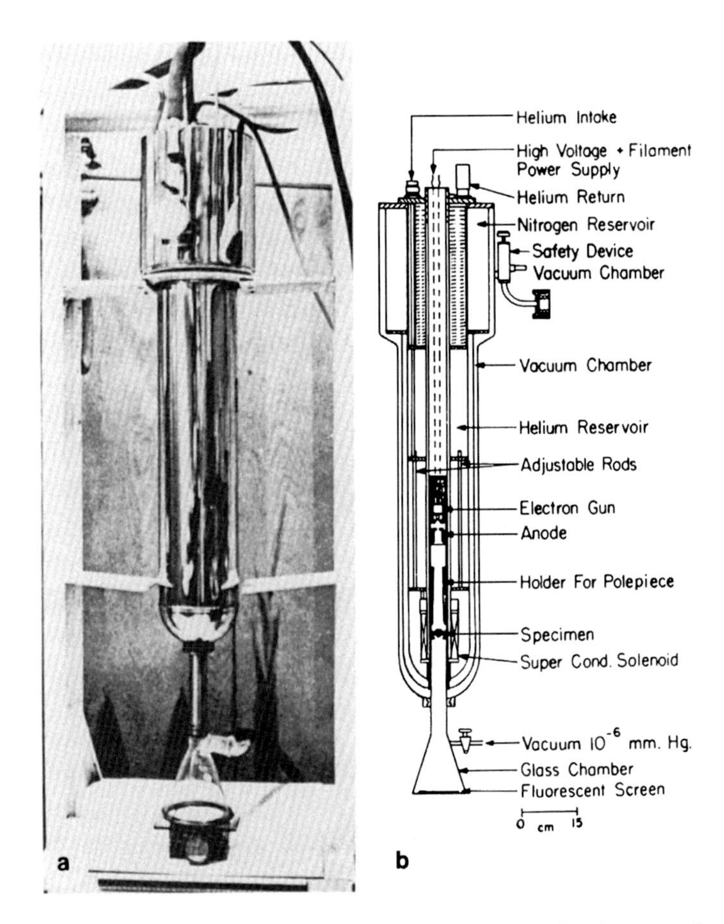

Figure 31 (a) Photograph and (b) diagram of basic equipment for electron microscopy with high-field superconducting lens, comprising air-core liquid-helium Dewar with superconducting solenoid (operating at 32,200 G in persistent-current mode), and inserted experimental electron microscope. [H. Fernández-Morán, 1965 (Fernández-Morán, 1965a).]

6.2 Development of cryo-electron microscopes with superconducting lenses

Following our work on specimen preparation techniques at liquid-helium temperatures (Fernández-Morán, 1960c, 1961a), major developments in the generation of high-field superconducting magnets with solenoids of niobium-tin or niobium-zirconium alloys (Hulm et al., 1971; Laverick, 1966) suggested the design of new types of electron microscopes. These "cryo-electron microscopes" (Fig. 31) (Fernández-Morán, 1964,

1965a, 1965b, 1966a) operating at liquid-helium temperatures would embody the following significant advantages: (1) high-field superconducting lenses with unprecedented time stability when driven in the "persistent-current" mode; (2) operation at liquid-helium temperatures, resulting in the decisive features of ultrahigh cryogenic vacuum, minimized specimen contamination, reduced specimen radiation damage, and thermal noise; (3) optimum conditions for both low-voltage and high-voltage electron microscopy; (4) high-efficiency image viewing, electronic image intensifiers, and recording devices would make it possible to use high-speed cinematography and stroboscopic recording (e.g., obtained through pulsed T–F emission from pointed filaments) for attainment of high temporal resolution combined with high spatial resolution (Fernández-Morán, 1965a). Preliminary experiments with the first cryo-electron microscopes using high-field superconducting niobium-zirconium solenoid lenses demonstrated the exceptional time stability of the images when operating in the persistent-current mode at 32.2 kG (Fernández-Morán, 1965a, 1966a, 1966b).

The first electron micrographs of biological specimens were recorded at liquid-helium temperatures (4.2 K) by Fernández-Morán with a specially designed superconducting objective lens (Fernández-Morán, 1966a, 1966b, 1967). This superconducting lens type, which can be adapted to replace the objective lens in modified high-resolution microscopes (Fernández-Morán, 1966a, 1966b, 1970; Hulm et al., 1971), comprises niobium-zirconium coils equipped with a cold ferromagnetic circuit, used without or preferably with polepieces, and including a cryogenic specimen stage of exceptional specimen stability. The entire lens assembly is immersed in liquid helium and is fully shielded by several concentric cryogenic devices from the warmer parts of the microscope. The complete lens with its superconducting control circuitry was contained in a flat Dewar (Fernández-Morán, 1966a, 1966b; Hulm et al., 1971) which served as a prototype for subsequent work with superconducting lenses (see Fig. 32). An important feature of the superconducting regulating circuitry, including stigmators and correction coils, is that extremely fine field-strength adjustments (a few parts in 10^7–10^9) can be made without leaving the persistent mode. This is essential for achieving reproducible "superfine focusing," which is orders of magnitude better than conventional lens current regulation systems (Fernández-Morán, 1966a, 1966b, 1973, 1980, 1982). The high electrical and mechanical stability permitted long exposure times with low-intensity microbeam illumination, thus yielding reduced radiation damage with reproducible resolutions of 8 nm in catalase crystals and 1–2 nm in asbestos

Figure 32 Experimental high-resolution cryo-electron microscope with special superconducting objective lens assembly in flat cryostat, power supply with persistent-current switches and improved current-control devices, ultrahigh-vacuum ion pump system, magnetic shielding, image intensifier, and vibration-isolated base [H. Fernández-Morán, 1966–1970 (Fernández-Morán, 1966a, 1966b, 1971; Hulm et al., 1971; Laverick, 1966).]

(see Fig. 33) (Fernández-Morán, 1966a, 1966b, 1967, 1971, 1972, 1973, 1980; Lefranc et al., 1982).

Systematic efforts by several international research groups working independently during the past two decades have contributed to the development of cryo-electron microscopy (Boersch et al., 1966; Bonjour & Septier, 1968; Dietrich, 1976, 1978; Dietrich et al., 1967, 1969, 1973; Dietrich, Fox, et al., 1975; Dietrich et al., 1977, 1978, 1979, 1980; Dubochet, Knapek, et al., 1981; Fernández-Morán, 1980; Formanek & Knapek, 1979; Fox et al., 1978; Knapek & Dubochet, 1980; Laberrigue & Levin-

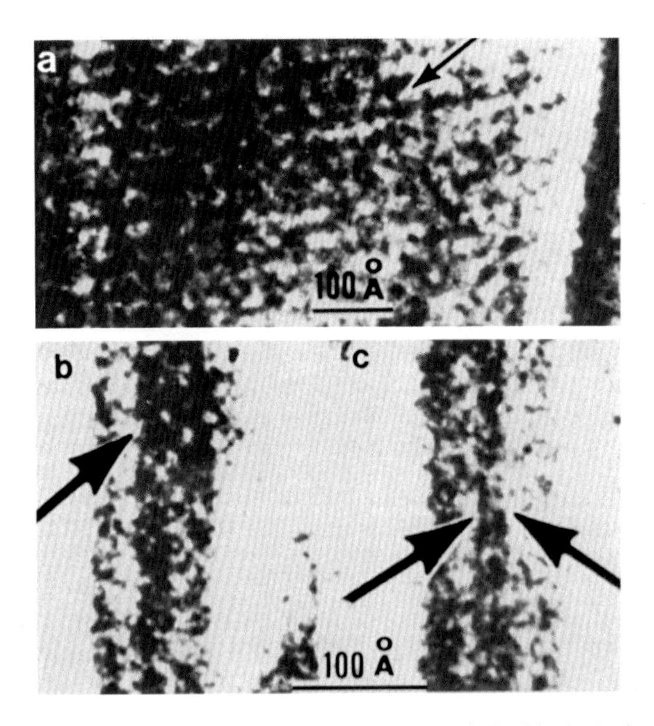

Figure 33 First high-resolution "cryo-electron micrographs" of biological specimens: (a) catalase crystals, and (b and c) asbestos specimens recorded at liquid-helium temperature (4.2 K) with cryo-electron microscope using superconducting objective lens in persistent-current mode and low-dose, microbeam illumination. [H. Fernández-Morán, 1966–1967 (Fernández-Morán, 1966a, 1966b, 1967).]

son, 1964; Trinquier & Balladore, 1968). The development and application of the shielding lens by Dietrich and co-workers during the past 16 years deserves special mention because of its outstanding performance (Knapek, 1982; Lefranc et al., 1982; Zeitler, 1982).

6.3 The Collins closed-cycle superfluid-helium refrigerator

The operational conditions of the cryogenic system must be optimized to obtain reproducible performance at a reasonable cost. Thus, in order to reduce the vibrations introduced by boiling helium at 4.2 K we soon found it necessary to use superfluid helium at 1.8 K (Fernández-Morán, 1970, 1971, 1973, 1980). The unique nonviscous flow properties of superfluid helium and its singularly high thermal conductivity yield distinct advantages for operation of the superconducting lenses, improvement of the performance

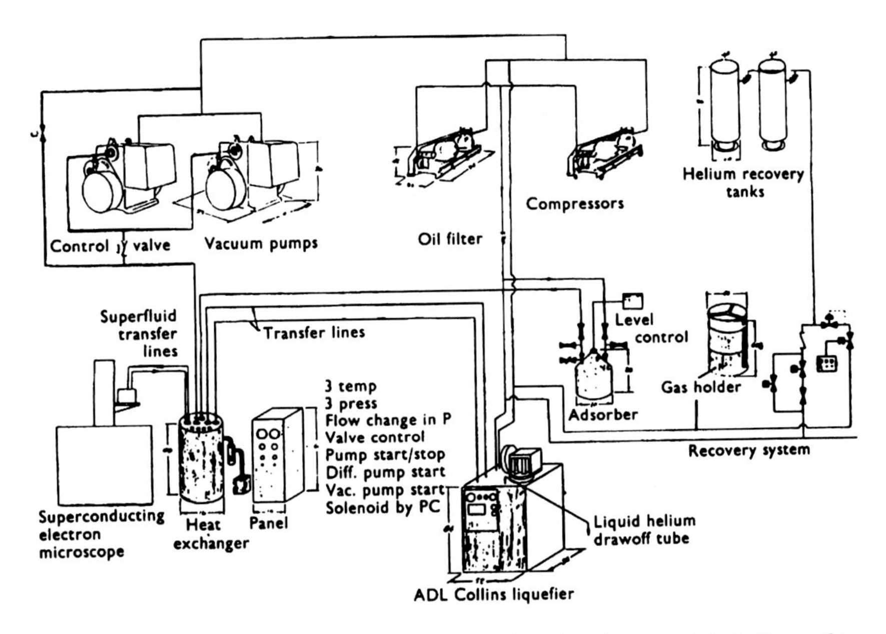

Figure 34 Schematic diagram of special Collins closed-cycle superfluid-helium refrigerator system (20-W capacity) supplying through long transfer lines an integrated superconducting cryo-electron microscope with a continuous flow of He II (1.85 K) at a rate of 8 liters/hr. [H. Fernández-Morán, 1969–1978 (Fernández-Morán, 1970, 1971, 1973, 1980).]

of the specimen stage at a lower temperature (1.8 K), and considerable reduction of vibrations in the cryostat.

Professor Samuel C. Collins and his associates (Fernández-Morán, 1971, 1973, 1980) have made large quantities of superfluid helium available thanks to a special closed-cycle helium II refrigerator featuring a novel heat exchanger developed and tested by them. With the assistance of Professor Collins and M. Streeter the large-scale, closed-cycle superfluid-helium refrigerator (Fig. 34) was installed and fully integrated with the modified 200-kV cryo-electron microscope in our laboratory. After elimination of superleaks the centrally located unit with its long transfer lines demonstrated for the first time the feasibility of routinely supplying superfluid helium at 1.8–1.9 K with a 20-W refrigeration capacity at a rate of 5–8 liters/hr. Thanks to the extraordinary longterm support of Dr. Orr Reynolds and Dr. George Jacobs of NASA's Life Sciences Division this entire project was funded and brought to fruition during the critical 10-year period of 1964–1974.

With liquid-nitrogen precooling, the rate of liquification approaches 216 liters of helium II per day. During several years of flawless operation more than 190 successful experiments were performed, permitting prolonged vibration-free examination of different types of organic and biological specimens under conditions of significantly reduced radiation damage and thermal noise (Fernández-Morán, 1971, 1973, 1980, 1982). The closed-cycle refrigerator can be run continuously or intermittently, since a prolonged maintenance of liquid-helium temperatures has been observed in the cryogenic stage of the microscope connected with the superfluid-helium system for periods of up to 29 min after all of the refrigeration equipment had been turned off. In fact, the capacity of this refrigerator is so vast that very large Dewars with improved superinsulation could be filled with several hundred liters of helium II, adequate to provide individual cryo-electron microscopes for weeks at a time. This mode of operation, which would still retain the distinctive advantages of a closed-cycle system, may be considered for a facility containing several instruments. Resolutions of 0.8–1.6 nm have been consistently achieved, mainly at 1.8 K, in the study of biological specimens, including supercooled cells in vitreous ice within special "wet chambers" or vacuum-tight "cryo-embedding cells," provided with thin, single-crystal graphite windows (Fernández-Morán, 1960b, 1973, 1980). New classes of layered transition-metal dichalcogenide intercalation complexes with unique superconducting properties have also been examined in correlated cryo-electron microscopy and diffraction studies (Fernández-Morán, 1973, 1980).

6.4 The cryomicroscope with superconducting lenses at Munich

During the past decade major advances have been made in high-resolution imaging of organic and biological specimens with considerable reduction of radiation damage using a cryo-electron microscope with superconducting lenses developed and applied by Dietrich, Weyl, Knapek, and co-workers at the Research Laboratories of Siemens AG in Munich (Dietrich, 1978; Dietrich et al., 1967, 1969, 1973; Dietrich, Fox, et al., 1975; Dietrich et al., 1977, 1978, 1979, 1980; Dubochet, 1975; Dubochet, Booy, et al., 1981; Dubochet, Knapek, et al., 1981; Fernández-Morán, 1980; Formanek & Knapek, 1979; Fox et al., 1978; Knapek & Dubochet, 1980) in collaboration with Dubochet and Knapek (Dubochet, Knapek, et al., 1981; Knapek & Dubochet, 1980). Salient aspects of this exemplary work are summarized here to illustrate the great potential of this field.

The cryomicroscope at Munich is a fixed-beam 220-kV electron microscope equipped with a superconducting lens system, including an objective lens of the shielding type developed by Dietrich et al. (Dietrich et al., 1967, 1977; Weyl et al., 1972), contained in a cryostat cooled by liquid helium at 4 K. This instrument has exceptional mechanical and electrical stability and absence of contamination, yielding a point resolution of 0.15 nm, a value which approaches the theoretical resolving power of the electron-optical system. This resolution is better than that of conventional lenses, and it permits bright-field imaging of single heavy atoms in an organometallic compound (Formanek & Knapek, 1979). Since the specimen is effectively cooled to 4 K, considerable reduction of radiation damage has been observed in a wide variety of specimens. Although the gain in radiation protection (Cosslett, 1978; Taylor & Glaeser, 1976) is still under discussion, the available evidence is consistent with the assumption that cooling the specimen to 4 K reduces the beam damage in organic specimens by more than an order of magnitude, provided that the specimen is not heated by the electron beam (Dietrich et al., 1982; Dubochet, Knapek, et al., 1981; Knapek, 1982; Knapek et al., 1981; Zeitler, 1982).

6.5 Structure determination of an organic complex with a superconducting electron microscope

Recently, Dietrich, Formanek, von Gentzkow, and Knapek (1982) have demonstrated that the steric structure of a radiation-sensitive organic copper complex can be determined with a superconducting electron microscope when the specimen, consisting of very small crystals, is kept at a temperature of 4 K, so that radiation damage is drastically reduced. Due to this effective cryoprotection, and the exceptional performance of the superconducting electron microscope (point-to-point resolution of 0.15 nm, and specimen drift below 0.03 nm/min), direct high-resolution images could be recorded in which the 1.1- and 0.9-nm lattice lines are visible. A structure model as derived from the single-crystal electron diffraction pattern, the direct images, and the results of chemical experiments is demonstrated. *N*, *N*′-Bis-salicyloyl-hydrazine (BSH) is one of the best copper deactivators, and since these organic chelates deactivate the metal, which drastically reduces the service life of low-voltage cables, it was essential to investigate the steric structure of the BSH–copper complex. A complete structure analysis of the microcrystals obtained from this complex is not possible with X-ray methods. In principle, the electron microscope is a very efficient tool for this purpose, provided that the radiation damage can be drastically reduced. The importance of this work for structure determination of biological complexes, which can only be obtained as microcrystals unsuitable for X-ray analysis, is evident, since it would open up a whole new field of biological ultrastructure analysis, yielding new data of considerable significance in molecular biology and related disciplines (Scherzer, 1979; Zeitler, 1982).

7. Reflections and outlook

Having covered such a broad field in these reminiscences it may be opportune to offer some reflections which could adumbrate future trends. Some specific approaches which appear promising, particularly in biology and medicine (Fernández-Morán, 1982), would include the following developments:

1. Improved instrumentation in cryomicroscopy, encompassing all types of microscopy and making full use of advances in superconducting electron optics, cryogenics, and radiation chemistry (Hüttermann, 1982; Zeitler, 1982), with especial emphasis on synergistic integration with advanced computer technology for processing data, to assist interpretation, and to infer and evaluate conceptual models (Heide, 1982; Hüttermann, 1982; Scherzer, 1979; Zeitler, 1982).
2. Systematic development of the next generation of high-voltage superconducting cryo-electron microscopes, following the proposed combination of a superconducting lens system with a compact superconducting linear accelerator, providing optimum cryogenic conditions by cooling with helium II (Dietrich, Herrmann, et al., 1975; Fernández-Morán, 1971).
3. Improvement and application of the cryoultramicrotome with diamond knife designed to operate at liquid-helium temperatures (Fernández-Morán, 1973; Zeitler, 1982) (Fig. 35) and used primarily to prepare ultrathin sections of rapidly frozen biological specimens for direct examination by cryomicroscopy, but including also thicker sections for histochemistry (Sitte, 1982).
4. Application of improved diamond knives (Fig. 36), which are prepared by new precision microirradiation techniques resulting in glazing and annealing for enhanced performance in cryomicrotomy.

One could add many other interesting approaches, but would have to admit that all of these suggestions refer to improvements in techniques and

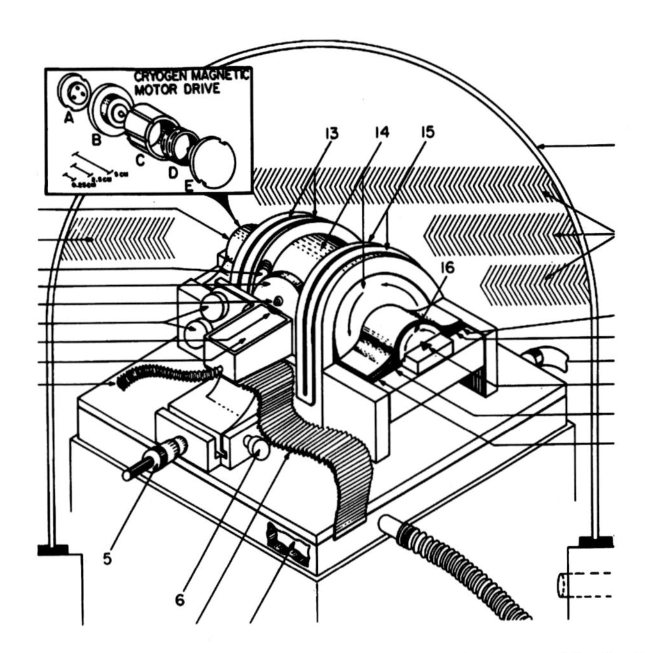

Figure 35 Schematic drawing of cryoultramicrotome with diamond knife designed to operate mainly at liquid-helium temperatures (1.8–4.2 K) within a specially shielded cryostat using superconducting motor drive. Ultrathin serial sections of rapidly frozen biological specimens can be prepared for direct examination by cryo-electron microscopy and cryo-electron diffraction. [H. Fernández-Morán, 1968–1982 (Fernández-Morán, 1968, 1973, 1980, 1982).]

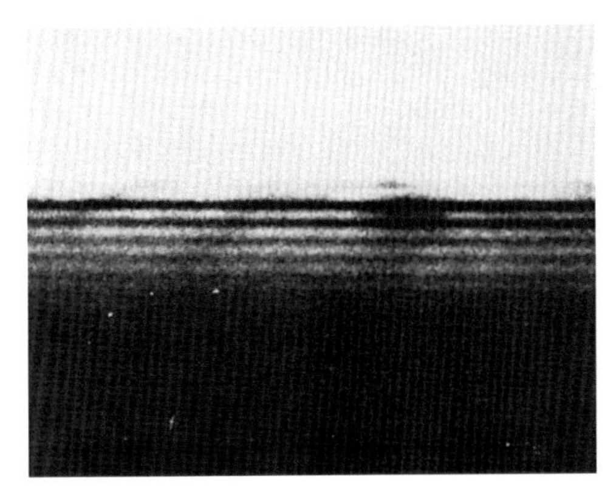

Figure 36 Transmission electron micrograph of improved diamond knife showing very uniform and extremely sharp cutting edge, and a characteristic periodic structure of the crystalline diamond facet. These ultrasharp and resilient diamond knives are produced by new precision microirradiation techniques, especially developed for cryomicrotomy of frozen biological specimens. [H. Fernández-Morán, 1980–1982 (Fernández-Morán, 1982).]

methodology which would remain sterile contributions if they are not applied, in correlated studies, to the solution of important problems in the life sciences.

Yet, by far the most important lesson to be derived from past experience is that electron microscopy, however interesting and novel its revelations, is only a tool, one which vastly extends man's visualization but remains restricted to morphology until it is linked to function and correlated with living systems. Electron microscopy has invariably proved most fruitful when it has been conducted in an environment of scholarly endeavor, bringing together the skills and experience of experts in physiology, biochemistry, polarized-light microscopy, X-ray diffraction, and many other disciplines, all devoted in a stimulating intellectual environment to the investigation of a specific problem.

Therefore, there is now an urgent need to establish international centers where electron microscopy in its broadest sense and scientists from different countries, each an expert in an important discipline, can work together and "think" together, sustaining an intense but spontaneous intellectual interaction continuously, and over a prolonged period of time (Wilson, 1980). In such an environment electron microscopy, correlated with biochemical studies, with X-ray diffraction, and with other disciplines, can contribute most effectively to molecular biology and molecular genetics, to biotechnology and the present revolution in applied biology. It will also serve a vital training function by attracting imaginative and creative young scientists, engineers, and technologists. Evoking fond memories of a unique anniversary I can still state "Thus, in a deeper sense we realize that electron microscopy represents the ultimate direct extension of eye and hand. Man may hereby be granted strange new freedoms and capabilities of 'seeing and doing' in those pivotal microcosmic domains of life which are still beyond mind's grasp" (Fernández-Morán, 1967).

Acknowledgments

The work carried out in Sweden (1946–1954) was mainly supported by the Luis Fernández-Morán Fund. Economic support in the latter part was also granted by the Nobel Fund through the Karolinska Institutet, and from research grants by the Wallenberg Foundation, the Therese and Johan Anderson Foundation, and the Ministry of Education of the Venezuelan Government. The research work in cryo-electron microscopy has been supported by NASA Grants NSG-441-63, NGL 14-001-012, NGR 14-001-166, and NSG 7303; by NIH Grant B-2460, NB-04267; and by the L. Block Fund of the University of Chicago, the Pritzker Fund, and the Spastic Paralysis Research Foundation. During the last few years special research grants were kindly made available by the Luis Fernández-Morán and the C. Jiménez-Macías Funds of Venezuela.

The work reported here encompasses salient aspects of a continuous 38-year effort in the development and application of electron microscopy, particularly cryo-electron microscopy, diamond-knife ultramicrotomy, and improved preparation techniques for the study of nerve ultrastructure and the submicroscopic organization of biological systems. This effort is international in scope, and it was carried out thanks to the assistance granted by distinguished colleagues and mentors. Many of the concepts and experimental approaches stem from the period of apprenticeship in electron microscopy which the author spent at the Nobel Institute for Physics in Stockholm, Sweden. I wish to reiterate my profound gratitude to the late Professor Manne Siegbahn for his generous hospitality, kind support, and encouragement at all times. It is a pleasure to thank Professor Torbjörn Caspersson, former director of the Institute for Cell Research, Karolinska Institutet, for his thoughtful help and advice.

The author is especially indebted to Professor Samuel C. Collins, former director of the Cryogenic Engineering Laboratory at the Massachusetts Institute of Technology, for his guidance and support by placing the facilities of the Laboratory at his disposal, and granting him the benefits of his valuable experience throughout all stages of the superfluid-helium experiments over a period of two decades.

The author has been privileged and grateful to work with Professor Francis O. Schmitt of the Massachusetts Institute of Technology and Professor William H. Sweet of the Massachusetts General Hospital and the Harvard Medical School. The resourceful assistance of Dr. Frederick B. Merk during that time is sincerely appreciated.

The kind help of Professor H. Stanley Bennett, of the late Professor William Bloom, and of Professor Raymond Zirkle was essential during the initial stages of the cryo-electron microscopy work at the University of Chicago. I also wish to thank Dr. Charles Laverick of Argonne National Laboratory; Milton Streeter and R. Osburn of Cryogenic Technology Inc.; and F. Lins, A. Peuron, D. Kasun, and Dr. S. Autler of Westinghouse Cryogenics Division. The author wishes to express his deep appreciation to Dr. Orr Reynolds, Director of the Biosciences Program, and to Dr. George J. Jacobs, Chief, Physical Biology of NASA, for their unfailing support and encouragement during the entire development of the cryo-electron microscopes and superfluid-helium system. The author is also indebted to Drs. F. R. Gamble and T. Geballe of Stanford University and F. J. DiSalvo and J. A. Wilson of Bell Laboratories for discussions and suggestions relating to intercalated compounds in the course of cooperative research programs. The stimulating discussions with Professor Gaston Dupouy of the Laboratoire d'Optique Electronique, Toulouse, France (Dupouy & Perrier, 1964); with Dr. Isolde Dietrich of the Research Laboratories of Siemens AG in München, Federal Republic of Germany; with Professor Elmar Zeitler, director of the Electron Microscope Department of the Fritz-Haber Institute, Max-Planck-Gesellschaft in Berlin; and with Dr. Gunter F. Bahr, Department of Cellular Pathology of the Armed Forces Institute of Pathology in Washington, D.C. are all gratefully acknowledged. Throughout the past decades the generous encouragement and valuable suggestions of Dr. W. O. Milligan, former Director of Research of the Robert A. Welch Foundation in Houston, Texas, were deeply appreciated. The major part of the cryo-electron microscopy development and research has been carried out in a systematic team effort by the members of an interdisciplinary group, particularly by Helmut Krebs, Charles Hough, the late G. Arcuri, G. Bowie, and C. Weber. All of the special equipment was made and assembled by expert co-workers of the Physical Sciences Development Workshop.

We are much indebted to D. Meddoff, V. Manrique, A. Hibino, M. Ohtsuki, and A. Gustafson for their skillful technical assistance in specimen preparation and electron microscopy.

The author wishes to thank all the colleagues and co-workers who contributed decisively to the foundation and research at The Venezuelan Institute for Neurology and Brain Research, IVNIC, near Caracas, Venezuela, during the years 1954–1958. Foremost is his profound debt of gratitude to the late Dr. Pedro A. Gutíerrez Alfaro, former Minister of Public Health, who helped found and support IVNIC throughout its existence. Among the many who participated in this pioneering project, Don Pedro Arvelo Crespo, Don Andrés Parra Pulido, Dr. Arturo Kan, Engineer Josef Weibel, and Professor Pierre Denis deserve special recognition, and we are all greatly indebted to them. It is also a pleasure to thank Professor Geoffrey H. Bourne for his stimulating encouragement during all phases of this project.

The generous support of Dr. Luis V. Amador, Mr. George Scharringhausen, Jr., and their colleagues of the Spastic Paralysis Research Foundation during the past decade is gratefully acknowledged.

I wish to express my deep gratitude to the late Professor Hans Selye and to the late Professor David E. Green for sharing the wisdom of their experience, and for the bequest of their inspiring sense of commitment.

The author thankfully acknowledges the encouraging support of Dr. Cipriano Jiménez-Macias, of Dr. Alberto Olivares, of the late Dr. Francisco J. Duarte, of Dr. Marcel Granier D., and of the late Dr. Miguel Parra Leon, President Emeritus of the Venezuelan Academy of Sciences in Caracas, Venezuela.

8. Afterword by Peter Hawkes

I can add little to Fernández-Morán's own account of his tumultuous life and times as well as his scientific career. He was born on 18 February 1924 in Maracaibo and died on 17 March 1999 in Stockholm.

References[1]

Bahr, G. F., & Zeitler, E. H. (Eds.). (1965). *Quantitative electron microscopy*. Baltimore, Maryland: Williams & Wilkins.

Boersch, H., Bostanjoglo, O., & Lischke, B. (1966). *Optik*, *24*, 460.

Bonjour, P., & Septier, A. (1968). *Rome, 1968* Vol. 1. (p. 189).

Bourne, G. H. (1955). *Nature (London)*, *176*, 1049.

Bourne, G. H. (1957). *Nature (London)*, *179*, 1296.

[1] The following background references are not cited in the text: Box (1977), Cosslett (1951), Frey-Wyssling (1948), Hodge and Schmitt (1960), Lonsdale (1942), Palade (1952), Prebus and Hillier (1939) and Sjöstrand (1949).

Box, H. C. (1977). *Radiation effects: ESR and ENDOR analysis*. New York: Academic Press.
Burton, E. F., & Kohl, W. H. (1942). *The electron microscope*. New York: Reinhold.
Caspersson, T. (1937). *Protoplasma, 27*, 463.
Caspersson, T. (1950). *Cell growth and cell function*. New York: Norton.
Collins, S. C. (1960-1982). private communications.
Cosslett, V. E. (1951). *Practical electron microscopy*. London: Butterworth.
Cosslett, V. E. (1978). *Journal of Microscopy (Oxford), 113*, 113.
Dietrich, I. (1976). *Superconducting electron optic devices*. New York: Plenum.
Dietrich, I. (1978). *Toronto, 1978* Vol. 3. (p. 173).
Dietrich, I., Weyl, R., & Zerbst, H. (1967). *Cryogenics*, 7, 178.
Dietrich, I., Pfisterer, H., & Weyl, R. (1969). *Zeitschrift für Angewandte Physik, 28*, 35.
Dietrich, I., Lefranc, G., Weyl, R., & Zerbst, H. (1973). *Optik, 38*, 449.
Dietrich, I., Fox, F., Knapek, E., Lefranc, G., Weyl, R., & Zerbst, H. (1975). *Canberra, 1974* Vol. 1. (p. 146).
Dietrich, I., Herrmann, K. H., & Passow, C. (1975). *Optik, 42*, 439.
Dietrich, I., Fox, F., Knapek, E., Lefranc, G., Nachtrieb, K., Weyl, R., & Zerbst, H. (1977). *Ultramicroscopy, 2*, 241.
Dietrich, I., Fox, F., Heide, H. G., Knapek, E., & Weyl, R. (1978). *Ultramicroscopy, 3*, 185.
Dietrich, I., Formanek, F., Fox, F., Knapek, E., & Weyl, R. (1979). *Nature (London), 277*, 380.
Dietrich, I., Formanek, H., von Gentzkow, W., & Knapek, E. (1980). *The Hague, 1980* Vol. 1. (p. 116).
Dietrich, I., Formanek, H., von Gentzkow, W., & Knapek, I. (1982). *Ultramicroscopy, 9*, 75.
Dubochet, J. (1975). *Journal of Ultrastructure Research, 52*, 276.
Dubochet, J., Booy, F. P., Freeman, R., Jones, A. V., & Walter, C. A. (1981). *Annual Review of Biophysics and Bioengineering, 10*, 133.
Dubochet, J., Knapek, E., & Dietrich, I. (1981). *Ultramicroscopy, 6*, 11.
Dupouy, G., & Perrier, F. (1964). *Journal de Microscopie (Paris), 3*, 233.
Durham, D. G. (1966). *Delaware Medical Journal, 38*, 202.
Ehrenberg, C. G. (1849). *Monatsberichte Preussischen Akademie der Wissenschaften, 64*, 73.
Elvers, I. (1943). *Acta Horti Bergiani, 13*(5). *Bot. Tidskr., 37*, 331 (1943).
Fernández-Morán, H. (1944). *Zeitschrift für Zellforschung und Mikroskopische Anatomie, 33*, 225.
Fernández-Morán, H. (1948). *Arkiv för Zoologi, 40A*(6).
Fernández-Morán, H. (1950a). *Experimental Cell Research, 1*, 143.
Fernández-Morán, H. (1950b). *Experimental Cell Research, 1*, 309.
Fernández-Morán, H. (1952a). *Experimental Cell Research, 3*, 282.
Fernández-Morán, H. (1952b). *Arkiv för Fysik, 4*, 471.
Fernández-Morán, H. (1953a). *Experimental Cell Research, 5*, 255.
Fernández-Morán, H. (1953b). *Boletín de la Academia de Ciencias Físicas, Matemáticas y Naturales (Caracas), 51*, 1–162.
Fernández-Morán, H. (1954). *Progress in Biophysics, 4*, 112.
Fernández-Morán, H. (1956a). *Industrial Diamond Review, 16*, 128.
Fernández-Morán, H. (1956b). *Nature (London), 177*, 742.
Fernández-Morán, H. (1956c). *The Journal of Biophysical and Biochemical Cytology, 4*(Suppl. 2), 29.
Fernández-Morán, H. (1957). In *Proc. int. neurochem. symp., 2nd, 1956*.
Fernández-Morán, H. (1958). *Experimental Cell Research*, (Suppl. 5), 586.
Fernández-Morán, H. (1959a). *Reviews of Modern Physics, 31*, 319.
Fernández-Morán, H. (1959b). *Journal of Applied Physics, 30*, 2038.
Fernández-Morán, H. (1960a). In F. O. Schmitt (Ed.), *Fast fundamental transfer processes in aqueous biomolecular systems* (p. 26). Cambridge, Massachusetts: M.I.T. Press. See also p. 33.

Fernández-Morán, H. (1960b). *Journal of Applied Physics, 31*, 1840.
Fernández-Morán, H. (1960c). *Annals of the New York Academy of Sciences, 85*, 689.
Fernández-Morán, H. (1961a). In M. V. Edds (Ed.), *Macromolecular complexes* (p. 113). New York: Ronald Press.
Fernández-Morán, H. (1961b). In G. K. Smelser (Ed.), *The structure of the eye* (p. 521). New York: Academic Press.
Fernández-Morán, H. (1962a). In S. Korey (Series Ed.), *Association for Research in Nervous and Mental Disease (A.R.N.M.D.) ser.: Vol. 40. Ultrastructure and metabolism of the nervous system* (p. 338). Baltimore, Maryland: Williams & Wilkins.
Fernández-Morán, H. (1962b). In R. J. C. Harris (Ed.), *The interpretation of ultrastructure: Vol. 1* (p. 411). New York: Academic Press.
Fernández-Morán, H. (1964). *Journal of the Royal Microscopical Society, 83*, 183.
Fernández-Morán, H. (1965a). *Proceedings of the National Academy of Sciences of the United States of America, 53*, 445.
Fernández-Morán, H. (1965b). *Annals of the New York Academy of Sciences, 125*, 739.
Fernández-Morán, H. (1966a). *Proceedings of the National Academy of Sciences of the United States of America, 56*, 801.
Fernández-Morán, H. (1966b). *Kyoto, 1966* Vol. 1. (p. 147).
Fernández-Morán, H. (1967). *Proceedings – Annual Meeting, Electron Microscopy Society of America, 25*, 10.
Fernández-Morán, H. (1968). In *Science year book* (p. 216). Chicago: Field Enterprises.
Fernández-Morán, H. (1970). *Experimental Cell Research, 62*, 90.
Fernández-Morán, H. (1971). *Grenoble, 1970* Vol. 2. (p. 91).
Fernández-Morán, H. (1972). *Annals of the New York Academy of Sciences, 195*, 376.
Fernández-Morán, H. (1973). *Applications of Cryogenic Technology, 5*, 153.
Fernández-Morán, H. (1980). *Proceedings of the Robert A. Welch Foundation Conferences on Chemical Research, 23*, 315.
Fernández-Morán, H. (1982). *Hamburg, 1982* Vol. 1. (p. 751).
Fernández-Morán, H., & Brown, R. (Eds.). (1958). *The submicroscopic organization and function of nerve cells, Exp. Cell Res., Suppl. 5*. New York: Academic Press.
Fernández-Morán, H., & Dahl, A. O. (1952). *Science, 116*, 465.
Fernández-Morán, H., & Engström, A. (1956a). *Nature (London), 173*, 494.
Fernández-Morán, H., & Engström, A. (1956b). *Biochimica et Biophysica Acta, 23*, 260.
Fernández-Morán, H., & Finean, J. P. (1957). *The Journal of Biophysical and Biochemical Cytology, 3*, 725.
Fernández-Morán, H., & Luft, R. (1949). *Acta Endocrinologica (Copenhagen), 2*, 199.
Fernández-Morán, H., & Schramm, G. (1958). *Zeitschrift für Naturforschung, 13B*, 68.
Fernández-Morán, H., Zinn, W. H., Cerutti, B. C., & Lang, C. (1957). In *First interamerican symposium on nuclear energy* (p. 1). Brookhaven National Laboratory.
Fernández-Morán, H., Oda, T., Blair, P. V., & Green, D. E. (1964a). *The Journal of Cell Biology, 22*, 63.
Fernández-Morán, H., Reed, L. J., Koike, M., & Willms, R. R. (1964b). *Science, 145*, 930.
Formanek, H., & Knapek, E. (1979). *Ultramicroscopy, 4*, 77.
Fox, F., Knapek, E., & Weyl, R. (1978). *Toronto, 1978* Vol. 2. (p. 342).
Frey-Wyssling, A. (1948). *Submicroscopic morphology of protoplasm and its derivatives*. New York.
Gabor, D. (1975). *Canberra, 1974* Vol. 1. (p. 6).
Gargiulo, E. P. (1969). In *Proc. int. indust. diamond conf. 1969* (p. 145).
Hall, C. E. (1953). *Introduction to electron microscopy*. New York: McGraw-Hill. (2nd ed.) (1966).
Hall, C. E. (1955). *The Journal of Biophysical and Biochemical Cytology, 1*, 1.
Hall, C. E. (1956). *The Journal of Biophysical and Biochemical Cytology, 2*, 625.
Hall, C. E., Jakus, M., & Schmitt, F. O. (1945). *Journal of Applied Physics, 16*, 459.

Hast, N. (1947). *Nature (London), 159*, 370.
Heide, H. G. (1982). *Ultramicroscopy, 10*, 125.
Hillier, J., & Ramberg, E. G. (1947). *Journal of Applied Physics, 18*, 48.
Hodge, A. J., & Schmitt, F. O. (1960). *Proceedings of the National Academy of Sciences of the United States of America, 46*, 186.
Hulm, J. K., Kasun, D. J., & Mullan, E. (1971). *Physics Today, 24*, 48.
Hüttermann, J. (1982). *Ultramicroscopy, 10*, 7.
Johnson, F. E., Curcio, M. E., Smith, D. C., & Lockett, L. E. (1981). *Laser Focus Magazine*, 36.
Kleinschmidt, A. K., & Zahn, R. K. (1959). *Zeitschrift für Naturforschung, 14B*, 770.
Knapek, E. (1982). *Ultramicroscopy, 10*, 71.
Knapek, E., & Dubochet, J. (1980). *Journal of Molecular Biology, 141*, 147.
Knapek, E., Formanek, H., Lefranc, G., & Dietrich, I. (1981). *Proceedings – Annual Meeting, Electron Microscopy Society of America, 39*, 22.
Knoll, M., & Ruska, E. (1932). *Zeitschrift für Physik, 78*, 318.
Laberrigue, A., & Levinson, P. (1964). *Comptes Rendus Hebdomadaires des Séances de L'Académie des Sciences, 259*, 530.
Latta, H., & Hartmann, J. F. (1950). *Proceedings - Society of Experimental Biology and Medicine, 74*, 436.
Laverick, C. (1966). Argonne Natl. Lab. [Rep.] ANL-72-75, 38.
Leduc, E. H., & Bernhard, W. (1962). In R. J. C. Harris (Ed.), *The interpretation of ultrastructure* (p. 21). New York: Academic Press.
Lefranc, G., Knapek, E., & Dietrich, I. (1982). *Ultramicroscopy, 10*, 111.
Lewis, T. G. (1962). *Tool Manufacturing Engineering, 49*, 65.
Lonsdale, K. (1942). *Proceedings of the Physical Society of London, 54*, 314.
Luria, S. E., & Anderson, T. F. (1942). *Proceedings of the National Academy of Sciences of the United States of America, 28*, 127.
Marton, L. (1934). *Physical Review, 527*.
Marton, L. (1968). *Early history of the electron microscope*. San Francisco, California: San Francisco Press.
Müller, E. W. (1937). *Zeitschrift für Physik, 106*, 541.
Müller, E. W. (1960). *Advances in Electronics and Electron Physics, 13*, 83.
Newman, S. B., Borysko, E., & Swerdlow, M. J. (1949). *Science, 110*, 66.
Olivecrona, H. (1946). *Svenska Laekartidn, 19*. H. Fernández-Morán, *Apuntes Biograficos sobre el Profesor Herbert Olivecrona*. Stockholm: Kugelbergs Boktryckeri (1947).
Palade, G. E. (1952). *The Journal of Experimental Medicine, 95*, 285.
Phillips, V. A. (1967). *Praktische Metallographie, 4*, 637.
Porter, K. R., Claude, A., & Fullam, E. F. (1945). *The Journal of Experimental Medicine, 81*, 233.
Prebus, A., & Hillier, J. (1939). *The Canadian Journal of Research. Section A, 17*, 49.
Runnström, V. (1928). *Protoplasma, 4*, 388.
Ruska, E. (1934). *Zeitschrift für Physik, 89*, 90.
Ruska, E. (1966). *Advances in Optical and Electron Microscopy, 1*, 115.
Ruska, E. (1980). *The early development of electron lenses and electron microscopy* (Transl. by T. Mulvey). Stuttgart: Hirzel Verlag.
Ruska, H. (1939). *Archiv für Experimentelle Zellforschung, 22*, 673.
Sayre, A. (1975). *Rosalind Franklin and DNA*. New York: Norton (p. 187).
Scherzer, O. (1979). In W. Hoppe, & R. Mason (Eds.), *Advances in structure research by diffraction methods: Vol. 7* (p. 101). Braunschweig: Vieweg.
Schmidt, W. J. (1937). *Die Doppelbrechung von Karyoplasma, Zytoplasma und Metaplasma*. Berlin.
Schmitt, F. O. (1950). *Biochimica et Biophysica Acta, 4*, 68.

Schmitt, F. O., Bear, R. S., & Clark, G. L. (1935). *Radiology, 25*, 131.
Siegbahn, K., & Ingelman, B. (1944). *Arkiv för Kemi, Mineral. Geol., 18B*(1).
Siegbahn, M. (1939). *Kungl. Svenska Vetenskapsakademiens Årsbok, 37*, 147.
Sitte, H. (1982). *Hamburg, 1982* Vol. 1. (p. 9).
Sjöstrand, F. S. (1949). *Journal of Cellular and Comparative Physiology, 33*, 383.
Taylor, K. A., & Glaeser, R. M. (1976). *Journal of Ultrastructure Research, 55*, 448.
Trinquier, J., & Balladore, J. L. (1968). *Rome, 1968* Vol. 1. (p. 191).
Valentine, R. C., & Horne, R. W. (1962). In R. J. C. Harris (Ed.), *The interpretation of ultrastructure* (p. 263). New York: Academic Press.
von Ardenne, M. (1940). *Elektronen Übermikroskopie*. Berlin and New York: Springer-Verlag.
von Borries, B., & Ruska, E. (1941). *Ergebnisse der Exakten Naturwissenschaften, 19*, 237.
von Helmholtz, H. (1873). *Monatsberichte der Deutschen Akademie der Wissenschaften zu Berlin*, 625.
Walter, F. (1958). *Zeitschrift für Instrumentenkunde, 66*, 246.
Weyl, R., Dietrich, I., & Zerbst, H. (1972). *Optik, 35*, 280.
Williams, R. C., & Wyckoff, R. W. G. (1944). *Journal of Applied Physics, 15*, 712.
Wilson, R. R. (1980). *Scientific American, 242*, 42.
Wischnitzer, S. (1970). *Introduction to electron microscopy*. Oxford: Pergamon.
Woodcock, C. L. F., & Fernández-Morán, H. (1968). *Journal of Molecular Biology, 31*, 627.
Zeitler, E. (1982). In E. Zeitler (Ed.), *Cryomicroscopy and radiation damage* (p. 1). Amsterdam: North-Holland Publ. See also p. 155.
Zworykin, V. K., Morton, G. A., Ramberg, E. G., Hillier, J., & Vance, A. W. (1945). *Electron optics and the electron microscope*. New York.

CHAPTER TEN

Dennis Gabor☆,☆☆

5 June 1900–9 February 1979, Elected F.R.S. 1956

T.E. Allibone FRS[a,b]

[a]Formerly Metropolitan–Vickers, Manchester, United Kingdom

[b]Formerly AEI, Aldermaston, United Kingdom

Contents

DENNIS GABOR, originally in Hungarian Gábor Dénes, scientist, engineer, inventor, humanist, Nobel Prize winner, was born in Hungary on 5 June 1900, came to England in 1934, married Marjorie Louise Butler in 1936 and, after a life of brilliant scientific and philosophical achievement coupled with great happiness, died in London on 9 February 1979. To use his own words he was "one of the lucky physicists who have been able to see one of their ideas grow into a sizeable chapter of physics", a chapter he continued to enlarge up to the onset of illness in 1974, and in addition he applied his penetrating intellect to some of the problems of man's survival, problems created by the advance of technology: in a series of books and discourses he—to use again his own words—"invented a future,—and a good one,—one that preserves the values of civilization and yet is in harmony with man's nature—based on hope and the love of life". Few scientists are masters of the 'two cultures', but this, Dennis Gabor certainly was.

☆ Reprinted from the Biographical Memoirs of Fellows of the Royal Society, **26** (1980) 107–147, by kind permission of the Royal Society.

☆☆ Corresponding author: Peter Hawkes. e-mail address: hawkes@cemes.fr.

Advances in Imaging and Electron Physics, Volume 220
ISSN 1076-5670
https://doi.org/10.1016/bs.aiep.2021.08.010

1. Early life

He was the oldest of a family of three boys; he was followed by George who died in 1935 from pancytopaenia, and by André born in 1903 who also came to England and is in the Economics Department of the University of Nottingham; André too has earned international recognition in his own field, and collaborated with Dennis in two papers to which reference will be made later. He has given me very great help in correcting some details in the excellent personal account left by Dennis with the Royal Society and by adding many other interesting facets of his brother's life. In what follows, direct quotations from Dennis's own writings are always given between double inverted commas.

Their father, Bertalan (or Bartholomew) was born at Eger in Hungary in 1867 with the name of Günsberg and changed his name to Gábor in 1899. Dennis knew his paternal grandfather who was born in 1832 of parents who had settled in Hungary at the end of the 18th century having come from Russia. The family were tall, fair, blue-eyed people, presumably descendants of one of the Russian tribes, the Cerims or the Kuzri who took up the Jewish faith centuries earlier.

Their mother, Adrienne, née Kálmán, born in 1879 was originally an actress but gave up the stage at 20 when she married. She lived through the terrible siege of Budapest in World War II and shortly afterwards came in 1946 to England living first with Dennis and later with André; she died in 1967. Her father was a highly skilled watchmaker, the son of an excellent tailor, but Dennis knew very little of his mother's remote ancestry; they were probably descendants of Spanish Jews who settled in Hungary in the 18th century.

Latent talent in the Gábor family came out in engineering. Bertalan was a gifted and ambitious child who hoped to go to university and qualify as an engineer. One of Dennis's uncles was a mechanical engineer, one cousin was a civil engineer, one an architect, and brother George became an excellent inventive engineer. Unfortunately Bertalan's father's business failed when Bertalan was only 17 years old and he had to leave school and take a job as a clerk: he prospered well enough to marry at 31 and then rose to be the director of the largest industrial enterprise in Hungary, the 'MÁK', the Hungarian General Coalmines.

Probably because Bertalan's scholastic career had been broken he always considered himself a frustrated engineer and inventor (though neither Dennis nor his father could differentiate between those two for many years). He read Jules Verne to the boys and was an admirer of the great inventor Thomas Edison whose life story he read to Dennis at the age of 6. He took Dennis to the Deutsches Museum in Munich at the age of 11 and his interest in science was unabated throughout his life; he died in 1942. His son's early desire to be an engineer afforded him great happiness.

The Gábor home provided the perfect background for a family of three boys. Both parents had excellent taste, collected books, fine pictures and *objets d'art*; both parents spoke German fluently and the children had German, French, and English governesses in succession "so there was no need to listen to the poor language courses in school". In addition, in their early teens the boys had a highly gifted tutor who stimulated their intellects. "But perhaps the most powerful influence in my general education was that of two highly educated friends of my father, a doctor and a lawyer. In Hungary, as also elsewhere on the Continent, there was no need for an educated man to be quiet if he finds himself in a less educated company. My parents, our friends, and we children listened avidly to the conversations of these two learned men on history, literature, the future of society, etc. and whenever an author or a book was mentioned which I did not know, I ran to read it." Dennis was a voracious reader (indeed he was throughout the whole of his life) and had a prodigious memory. At about 12 his father offered him a prize if he could memorize Schiller's long (430 lines) epic poem 'Das Lied von der Glocke'; he won the prize and remembered the whole poem for many years. He amused himself by translating Hungarian poems into very good German, using the same meter as the original.

After 4 years in the elementary school Dennis went to the 'Reáliskola' Toldi Miklós Állami Föreáliskola or grammar school in which modern languages were taught and scope was given to mathematics and the sciences. After some months the form master complained that he had a neurotic child on his hands and suggested that Dennis be sent to a special institution catering for unmanageable children;—"The influence of my family was so strong, that in comparison, the teachers had little influence on my development. Besides I was almost always ahead of the curriculum as I was a voracious reader and had not overmuch respect for my teachers." Father Bertalan recognized that his son's fidgetiness was due to impatience with the low standard in the form and persuaded the master to keep him. The master had the pleasure of seeing Dennis gain top marks throughout his school life, and in due course master and pupil were elected to the Hungarian Academy of Sciences.

The physics master was an extremely conscientious man who constructed teaching aids not provided by the school. Dennis irritated him too, since Dennis's knowledge of physics even at that early age was superior to that of his teacher, but later in life they became good friends. His superior knowledge was due to the fact that his father allowed him to buy almost any book he desired; he found German textbooks of advanced mathematics and physics and "before I went to the University I knew all about the mathematics I was to learn there, and more electromagnetic theory than I ever learned at the Technische Hochschule in Berlin". By the age of 16 he had mastered the two subjects up to degree standard. "I remember how fascinated I was by Abbé's theory of the microscope and by Gabrial Lippmann's method of color photography which played such a great part in my work 30 years later". Although he never took a degree in mathematics he was an extremely competent mathematician as is obvious from even a casual study of his life's work and all his inventive ideas were based on full mathematical analyses of the details they embraced.

André provides a story about his brothers: 'In 1915 the Gábor family moved to a large flat, the spare room of which became a laboratory. We had a carpenter's bench and a woodlathe, and father allowed us to buy whatever we needed. Dennis and George started serious experiments in physics and built intricate apparatus partly for their school and partly for their own amusement and edification. The former included a clever device for the demonstration of the law of falling bodies, and an epidiascope',—already, then, Dennis was turning his interest to optics—'and the latter included a large induction coil which produced sparks 12 inches (30 cm) long: they also built a Tesla coil and these items evoked the admiration of many experts. Father obtained an old X-ray tube and we took X-ray pictures; they also had a glass-blowing equipment'.

Throughout school days Bertalan subscribed to the popular scientific journals; the family were together at lunch and at dinner and, especially in later years, mealtimes were like the meetings of a discussion society. His interest in science went hand in hand with his appreciation of the arts. Again André contributes: 'In our home we had a fine selection of histories of the visual arts, with colored reproductions of the greatest masterpieces and these we knew long before our visits to the galleries of Vienna, Dresden, Berlin, Paris and London brought us face to face with the originals. Dennis also liked music, and, having a marvellous memory and an unerring instinct for style, he could at once identify the composer of a piece of music and also the artist of practically any item in a gallery.'

Physically Dennis was a delicate child but later he developed a good physique becoming one of the best jumpers and runners in his class, and in later life played a very good game of tennis. In 1917 he passed the medical examination preparatory to being called up for military service on 15 March 1918. (The very moment that Hindenberg launched his offensive on the Western Front.) He was allowed to take the Matura examination, the qualification necessary for entry into University, in February 1918 and then joined the O.T.C., training in artillery and horsemanship in Lugos (now Lugoj in Romania). Here he met Tiber Reiter, a nephew of his father's friend, and they became friends, and later collaborators, in Berlin—of this, more anon. In the early autumn of 1918 he was transferred for the final part of his military training to Northern Italy, then occupied by the Hungarian forces. After his return to Budapest he took lessons in Italian, memorizing 200 words a day and thus, all his life, he had command of four foreign languages.

At the end of the war the whole Gábor family adopted the Lutheran faith, the Church of which is in communion with the Church of England and Dennis remained in that faith all his life.

2. College life

With the collapse of the Central Powers he returned home on the last day of October 1918 and at once registered into the Magyar Királyi József Müegyetem (the Technical University in Budapest) as a student of a 4-year course in mechanical engineering (Dennis had an engineering drawing-board in his office all his life). André explains his brother's choice of subject as follows: 'As regards the attitude in Hungary to pure science and engineering, the people who studied pure science at the University had only two outlets, one of which was to go into schoolteaching, the other to emigrate, and it was engineering rather than pure science that attracted people of talent and ambition'. Von Karman noted 'The scientist describes what is; the engineer creates what never was.' At the end of his first year, Dennis won the Prize in Mechanics (Theory of the spherical pendulum) but during his third year he received an order to register for military service; as he disagreed with the policies of the reactionary government he left to continue his studies in Berlin and registered in the electrical engineering department of the Technische Hochschule, Charlottenburg. He noted: "Though still at its height at the time, I could not call the T.H. Berlin an ideal institution. There were far too many students and there was hardly any personal contact between students and teachers. It was a sort of slot machine into which one had to throw' no end of machine designs, essays and papers and out came a diploma (1924) in the end. But it certainly made one get used to hard work! My real education, and the memory which I cherish most at that time, was the Physical Colloquium at the University, every Tuesday, and the unforgettable Seminar on Statistical Mechanics which I had under Einstein's guidance in 1921–22." Two options were open, heavy currents and electronics but the latter was not available to foreign students.

The great German inflation of this period forced many impoverished people to let rooms, preferably to foreign students; thanks to his father's generosity Dennis was never short of money but he considered it a moral duty to share the deprivations and live modestly.

3. Research in Germany and Hungary

"When at last I had my Diploma I asked my father for permission to live on his money for another 2–3 years to get my Dr.-Ing., and—he thanked me for it! (This veneration of learning in the Jewish middle-class of Budapest is probably the main reason why there are now so many Hungarians in science.) My professor, Orlich, was a gentleman of whom I have the kindest memory but he had no scientific imagination and could give no worthwhile problem. Fortunately I found one myself. Electrical transients were a serious problem in high voltage networks; looking for methods of recording them I came upon the Braun tube and I was one of the first to turn it into a high-speed oscillograph with internal photography. (Professor Rogowski and his pupils in Aachen worked independently of me and he recorded the first rapid transient actually a few weeks before me in 1925.) I was the first to make transients photograph themselves with what is now known as a bistable system, which I called a Kipprelais." This work is fully described in his thesis, (Gabor, 1927b),[1] which was preceded by three papers in the *Archiv für Electrotechnik*, (Gabor, 1925, 1926a, 1927a). The distinguishing feature of this work in Berlin and in Aachen was that the voltage on the oscillograph was far higher than on the Braun tube, that the oscillograph was continuously pumped, and that the photographic plate was actually introduced *into* the oscillograph to record the transients, so the sensitivity was increased by orders of magnitude. But Gabor was the first to shroud with iron the short solenoids used to focus the electron beams and thus concentrate their lens-effect. His kipprelais held the beam to one side until the arrival of the transient. Oscillographs all over the world copied the Aachen-Berlin models.

Gabor once asked Professor Orlich if it was possible to learn how to invent, to which Orlich replied 'don't waste your time trying to invent, because in these days, all inventions are useless; either they will not work, or someone else has already made the invention'; Dennis in fact took out more than 100 patents and sold some at good profit! During his third and last year at college his research concerning the recording of transients on a 100 kV transmission line was supported by the German High-Voltage Research Association, (Gabor, 1926b, 1927b).

Gabor again met Reiter, now a medical student, who had come across the publication by A. Gurwitsch who claimed[2] that growing onion

[1] A list of all his publications that I have been able to trace is appended.
[2] Das Problem der Zellteilung physiologisch betrachtet, 1926, Berlin (in German).

roots emitted a radiation which enhanced mitosis in roots very close to them,—mitogenetic rays. In spare time Reiter and Dennis worked on this subject and found there was a radiation of $\lambda = 3400$ Å. They established that mitosis was indeed 'induced' in an onion shoot by a close-spaced growing tip of another onion root (a mature root did not emit a radiation), that the radiation, although very weak, could affect a photographic emulsion, and that sunlight of this same wavelength could cause mutations, i.e. had an 'induction effect', though an excess of the radiation had a deleterious effect. André Gabor joined them in conducting these experiments which were done in a room just above the morgue of the Charité Hospital where Reiter was a student, and André reports that 'the room was full of human specimens and smelled to heaven'! By 1927, Reiter had taken his medical degree and Dennis his senior research degree: Dennis described some of the work to the Director of one of the Siemens group of companies and he in turn introduced Dennis to the head of the Physics Laboratory of Siemens and Halske Co. at Siemensstadt. Both graduates were offered employment and continued some further experiments on the phenomenon, publishing a paper, (Gabor, 1928a), and a book, (Gabor & Reiter, 1928), in 1928.

Over 100 experiments were described and from a moderately careful study of the book one would conclude that the work was authentic. In his autobiographical Notes Dennis wrote: "The results ... seemed to support ... the hypothesis of some radiating agency; on the other hand all experiments for proving the radiation by physical means have failed. To this day (1956) nobody knows what these experiments really mean." This was not quite his attitude at the time. Clearly a mystery surrounds this work and although I asked Dennis about it on several occasions I never penetrated the mystery.

There was some indication that a health lamp strong in the part of the spectrum needed for mitosis might be more effective than a mercury lamp and Dennis found that the addition of cadmium enriched the mercury vapor spectrum in the u.v. but the life of these lamps was short and he spent time trying to find a way to eliminate this shortcoming. In so doing he came up against the Langmuir paradox—of which more anon—and published papers, (Gabor, 1930, 1932, 1933a). It was in the course of this work that he invented a thin molybdenum strip seal to carry high currents into the quartz vessel, a seal in general use even today. Concerning paper (Gabor, 1930), Professor Franke formerly of Siemens writes 'Gabor was not a member of our section of the laboratory but he stepped in at least once a day.... We were developing loudspeakers and microphones ... all problems of our work were discussed with him especially the theoretical and mathematical ones; he did all the design of magnetic fields and coils and transformers for us. When I next met him 25 years later, he immediately recognized me and gave proof of his marvellous memory by telling me the title of my doctor's paper.' In later years the Gabors, visiting Berlin, always went to see Dr Franke.

A few weeks after Hitler came to power Dennis's contract of service ran out and Siemens did not renew it, so he returned to Hungary and approached Mr Aschner the Chairman of the Tungsram Electron Tube Research Institute in Budapest, asking if he could work on his new invention, the plasma lamp, without a salary. This lamp contained a perforated disk through which strands of mercury discharge plasma emerged. He developed a close link with Dr Andov Budincsevits, who has written to me 'Gabor at that time was not only an eminent theoretician but also an experimental physicist, determined to develop the practical application of his concept.... We produced a plasma lamp conforming to Gabor's original idea and it had a life span of some hundreds of hours ... but the yellow light emitted would probably restrict its applicability.' Gabor demonstrated the lamp to Mr Aschner offering the patent to the Company but the Chairman did not like to create a precedent by paying for an invention and, instead, offered Gabor a position on the staff. Gabor refused this offer and decided to try to sell his patent in England. He might have been influenced by the fact that, owing to the deteriorating political situation other members of the staff (including Dr E. Orowan—later F.R.S.) 'also intended to emigrate to England'. Paper (Gabor, 1933b), written in Budapest, is the first exhaustive analysis of the Langmuir plasma. Dr Budincsevits adds that 'our friendship was renewed in 1962 when Dennis first returned to Hungary: he became a faithful citizen of his new country but he never denied the ties which connected him with Hungary: he liked to sing Hungarian songs, he never cut his roots'.

In the summer of 1928 I went to Germany to discuss oscillographs with Professor Rogowski and Dr Gabor, and benefited greatly from these meetings, I was the first British scientist he had met. Thus it came about that Dennis wrote to me (1934), in the High Voltage Laboratory of the Metropolitan-Vickers Co., Manchester, asking if M.-V. could offer him employment. The M.-V. Company was the heavy-current half of the parent A.E.I. Company, the B.T.H. being primarily the light-current half of the A.E.I. My research director suggested that I wrote to the B.T.H. research director, Mr H. Warren, recommending Gabor. This I did and Dennis was duly invited to Rugby where he spent some 15 years, 1934–48.

Some time in 1934 Dennis wrote to ask me whether his name should be spelt with one n or two. I replied that both forms were in use in England and as his Hungarian name had contained only one he could Anglicize it into Denis. This he used for a while but later added the second n.

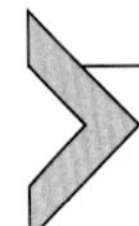

4. Research in the British Thomson-Houston Company, 1934–48

England was very sympathetic to the refugees from Nazism, and Dennis must have been given a friendly welcome in Rugby (indeed Mr Warren was a very kind man), for his brother André writes: 'After our experiences in Germany we were most agreeably surprised by the fact that the English, who had a reputation for being insular, cold and inaccessible, were exactly the opposite. Soon after Dennis came to this country he said in his letters that he had already made a number of friends, had been the guest of several families, and that he had only just realized how unfriendly the Germans were: after 12 years in their country Dennis had made fewer friends than in England after a few weeks.' And within two years he had found a wife. Marjorie was the eldest daughter of Mr J. K. Butler and his wife Louise, and was at that time working in the B.T.H. Co. She had great musical interests and the Amateur Dramatic Society (with Gilbert and Sullivan operas) made a good meeting ground. Dennis played no instrument but he sang well, and all his life—with his fine memory—he could entertain with songs from very many operas. They were married on 8 August 1936 (and late in life, Dennis wrote "and lived happily ever after". They did).

Gabor's appointment with B.T.H. began as a patent agreement to develop further the plasma lamp on which he had worked in Hungary. B.T.H. were the part of A.E.I. dealing with the lamp and lighting business so a fluorescent lamp needing no choke would be a great asset, but two further years of work failed to overcome the low life of the lamp, no material then available could withstand the ion bombardment of the cathode, and work finally ceased, and he was then given a staff appointment.

Dennis dismisses the following years spent in Rugby, 1937–48, as very sterile, comparing badly with his postgraduate work in Berlin and with his research after leaving industry but I and his few colleagues now alive would not accept this verdict. It is true that, had he not been an 'enemy alien' he would have been more fully engaged on war work and might well have contributed some brilliant ideas to radar research, but his scientific output was considerable, was excellent, and has stood the test of time, and at the end of this phase he invented holography and thereby won a Nobel Prize—hardly 'sterile.'

Unfortunately we do not have a record of all the reports he may have written, but a few of his colleagues (Emeritus Professor G. Carter of Leeds, Emeritus Professor C. J. Milner of Sydney, N.S.W., Dr J. Dyson F.R.S. and Mr Ivor Williams who made most of the apparatus) have provided contributions from which a history of those years can be roughly reconstructed.

Electron optics dominated his attention though he was given other tasks to do. It will be recalled that he left the Technische Hochschule, Berlin in 1927 to work on transient electrical phenomena for a year and then went to Siemens. Just after he left the Hochschule, Busch[3] proved, in an important theoretical paper, that the magnetic field of a short cylindrical coil of the type Gabor had used in his oscillograph acted as a lens, obeying the lens law of optics relating object/image distances to focal length: this at once meant that magnification could be achieved. "The paper was more than an eye-opener; it was a spark in an explosive mixture," and even with parts of Dennis's old oscillograph, Knoll & Ruska in the Hochschule began work on the world's first electron microscope. Dennis "could have kicked myself", but his mind was set on electron optics and remained so for the rest of his life.

His first task was to try to increase the angle of scan of a cathode-ray tube by shaping the electrostatic field by means of high-resistance deflecting electrodes which gave more freedom in defining the boundary conditions (Dr Dyson recalls this). He first built an electrolytic-tank model of the electrode system and, with probes mounted on a tricycle carriage, measured the potential and field, (Gabor, 1937a). This was to be shown at the Physical Society's Exhibition and Dr Milner (as he then was) was selected by Mr Warren to be the alternate demonstrator, Milner being selected as he was a physicist. Dennis never liked being called a physicist. "No, Mr Warren" he retorted "I am an engineer. I insist, I show you my Diploma!" An almost identical electrolytic tank was independently developed in America, and Dennis showed the account of it to Dr Milner "You see it shows the American passion for the automobile, it has four wheels, not three, and is motor-propelled." Experimental tubes were made with wide-angle deflexion but work on these ceased because electrostatic deflexion became replaced by magnetic deflexion.

[3] H. Busch *Arch für Electrotechnik lS*, 583–591, 1927.

The advent of television in 1937 sent shock-waves through the cinema industry and in 1938 Oscar Deutsch (of the Odeon Cinemas chain which purchased sound-film projectors from B.T.H.) sought to encourage B.T.H. to develop a system for cinemas in which the viewers would see three-dimensional images without the aid of any special spectacles or other appliances. Dennis reflected on this and conceived an ingenious solution of the problem, taking out two lengthy patents on 3 May 1940 (541,751 and 541,752), the first containing rigid mathematical formulations of the conditions to be fulfilled. No experiments could, however, be made as, by now, the B.T.H. Research Laboratory was actively engaged on war work; the experimental work done with Mr Ivor Williams after the war will be reported later.

He gave an excellent lecture to the Rugby Engineering Society in 1938, (Gabor, 1939), on the many discharge devices which operated at low gas pressure and concluded "the theory of gas discharges has, up to now, only been able to give a lead in a few cases; in most instances it has lagged behind invention and the best it could do was to help to improve the devices invented by experimenters" (Lord Rutherford would have enjoyed this).

Professor Carter recalls an article Dennis wrote, (Gabor, 1938), on the menacing situation of 1938–39. Gabor commented on President Wilson's 1919 'Doctrine of national self-determination'; this idealism had a wide appeal in those days and the rightness of the President's ideas was taken too much for granted until Gabor pungently described them as "universal principles so evidently right that people do not see what nonsense they are"; at a time when barriers between nations were being lowered, it was nonsense to invoke the principle of selfdetermination to erect new ones, not least between the component parts of the Austro-Hungarian empire.

André Gabor paid a visit to England in December 1938 with the intention of staying for a fortnight but Dennis, being convinced that war was imminent, would not let him return and supported him financially until he found employment in the service of the Ministry of Agriculture. In August 1939 Dennis arranged for his parents to come to England and tried to persuade them to remain but they returned home just before Hitler assaulted Poland. He wrote a monthly letter to them via the Red Cross or the Apostolic Delegation; his father died in 1942 but, as already mentioned, his mother survived the war and came to the Gabor's home in 1946. When war was declared Dennis volunteered for military service but was rejected and his name was placed on the Register of Aliens with Special Qualifications and it was left to B.T.H. to see that he had no access to classified information. For part of the time he worked at home, and later a capacious hut was erected for him outside the security fence of the Company. He invented an improved version of the Schmidt lens system, an objective for photography or for projection, with large apertures and good chromatic properties, consisting of a spherical mirror in combination with one or more lenses having spherical surfaces: the original Schmidt system employed non-spherical surfaces not readily made in those days. Patent 544,694 of January 1941 gives full details.

Since Dennis was not informed of the secret work aimed at detecting aeroplanes at night by radar he invented his own solution, detecting a plane by the heat from its engines. To render the infrared of the spectrum visible he devised what he called a "relay screen" consisting of a transparent plate with optically smooth surfaces on one face of which were cemented parallel rows of thin strips. Over these ridges was a thin clear plastic film incorporating a fluorescent dye, and coated with a very thin infrared absorbing layer. When an infrared real image was focused on the absorbent layer the film expanded and, between each ridge, it sagged closer to the baseplate. Light striking the remote side of the transparent plate at 45° was totally internally reflected except at the parallel rows of thin strips and at the mid-span of the thin plastic film wherever it had been heated by the infrared: here the light passed into the thin film causing fluorescence. News of this relay screen reached military ears and experimental work was requested by Government. Professor Milner writes 'As Gabor was still proscribed from the Laboratory, engineer W. D. Sinclair, who had superb manipulative skill, and I visited Gabor at home once weekly to discuss progress, Sinclair being the experimenter. He made screens 1 inch square, using microscopic-interferometry inspection techniques based on Newton's rings, but, though thermal sensitivity was observed, it was never more than 1% of theoretical. Perforated films were developed to facilitate motion of the thin plastic films. By mid-1943, security regulations were eased and Gabor continued the experiments with Sinclair in the hut. In spite of poor conversion efficiency the device reached a stage when it could detect a soldering iron on the other side of the room.'

Professor Milner writes about another aspect of Dennis's activities of those days, an aspect which is a foretaste of the great concern for the future of the human race which he showed in his books and lectures after retirement. He and Dr Milner were members of the Branch Committee of the Association of Scientific Workers in the 1939–41 period and were concerned that science should be used and not abused, but at that moment in our history they were convinced of the over-riding importance of all scientists working together to win the war. In 1941, before Hitler attacked Russia, the Central Executive of the A.Sc.W. (containing to my knowledge several avowed Communists—T.E.A.) completely opposed the Rugby views and the Branch Committee resigned *en bloc.* Dennis had prepared for them "a kind of Hippocratic Oath", a fascinating document which, as far as I am aware, has never been published. He believed that the new world must contain a stronger element of reason in the management of human affairs and here the scientist would come into his own: "We represent the greatest body of men and women schooled in thinking ... in a field in which lies do not exist, no amount of shouting will make an incorrect mathematical error right. We are accustomed to be critical of ourselves as we work under a unbribable critic, unmoved by any flow of eloquence, but this very fact is also our greatest weakness. We may be masters of all kinds of engineering and yet ignorant in regard to the engineering of human consent." He offered a 10-point advice to scientists as to how best to equip themselves to play a full part in society. Space does not permit me to deal with these, but, writing to Milner in 1966, Gabor added: "I well remember writing those 10 points and I still subscribe to them; we anticipated by 5 years the shock-wave which hit physicists after Hiroshima when they felt they had to take a stand in a world which had become too dangerous by their own work."

In 1942 he gave a lecture on Electron Optics, (Gabor, 1942, 1942–43), in which the question behind his Nobel Prize is asked: "Can we make electron-lenses as perfect as the highly corrected lenses of modern optics?" The answer in 1942, 15 years after Busch, was "No"; for Scherzer had shown that neither spherical nor chromatic aberrations could be eliminated, though by using high accelerating voltages the RCA microscope produced electrons constant in velocity to 0.004% and with high intensities the numerical aperture of the electromagnet could be cut down to 1/1000 thus reducing aberrations to very low values. But Dennis was a perfectionist, he wanted to know *how* to arrive at ideal stigmatic lenses even if they could never be produced: he had labored for years on this, but now he had solved the problem using Hamilton's century-old analogy of mechanics and optics; "for 3 years during which time I have not worked in electron-optics I must have been unconsciously digesting Hamiltonian methods and now, effortlessly, I have solved the trajectories and want my experience to benefit others". (He even won the Nobel 'effortlessly' as we shall see.) The full paper, (Gabor, 1942–43), is a joy to read and a portent of the future.

He continued to report the applications of Hamilton's formulations of dynamics to dynamical problems in electronic devices such as magnetrons in another lecture in 1944, (Gabor, 1945a), "It may be added that their usefulness probably does not extend much further, as they seem to give little help in the problems of space-charge and noise, but the field they cover is an important one and their applications may save the specialist many hours of headaches."

In the midst of this work he found time in 1943 to write an I.E.E. paper, (Gabor, 1944a), on the fundamental processes by which steady electromagnetic oscillations can be produced in electronic devices. Regarding an oscillator as a machine for converting d.c. to a.c. with electrons as the moving parts he listed the only six possible processes and gave examples of each. Professor Carter, a colleague in B.T.H. at the time who contributed to the I.E.E. Discussions, writes of this classification and of his work on the theory of communication, (Gabor, 1946b), that 'one could not but be amazed at the brilliance of the analyses; he was gifted in *identifying* a seminal idea, whether from the writings of others or from his own mind; and when he transmitted such an idea to someone else the effect on the recipient's mind was permanent'. He was awarded the Duddell Prize of the I.E.E. for his paper, (Gabor, 1944a), and of paper (Gabor, 1946b) he wrote "This theory was little understood at the time but it received wide recognition when American workers, a few years later, put communication theory on the map"; further reference will be made to this.

That he was still pondering on the problem of the perfect electron-lens is shown by the next paper—to the Royal Society, (Gabor, 1945b), written in 1944—"The theory of the motion of electron assemblies in electric and magnetic fields is of interest in the theory of the magnetron and in electron-optics where lens correction is impossible in the absence of space-charge". He produced a complete theory of stationary swarms of electrons around an axially symmetrical magnetic field, and showed that electrons injected with small velocities would distribute themselves with a density inversely proportional to their radial distance from the axis, thus offering the possibility of realizing a dispersing electron lens. In 1946 he offered practical details of how such a zonally corrected lens might be made, (Gabor, 1946a). The electron beam would have to be fired off-axis by tilting the gun to miss the supports of the central wire supplying the electron-swarm and he calculated that spherical aberration would then be corrected in a zone of about twice the wire diameter. Dr M. E. Haine who was responsible for the investigations on electron microscopes, first in the Metropolitan-Vickers part of

A.E.I. and later in the A.E.I. Research Laboratory, has reported that he discussed this space-charge correcting lens system with Gabor and expressed anxiety about the extreme accuracy required for positioning of the central wire electrode. As a result of this, Dennis carried out extensive computations on the permissible tolerances and found them to be impracticably tight so this correcting system was never developed. He considered that the theory of electron swarms provided a better basis for magnetron theory than the official theory held during the war.

Two more short notes, (Gabor, 1947b, 1947c), deal with focusing of positive ions of very high energy.

Dennis has said that his paper on Communication Theory, (Gabor, 1946b), was one of the few he ever wrote as a scientist interested in understanding things rather than using them to a purpose, and the analyses he made so impressed his friends that, more than once, they advised him to do analytical work rather than to continue trying to invent. But his collaborator, Ivor Williams, puts a different complexion on the matter: he wrote to me 'The Transatlantic cable was running out of space and the Post Office was working on the American 1939 Vocoder', so Dennis's analyses of the signals necessary to convey the information (part I the I.E.E. paper (Gabor, 1946b)), and of the minimum requirements of the ear for intelligibility (part II), were really *essential* preliminaries to the building by Mr. I Williams of simple apparatus described in part III of this paper to compress speech into a narrow frequency band without losing its intelligibility. Paper (Gabor, 1947d) written in 1946 is a sequel to this, describing a device also made by Mr Williams, to compress and then to re-expand speech, and the paper includes an extensive analysis of possible variants. Demonstration of the device was not very satisfactory but it had only reached the early stages of development when the project was abandoned.

The penultimate activity of this B.T.H. phase of his life was the development of the 3-D projection system which he had patented in 1940; a Russian named Ivanov had, independently, patented a similar 3-D system. Mr Ivor Williams did the experimental work with him. It was based on Lippmann's lenticular screen of 1896, but by now there were Perspex lenticular plates available. Work started after the war with the use of two projectors correctly spaced, focused on to the back plane of the screen which had a diffusely reflecting surface. Reflected light passing through the lenticular front face forms a pair of beams nearly parallel, so that a person in a row of specified positions will see the left picture with the left eye and the right with the right eye. But the viewer must not move his head. In

Moscow a small stereo-theatre was built but, said Dennis, "only the Russians could be so disciplined to sit so still". Work was abandoned in 1948 and no account was written. However by 1960 Dennis had had further thoughts embracing ways to avoid the necessity of keeping still, (Gabor, 1960b), but he realized that the scheme was practicable only with *small* screens and the advent of the large cinema would, in all probability, have killed it.

The foregoing account shows how his mind was constantly returning to the problem of aberrations in magnetic lenses. In his personal account he wrote "I realized that I ought to have stuck to my original line in 1927 when electron optics was just discovered, but I hoped it was not yet too late to make a comeback by improving the electron microscope beyond the limits set by aberrations". His book *The electron microscope*, (Gabor, 1948a), published in 1944 re-analyzed the problems and concluded with a chapter on The Ultimate Limit of Electron Microscopy, searching hopefully to 'see' single atoms. He gave an illustration of Bragg's 'indirect' method of structure analysis obtained by interference of diffracted beams from millions of unit cells and he also had admired Zernike's use of a coherent background of waves to show up aberrations in optical lenses. It is therefore not surprising that while awaiting his turn for a game of tennis during the 1947 Easter holiday his subliminal ego presented him with the solution to this overriding problem *effortlessly*, "Why not take an electron picture, one which contains the *whole* information and correct it by optical means. To capture the whole information, *including the phase*, the coherent background must be supplied by the same electron beam, which will therefore produce interference fringes: photograph these, and then illuminate this photograph with light and focus it onto a photographic plate." He called the electron diffraction pattern a hologram, because it contained the whole information, amplitude and phase, and the magnification gain would be the ratio of the optical to the electron wavelength. To try out his idea he decided to produce first an optical hologram and reconstruct the object optically. Once again his assistant was Mr Ivor Williams: they produced an extremely fine beam of light by illuminating a pinhole of 3 μm diameter with monochromatic light: this light fell upon a microphotograph 1 mm diameter, and at a distance, a photographic plate recorded the interference patterns produced by waves from the details on the micrograph and waves by-passing them. Williams recalls the secrecy in which this idea was tested, and the interminable exposures on account of the very feeble illumination, but between Easter 1947 and May 1948 a beautiful reconstruction had been produced,

(Gabor, 1948c), followed up a year later, (Gabor, 1949), by a full account of 'microscopy by reconstructed wave-fronts', and a forecast was made that by this method a resolution of 1 Å might be achieved.

By 1948 the new A.E.I. Research Laboratory at Aldermaston had been created, one section of which was to include the study of the long-term problems of electron microscopy to back up the development of the electron microscope in the Metropolitan-Vickers Co. (the B.T.H. Co., the other half of A.E.I., was not engaged in this subject). Dennis asked me if he could join us and pursue his proposal in Aldermaston but he had other commitments in the B.T.H. Co. and they were not willing to share his time on a 50:50 basis. We did, however, decide to work on the subject with his help and guidance from a distance, and I applied to D.S.I.R. for financial assistance because, if we were successful in resolving atomic lattice spacings many of the sciences would benefit, but A.E.I. might not necessarily increase its sales of microscopes: D.S.I.R. accepted this argument and gave what, I believe, was the first-ever grant to an industrial laboratory. We ran into scientific difficulties which do not directly concern this biography, except in so far as they reacted upon Dennis. Again, Dr Haine says 'Probably our main contribution was to push Dennis repeatedly into thinking about the practical aspects of the method and to express himself in terms understandable to us and relevant to the practicabilities.' This in turn reacted upon Dr Haine and his team 'I remember how my understanding took a sudden jump forward and inspired me to realize the fact that this point-projection method was unnecessary and that a transmission method ... was free of the chromatic defects which made the point projection method impracticable'. As a result Gabor re-examined and also extended his theory, (Gabor, 1951c), but to cut out unwanted interference, a Zernicke phase-plate was needed and this was too difficult to make: the attempt to produce micrograms by his process was abandoned, but the effort had been of great value in focusing attention on other features of the electron microscope, and, as a result, resolving powers of 5 Å were obtained. Later in 1948 he was appointed Reader in Electron Physics in Imperial College and left the B.T.H. Co.

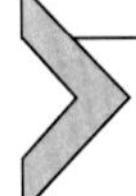

5. Mullard readership in electronics: Imperial College 1949–58

In his 'personal account' he wrote in 1956 that the preceding 6–7 years had been the happiest in his scientific career: "at last I was my own master and could work with young research students on problems of my own". Academics seldom realize that for those of us whose lives have been spent in industry the majority of our activities have been set by the requirements of our employers and only, almost as a side-line, have we been able to pursue ideas of a more academic character. Dennis had spent many months in 1947 on paper (Gabor, 1948c) but had the Metropolitan-Vickers Co. not been making electron microscopes it is *very* doubtful whether he would have been allowed the time in B.T.H. to proceed to paper (Gabor, 1949). Now he could please himself. I asked him immediately if he would consent to be a consultant of A.E.I. and we enjoyed his visits to Aldermaston for many years: he was a great favorite with the staff, always provocative, always inspiring. He was also a consultant to the Post Office.

In a lecture he gave at Delft, (Gabor, 1950b), he revealed some of the difficulties being encountered in the A.E.I. Research Laboratory and in (Gabor, 1951a) he proposed a complicated electron-optical system to avoid chromatic aberrations of electron lenses: as already stated, he also fully analyzed, (Gabor, 1951c), a proposal made by the Aldermaston group which had advantages over his original scheme for making electron holograms, but with the closure of this work his concentration on holography almost ceased, and apart from reviews contributed in French, (Gabor, 1953a, 1953b), and German, (Gabor, 1953c), and a final review of theoretical and experimental work completed in the A.E.I. Laboratory, (Gabor, 1954c), he left the subject until it was dramatically revived in 1963[4] when a laser beam was used to produce an optical hologram.

Although the experiments on electron microscopy by holography were not forging ahead rapidly, he realized from the optical holography done in Rugby that the optical microscope might benefit "by superimposing two strong coherent backgrounds on the images of an object having a phase shift of 90° between them; one can obtain a complete optical record of an object, containing amplitude and phase". His design for such an interference microscope was described in 1954, (Gabor, 1954c), after a research student, W. P. Goss, had begun to work on it and it was supported by the Paul Instrument Fund of the Royal Society for five years. However an account of the work done did not appear till 1966, (Gabor & Goss, 1966), It had achieved its goal; with it, three-dimensional objects could be seen though the separation in depth was not as good as had been anticipated.

[4] By E. Leith & J. Upatnieks of the University of Michigan. *J. Optical Soc. Am.* **53**, 1377–1381, 1963.

But by 1966 the laser was available and with such illumination it was hoped that the concept would be revitalized.

Dennis's interest in plasmas continued; he reviewed the experimental work and the theories of plasma oscillations, (Gabor, 1951d), produced a classical wave-theory of plasmas, (Gabor, 1952a), which upset his Budapest theory given in (Gabor, 1933b), and he suggested that there must exist strong internal oscillations in the arc plasma far exceeding what had been assumed. With two research assistants E. A. Ash and D. Dracott a very elegant piece of apparatus was devised with which they succeeded in 1955 in measuring such oscillations, (Gabor et al., 1955), and thus solved what Gabor had recognized, 25 years earlier, to have been a paradox in Langmuir's first observations on plasmas. A fuller account of this work is given in (Gabor, 1956e), and Ash's work on electron interactions in (Gabor & Ash, 1955).

He continued also to pursue his very erudite considerations of Communication Theory: since he had written paper (Gabor, 1946b) in 1944 his ideas had received wide recognition—as already stated—and he was invited to present a general report to the Symposium on Electronics and Television in Milan in 1954, (Gabor, 1954b), on 'communication theory and cybernetics'. He said he feared there was a danger that the subject would become abstract, even sterile and it was his "object to show that, far from being sterile, the theory is likely to become a powerful driving force in the future development of the techniques of communication and control". He dealt in great detail with two aspects of the theory, the analysis and use of all the *prior* information available—the structural part of the theory—, and the part which probability plays in the sending and receiving of information—the statistical part of the theory. He went on to propose a "learning filter for noise elimination, prediction, or recognition": he considered such a filter could be made but at that time was not prepared to forecast its practicability: we shall see how just such a device came into being a few years later. With his brother he read "An essay on the mathematical theory of freedom", (Gabor & Gabor, 1954), to the Royal Statistical Society; they "wanted to show that in the vast domain covered by this powerful but vague idea there exists a field, large and important in itself, which is amenable to quantitative discussion, the approach being based on the statistical concept of communication" already referred to. The mathematics of this is extremely complex to the layman, but, from the discussion it provoked, it had clearly been well appreciated by the statisticians present. The Gabors had been encouraged to attempt this formulation by the success of communication theory which enabled them "to make comparisons between different systems which otherwise would remain incommensurable". Four years later they presented a further paper on freedom, (Gabor & Gabor, 1958), confining themselves to statistical, and not to the philosophical, ideas of freedom. They laid down a number of axioms and then derived diversity and dependence coefficients of the variables. As an example they applied their results to the "choice of a profession", based on some Swedish statistics of (a) the occupation, and class, of the parent and (b) the IQ and the occupation of the son at age 24.

In the post-war years, applied electron-physics was becoming very popular, Ordinary and Higher National Certificate courses were being organized and in 1952 the British valve industry offered financial support to Imperial College to establish a post-graduate course in the subject and the U.G.C. gave help to re-equip the electronics laboratory. An account of the splendid new equipment was written in 1955, (Gabor, 1955c), equipment sufficient for making up almost any type of vacuum apparatus, glass lathes, hydrogen and muffle furnaces, induction coil heating, and exhaust pump stands, etc. The stage was thus set for Dennis to develop his inventive flair, and make working models. "In 1952 I had the idea of a flat, thin television tube for monochrome and for color ... I am now busily engaged on its development with three young assistants and I hope that this will at last satisfy my life-long craving to make a useful and successful invention in the field of electronics" (this was written in 1956). The thin flat tube hanging on the wall like a picture had been for years a standing feature of imaginative American advertisements and these gave Gabor the idea. He sought assistance from the National Research and Development Corporation and I was asked by them for an opinion. It was a very complicated device on paper, involving very wide angle sideways deflection, the bending of the beam first through 180° and then back through 90°, though the design for color appeared to be less demanding than the color concepts then being proposed. I, in turn, discussed the design with the specialists in the A.E.I. (Woolwich) who were making t.v. tubes, and finally gave qualified support, with a warning about putting so many complications *inside* the glass envelope, itself a very difficult thing to make in large sizes. The first description of its construction and operation was in 1958, (Gabor, 1958a, 1958b), Dr P. R. Stuart and Dr P. G. Kalman being co-authors, Ph.D. students at that time. The many difficult design features were calculated and developed with Dennis's usual brilliance. Dr Kalman, a fellow Hungarian, has given a fascinating insight into the way Dennis worked: 'He was, as you know, a very kind man but a hard taskmaster. Demanding the utmost from himself, he simply could not see how others could do otherwise, ... Because he had an uncommon insight into complex mechanics he could not see that lesser mortals found his example hard to follow. He would juggle with various would-be solutions bending the rules as he went and straightening out the mathematics afterwards; being apparently unmethodical, using no perceptible logic, but somewhow he knew the answer and was designing the hard-ware in the matter of a few breathless minutes. Once, seeing that I had been taken aback by his methods he burst out laughing, "Do you really think anything was ever invented in any other way? First you have to know the answer, logic comes afterwards".' This was indeed his way of working which I have seen on many occasions but I knew that he had wrestled with a concept for perhaps days or even years, as with holography, and what Kalman was describing was the outpouring of Gabor's subconscious mind which had already provided the answers. Kalman consulted him on the desirable length of the Ph.D. thesis he was about to write: "Well Kalman, that depends: Fermi wrote his on the back of a postcard but in your case I should make it a little longer."

Dennis had no lecturing duties but gave many specialist lectures. As he got more excited his speech became slurred and sibilant, losing intelligibility, though his style improved in later years. He disliked administrative duties and fortunately he was relieved of them. Dr Kalman thought that people like Gabor ought to live in a world where he would never have to think about money: that is precisely how General Electric of America treated Langmuir—he was given an open cheque account.

By 1958 the flat t.v. tube had reached the stage when all the electron-optical problems had been resolved; "a development of this sort cannot be wholly completed in an academic laboratory, but really in industry". But by this date the conventional cathode-ray tube had been greatly reduced in length and industry was not interested in a thin tube. "If I live long enough I may still be able to see some of my inventions through to success; my motto is that of G. de Taciturne: Il n'est pas nécessaire d'espérer pour entreprendre, ni de réussir pour persévérer!" Two other students, H. A. W. Tothill and J. E. Smith-Whittington spent more years on the flat tube, taking out some of the complications and in 1968 paper (Gabor et al., 1968) reports "We have now realized, in vacuum-tank experiments, a thin tube ... have indicated how parts could be further improved—but a full development would have exceeded our time and resources." A patent only last 16 years, and 16 years had elapsed since work began. By a quirk of fate, even as I prepare this memoir a very small, flat television tube probably ready for the market has been announced, within 4 months of the death of the inventor of the original flat t.v. tube.

Two other inventions in this (1958) period might be mentioned. In a review of the theory of electron interference experiments, (Gabor, 1956f), he showed that two new instruments could be constructed, a phase-marking diffraction camera, and an electron interference microscope. The other concerned a Wilson cloud chamber, (Gabor & Hampton, 1957); he proposed turning it into a microwave cavity operating so near to breakdown that multiplication of secondary electrons would take place on the α-particle tracks at each antinode, thus giving time-markings at, say, every 10^{-10} s. With B. Hampton the chamber was made and the marks were well recorded; the Royal Society had again provided help from the Paul Instrument Fund.

6. Professorship at Imperial College, 1958–67

He was appointed to a personal Chair of Applied Electron Physics in 1958, his Inaugural Address, (Gabor, 1959a), being on "Electronic Inventions and their Impact on Civilization". He reviewed in part I the work on three projects in the laboratory, on the t.v. tube, on an early attempt to raise the temperature of a plasma, aimed at thermonuclear fusion, and on the 'learning filter', to which I have already referred, Paper (Gabor, 1954b). Then in part II, Inventions and Civilization, he speculated on the part which sophisticated electronic machines might be expected to play and asked "Do we need inventors; do we need more inventions?" Sir George Thomson's *The foreseeable future* had just been published and Dennis concluded "for the first time in history we are faced with the possibility of a world in which only a minority need work to keep the great majority in idle luxury—a nightmare of a leisured world for which we are socially and psychologically unprepared". "Who has left Mankind without a Vision? The predictable part of the future may be a job for technology but for the part which is a matter of free human choice it ought to be, as it was in the great epochs of the past, the prerogative of the inspired humanists". And this is what Dennis set about to be for the rest of his valuable life. An abridged version, (Gabor, 1959b), was printed by *Encounter* in 1960 under the title "Inventing the Future"; a title proposed by the Editor. The German Merkur printed part II of the lecture, (Gabor, 1961f), and it also formed the basis for his Joseph Wunsch Lecture delivered in Haifa on 30 May 1965, (Gabor, 1965a).

I have already dealt with the demise of the flat t.v. tube. The other two projects unfortunately suffered similar fates. The 'learning filter', (Gabor, 1959c), and (Gabor, 1960c, 1961a), with his students Wilby and Woodcock as collaborators, was "an analogue computer which *learned* to do its job: one had to teach it by putting into it a magnetic tape which had a record on it which had to be recognized or predicted, together with another tape which had the solution on it. By a feedback process the non-linear filter modified its parameters until it optimized itself and did the job. The linear part worked beautifully as a predictor and simulator, but in the end, after 6 years work, the project was abandoned."

The thermonuclear fusion project had to be abandoned for different reasons. Gabor had started on a most ambitious project after the thermonuclear work in U.S.A. and Britain had been declassified in 1958. D.S.I.R. referred the proposal to me and I had commented that the project appeared to be too difficult for a small university group to tackle, but a grant of £67 000 was given. Work went on, indifferently, for 2–3 years but illness intervened: "I am not inclined to nervous breakdowns, my nerves are tougher than my body, but the serious thrombosis-phlebitis which I got in my legs in America in 1961 was probably due to this disappointment." I visited Dennis in the St Mary Abbots Hospital on 14 December 1961 and told him to withdraw from the D.S.I.R. contract for the sake of his health; that the project might be tackled by E. O. Lawrence and the expert team in Berkeley, California, but was too much for his Electronics Laboratory. He withdrew—and recovered.

The reproduction by *Encounter* of the second half of the Inaugural Lecture, with its new title 'Inventing the Future' appealed to Gabor's friend, Frederic Warburg of the publishing house of Seeker and Warburg and he pressed Dennis to write a book; a year of Saturdays 1961–62 was put to extremely good use, and this popular book has been translated into many languages. The new title put a different complexion on his work and "it gave direction to my thinking". The Inaugural Lecture was concerned with what the world *is*, and how technology might develop in the foreseeable future (as G. P. Thomson had written too): the new book dealt more exhaustively of course with all this but went much further in inventing a plan for survival: "The future cannot be predicted, but futures can be invented. It was man's ability to invent which has made human society what it is. The first step of the inventor is to visualize, by an act of imagination, a thing or a state which does not yet exist and which to him appears in some way desirable. He can then start rationally arguing backwards and forwards until a way is found from the one to the other. For the social inventor the engineering of human consent is the most essential and the most difficult step." "With Moses, I have given forty years for the education of a generation which will deserve to live in the Age of Leisure."

The last experimental work pursued in the laboratory before Gabor retired held out, for a time, great promise so I *must* deal with it: it concerned his very old work in Siemens, passing plasmas through perforated disks, paper (Gabor, 1933b) in Hungary, and (Gabor, 1937b, 1944a) in Rugby. The original plasma device failed because ionic bombardment destroyed the electrode but his student, J. Nilson, made an important discovery and Gabor realized that the dark discharge coming through the perforated disc or gauze might be used for making an inert gas-filled controlled thermionic converter (the Nilson contribution does not appear to have been written up so that its exact nature cannot be described). Work was then continued by H. A. Fatmi with N.R.D.C. support and in 1961 Gabor could envisage the physical design of a thermionic converter, (Gabor, 1961c, 1965e), filled with argon/mercury mixture or with caesium from which a.c. or d.c. power might be extracted at quite low cost, the heat for the cathode being supplied at 1200–1800 °C from for example, the flames before they licked over the surfaces of the boiler tubes of a steam-generator (or even from the fuel elements of a nuclear reactor—if they could operate at such temperatures!) "Provided that these problems can be solved there is a wide field of potential applications for thermionic generators for supplying d.c. power—and for improving the efficiency of power plants"—"After $1\frac{1}{2}$ years of rather disappointing research we had a stroke of luck; I had the idea that an addition of copper to the electrode—but my collaborator, Michael Albert, out of habit added zirconium hydride—and there is now good hope that we shall make—thermionic converters." A theory was written, (Gabor, 1963a), and in 1966 Albert, Atta and Gabor described the various cathodes on which work had been done, (Gabor et al., 1967), but time ran out, and in 1979 these converters are not even on the horizon. The development caused a bit of a stir. N.R.D.C. interested Unimation Inc. of America and Mr J. F. Engelberger came over hoping to secure a patent licence; 'Well I remember my first encounter with Dennis. When I introduced myself he said, "Oh yes, I have only known one other Engelberger, that was Willy, the Austrian composer". I replied with dignity that that was my uncle, whereupon Dennis stretched himself to full height and sang the refrain from one of my uncle's famous songs, Wenn Die Letzte Blaue Geht': Engelberger got his licence!—and Dennis's friendship. Mr Engelberger concludes his 1979 letter to me 'Thermionics is still not a practical technology, it may just be that a Gabor idea needs a quarter century of gestation before an environment exists that allows it to blossom forth in full glory! He took me round St Paul's; he was far better than any walkie-talkie for he knew not only the statutes but even the history of the sculptors, and the peccadillos interred with the remains of famous men. The secret of being a futurist, he said, was to start predicting in your 60s and make your predictions for 50 years hence'.

The junior collaborator of (Gabor et al., 1967), Dr M. J. Albert, also writes in 1979 from Israel, 'Dennis was not only a brilliant theoretical scientist, he took an active interest in the minutest details; the work was governed by diaries and edicts—typewritten instructions handed out to collaborators—followed by memoranda of discussions, made several times daily: jigs, fixtures, apparatus, all was designed and drawn by him. But one did not feel suppressed or overpowered, one had the opportunity of exercising initiative. His starting point was the concept of the invention, then work was directed towards its realization; his attitude was that of a starry-eyed young tycoon hoping to make his first million; maybe it was good for physics that none of his inventions became 'hits' in the sense of making him a millionaire'. But had he become a millionaire this would not, in my opinion, have made any difference to his pursuit of knowledge. He regarded his research students as his most precious products, and his laboratory as his Heaven on Earth.

Gabor was elected an Honorary Member of the Hungarian Academy of Science in 1964, the first emigré to be elected. He was rather surprised "seeing that in my book *Inventing the future* I have expressed rather critical views on Communism. Welcome sign of liberalization." He was also invited by a fellow-countryman, Peter Goldmark, with whom he had worked in Berlin, to the laboratories of the Columbia Broadcasting System (C.B.S.). The C.B.S. had bought a number of his patents and in 1965 Goldmark, the director, offered him an appointment for life as a Staff Scientist, but he did not want to give up his Chair and instead, accepted a part-time consultancy. He gave a very challenging lecture to an Education Seminar, (Gabor, 1965k), at U.C.L.A. in April 1965 on 'Inventing Education for the Future', a future world which *must* be economically stagnant.

After Aldous Huxley died in 1963 a Memorial Volume was published containing tributes from many friends including one from Dennis, (Gabor, 1965j). Huxley's lifelong influence on him started in student days. The early novels "acted like a shock wave on me and on my fellow Central

Europeans; we found he could discuss intellectual matters so much better than we had imagined from reading other English novelists, but the sort of Central European who could read and enjoy Aldous Huxley was destined to emigration or the gas chamber.... His greatness lay in the unparalleled span of his mind, in his ability to compose grand contrapuntal symphonies of human life and to put them into perfect literary form." "I am not however concerned with his literary fame ... but his heritage to those who really care about the future of the human race, and in this respect I hope he will be remembered with Thomas More."

Here we see the growing concern for the future to which Dennis devoted so much of his later life. It began (so far as his published works provide the evidence) way back in the early days of the war when he framed his version of "a kind of Hippocratic Oath" to guide scientists' behavior. This contained thoughts which may well have been provoked by *Brave new world* (Huxley 1932).

I have already referred to the brilliant theme of the second part of his Inaugural Lecture, 1959, (Gabor, 1959a), from which *Inventing the future*, (Gabor, 1959b), sprang and here there is a direct connection with *Science, liberty and peace* (Huxley, 1947) in which "Huxley gave a closely reasoned analysis of the threat to individual liberty arising from the progress of science." His views on the possible development or invention of beneficial drugs were influenced by *Doors of perception* (Huxley, 1957); and finally "To me *Island* (Huxley, 1962) was the last great gift with which he enriched my life", and Dennis too has enriched the lives—maybe partly on account of Aldous Huxley,—of many who care about the future of mankind.

Holography had been dramatically revived by the invention of the laser; this is not the place in which to describe the eruption which occurred except insofar as it directly involved the inventor of holography. Using a laser beam produced by a helium-neon gas operating at 6328 Å Leith and Upatnieks produced optical holograms in 1963 by Gabor's original technique, (Gabor, 1948c, 1949), with remarkable clarity. Gabor lectured at the C.B.S. Laboratories, also in April 1965, (Gabor, 1965f), reviewing the dramatic change, a change made possible because the holograms could now record 10 000 interferences and thus make full use of the information capacity of fine-grain photographic plates.

In October 1965 he described what might turn out to be one of his most important inventions 'Character recognition by Holography', (Gabor, 1965i), an invention which, almost certainly, will not have to wait 16–20 years to fructify. He showed that holography could "solve one of the most urgent problems of computers and other data-processing devices; the recognition of characters with many variants. ...*A single hologram* may discriminate between *all* the numerals and the letters of the alphabet, each with 30 variants." His brilliant proposal involved the making of one hologram combining two coherent waves at an angle to one another, both falling on a photographic plate each proceeding from different objects, the first object being, say, a letter of the alphabet, written or printed, the second being a combination of point sources of that same letter arranged as a code which can be read by a machine. When light from the handwritten letter is presented to the hologram, the coded version flashes out and can be read by the data-processing device.

Lensless holography also appeared in December 1965 and with American colleagues, (Gabor, Stroke, Brumm, et al., 1965), he reverted to work done (with Goss, see (Gabor & Goss, 1966)) in London in 1951–56 but not, as yet, published on the reconstruction of phase objects by holography. All this was reviewed in January 1966, (Gabor, 1966a); "lensless holography has an intrinsic advantage over ordinary photography because there is no lens which can fully exploit the information of the photographic plate,—it has been put to good use in the photography of explosive waves" (shown in (Gabor, 1966c)), "it has been applied to the optical testing of surfaces and their deformations, and holography with X-rays is particularly attractive."

His last major paper, (Gabor & Stroke, 1968), before retirement was written with an American, Professor G. W. Stroke of the University of Michigan, who had made many contributions to holography using lasers. In Gabor's pioneer work the light wave from the object was frozen onto the photographic plate by the superposition of the 'background' or 'reference' wave which might, or might not, be the original wave or even a wave of the original wave-length. But with lasers, the Americans had shown that the reference wave can be transmitted at an angle to the wave striking the object, that it can even be a diffuse illumination so that a three-dimensional image may even be seen with two eyes, that the technique can be used with a modification of Lippmann's (1894) method of color photography (that is to say by letting the reference wave fall on the emulsion from the opposite side to the object wave, thus giving a deep or volume hologram), and finally that volume holograms could be made to reconstruct multi-color images upon illumination with ordinary white light. Gabor and Stroke wrote a comprehensive theory covering all these variants.

The first book on holography had been published by Professor Stroke in 1966 *Coherent optics and holography* and in its preface he wrote 'Because of the uniquely important place held by Professor Gabor's three original "wavefront-reconstruction" papers which laid the ground-work for the principles of wavefront reconstruction imaging in optics, and more particularly for the retrievable recording of phase and amplitude information in imaging light fields, the three original Gabor papers, (Gabor, 1948c, 1949, 1951c), are reprinted in their entirety in an Appendix'. No greater testimony to the value of this work done 18 years ago could have been paid, and this by one of the leading authorities in the world.

"I am no longer afraid of retirement because I have acquired a new hobby, writing on social matters. Now that 'my future is mostly behind me' I am passionately interested in the future which I shall never see, but I hope that my writings will contribute to a smooth passage into a *very* new epoch." This was the last entry he made in his 'personal record'. He retired in September 1967 and was made a Research Fellow of Imperial College and Professor Emeritus.

André writes 'In 1921 Dennis spent part of his holiday at Igls in the Tyrol and part travelling in Italy. From Rome he travelled along the coast south of Rome and it was then his wish to have a villa there sometime.' Now he purchased land at Lavinio Lido, a pleasant holiday resort near Anzio, and built a beautiful villa named after his wife, where he and his wife subsequently spent all of their summers.

7. Retirement

Gabor was immediately offered a Senior Professorial Fellowship which would have left him in command of his research laboratory but there had been very few graduates wanting to work on his special subjects, the students who had last worked on the flat television tube had completed their theses and had left College, he had never worked in collaboration with a more junior member of the staff, and he had had the invitation from the C.B.S. to devote more time to work in America than he had been able to spare during his active professorship. So he declined the College offer and accepted, instead, a Professorial Fellowship which gave him the use of secretarial services and allowed him to keep his old office.

The Italian house had been built and was ready for occupation in the summer of 1968 and the Gabors spent the Michaelmas and Lent terms of 1967–68 in Stamford, Connecticut.

The first year of retirement was exceptionally busy. The Americans, Kahn and Wiener,[5] had published a list of one hundred inventions and innovations placed in possible chronological order which might materialize before the year 2000 A.D. and abstained from all judgments about their value. Dennis was stirred to think more deeply; he took the list as a basis for analysis, discarded some of them and added many of his own; he consulted specialists in the various fields of endeavor and also expressed his own view of the desirability or otherwise of the 137 inventions listed. In two lectures to the Science of Science Foundation in January-February 1968 he presented these reflections (they are to be found in French translation in (Gabor, 1968d)) and then prepared an extensive review of them for his next book *Innovations*, (Gabor, 1970a). Since writing his earlier book *Inventing the future*, (Gabor, 1964b), he had become more and more anxious for the distant future he would never live to see: "The most important and urgent problems of the technology of today are no longer the satisfactions of primary needs nor of archetypal wishes, but the reparation of the evils and damages wrought by the technology of yesterday ... Fossil fuels are threatened by exhaustion, so we must have nuclear power; death control has upset the balance of population so we must have the pill; mechanization, rationalization, and automation have upset the balance of employment: what is it we must have? For the time being we have nothing better than Parkinson's Law and restrictive practices." So many of the most profitable advances carry inherent penalties, for example some of the new remarkably strong materials of extreme value to industry "might be used to produce the poor man's hydrogen bomb ... innovation must work towards a new harmony otherwise it will lead to an explosion," and the author's intention was to make us *think*. The book was published in 1970 and in the same year a German tract, "Prognosen des technischen Fortschrittes", (Gabor, 1970c), summarized the main theme.

The 60th birthday celebrations of Professor Scherzer gave Dennis an opportunity of dedicating to him an "Outlook for holography", (Gabor, 1968c), recalling that it was Scherzer's brilliant proof in 1936 of the theorem that the spherical aberrations can never be eliminated in any axially symmetrical electron-optical lens system "that induced me, in 1947, to look for a way around this fundamental difficulty". He repeated the story of the Aldermaston experiments of 1948–53 which ended in disappointment owing to a combination of defects; "I wish we could start now when

[5] Kahn H. and Wiener A. J., 'The next 33 years. A framework for speculation', *Daedalus*, 1967, pp. 705–732.

the patient work of German and Japanese electron microscopists has eliminated these defects and has reduced the resolving power to the 'Scherzer limit' of 3.6 Å". "There are now many ways in which holography is developing, and more speculatively it may record wave phenomena for which good imaging systems are not available, such as X-rays, radio waves, sound waves and earth waves."

In his 'Fifty-eighth May Lecture' to the Institute of Metals, (Gabor, 1968e), "The prospects of industrial civilization" he strongly attacked our present way of industrial life concentrating mainly on Britain's malaise. His central theme was the problem of 'mismatch', to use an electrical engineering term; the mismatch of the I.Q. of most of the population and the I.Q. demanded by industry of the future, the relatively few people needed, for example, in the earth-moving industry, the containerized freight industry, etc., the mismatch between the military arms we need, and the huge size of the armament industry which (even in U.S.S.R.) exports to all and sundry, and many other similar examples of mismatch, culminating between life and economics,—life being now so safe and rich we *could* do so many good things like improving education, clearing slums, etc., but we **don't** do these things because we are up against economics of balance of payments, of insoluble unemployment, just bogged down in day-to-day politics. Of the potentially greatest mismatch of the future, between Man's nature and the Age of Plenty; he asks "the huge increasing crime rate, the drug taking, the restlessness, etc., are these all the portents of the Age of Leisure?" Here, as he saw it, is the challenge of our times; it epitomized his pleading to every thinking man and woman, to the students who heard him in Japan (The World in 2000 AD, (Gabor, 1968f)) and in due course it became the theme of his next book *The mature society*, (Gabor, 1972a). He reflected "Because I shudder at the thought of brain washing or pacifying drugs I would rather take a leaf out of the book of the successful élites of the past, such as the English aristocracy and the religious orders. I don't expect you to like it. It is Hardship. The calculated discomfort of the English public school (cold water tubs and the like) may look like the modern version of the savage's initiation ceremony, but actually there is wisdom here in the recognition that man does not appreciate what he gets for nothing. The permissive society is approaching a dead end. Despite the lunatic fringe, rebellious youth also contains a potential élite. For those with a fighting spirit among them let life be hard and competitive. Everybody must be made aware that a civilization achieved by a hundred thousand years of hard work and human sacrifice is worth preserving."

Professor Stroke, now Head of Electro-Optical Center in the State University of New York at Stony Brook, had invited Dennis to be a Visiting Professor, and there they gave an account of the successes, to date, of holography (Gabor & Stroke, 1969). Most of the basic features have already been mentioned in this Memoir. Some of the new applications were the taking of shock-wave pictures, the examination below a water surface of micro-fauna, of aerosol and jets emerging at high velocities, of faults below' the surface,—of, for example,—motorcar tyres, and then, just ahead, maybe, lay the prospect of holographic portraiture.

Returning home from America and going on to his Italian home he went to a meeting of the Accademia dei Lincei in April 1968 where the 'Club of Rome' was founded. The members of the Club belonged to the world of culture, industry, art, and science and had most diverse backgrounds, but, whether economists, philosophers, scientists, technologists, humanists, or artists, they were a group interested in studying problems of natural resources, nutrition, environment, climate, the Third World, income distribution, etc. Dennis had been invited as a founder-member in consideration of his broad vision of the problems of the world which he was able to set out from both the scientific and the humanistic viewpoints. In due course he made great contributions to the Club's activities as will later be recounted.

In the autumn he joined the O.E.C.D. Working Party on long-range forecasting and planning convened at Bellagio in October and there presented his views on "Open-ended planning", (Gabor, 1969a), the sort of planning which does not unduly restrict the freedom of other, later, planners. He admitted, of course, that he spoke as an amateur though he backed his proposals for safety and freedom of action with mathematical analogues. In essence he proposed that, since *potential* productivity was so much greater than *actual*, certain vital parts of state planning could be financed on the stock market with bonds redeemable at a future date at the current industrial index value at that date, so that the investor would be safeguarded from inflation and the bonds would have an anti-inflationary effect on the Exchange (a proposal anticipating the present—though greatly restricted—'Granny Bonds'). Gabor gave some reflections at the end of the Symposium; "The main dangers which threaten our civilization are the decline of the Gospel of Work—and the meaninglessness of a world in which rational thought (modern science and technology) has brought to the fore the irrationality of Man": "the new planners must modify the social structure to provide a component of self-interest of individuals and of corporations towards long-term stability", quoting Abraham Lincoln's words "like the Patent System, adding the fuel of self-interest to the fire of genius"—words aptly quoted by an inventor of the highest calibre.

Earlier in the year at Florence he had presented to the Symposium on Applications of Coherent Light some thoughts on prospective applications of holography, (Gabor, 1969b), holographic panoramas, three-dimensional pictures in monochrome and color, and information coding and storage which for the moment were held up by 'laser speckle'. Some were probably for the next century, but transmission and storage by holographic means he regarded as likely to be successful if speckle could be eliminated. A further goal he envisaged would be the translation of the information received by the radio-astronomers into the optical domain by optical means. At another Symposium, the first on Acoustical Holography he summarized the proceedings, (Gabor, 1969f), "...how thrilled I have been, I am sure this is the beginning of greater things. The acoustic reconstructions are in the stage my optical were in 1947 and it took me 15 years before optical holograms gave good pictures; we shall not have to wait 15 years before you get good results—ultra-sound for medical purposes and (non-destructive testing), and underwater imaging for oceanography." As for using holography for seismic waves, "use coherent oscillations, measure the field at the earth's surface and conclude what is below; holography for this type of problem is complementary to sonar, sonar and radar are wonderful for looking sideways, the only thing that they cannot do is to look forward. And holography can do exactly that."

It had been suggested[6] in 1968 that the remarkable property of the human memory of recognizing and recalling long sequences of which, at first, only a small fraction is consciously remembered, could be simulated by a mathematica model which had been called a temporal analogue of holography. Gabor devised another mathematical model, (Gabor, 1968ba,b), which he considered produced a closer analogue of holography though he did not wish to suggest that it corresponded to the reality of the nervous system. Later in the year 1969 in America he showed in "Associative Holographic Memories", (Gabor, 1969g), that in holography a mere *part* of the hologram suffices to recall the whole object and commented on the interest shown by the neurophysiologists; the fact that large parts of the brain can be destroyed without wiping out a learned pattern of behavior has suggested that the brain might contain a holographic mechanism. "I do not suggest

[6] By H. C. Longuet-Higgins, *Nature, Lond.* **217**, 104, 1968.

that such processes *are* present in the nervous system—but the possibility cannot be excluded and the hypothesis deserves careful examination."

He wrote, with H. Hartley, the Royal Society's Biographical Memoir, (Gabor & Hartley, 1970), on Sir Thomas Merton during 1969. Merton was a man after Dennis's own heart, a wonderfully able inventor who worked by inner compulsion rather than under economic necessity, a man for whom inventing was a creative joy; Merton had said that the research laboratory was the only place where one swallow can make a summer. Like Dennis, Merton had been an art connoisseur and, apparently like Dennis who spent the six winter months in America and most of the summer in Italy, would quote with relish a saying by J. M. Keynes. "The avoidance of taxes is the only intellectual pursuit that still carries any reward."

Several universities conferred honorary degrees on Gabor, the first being Southampton 1970, followed in a few months by Delft (1971) where the Oration was given by Professor Le Poole who had first heard Dennis at the electron microscope conference in Manchester in 1946 and had been in close contact with him ever afterwards. He told a little-known anecdote about the Graduand;—'walking in the streets of Berlin with your friend Leo Szilard and discussing the wave-nature of electrons advanced by de Broglie;—summing up, you both agreed that here was an ultra-short-wave radiation readily available with lenses to go with it, so why not build an electron microscope which should be capable of giving much better resolution than the light microscope. But you both agreed it would serve no purpose'. In his reply, (Gabor, 1971a), Gabor spoke of Delft as the premier technical university of Europe and told how he had missed making the first electron microscope "out of the pieces of the cathode-ray oscillograph with which I had made my doctorate in Berlin and this never ceased to annoy me. I spent much of my spare time thinking of ways and means to perfect the instrument I had failed to invent. The super-resolution I aimed at in 1948—and failed to achieve—could be achieved today if somebody went at it with sufficient determination." He spoke of his failure also to invent the laser: "In 1921 I attended a seminar by Einstein on Statistical Mechanics and shall never forget his brilliant derivation of Planck's Law in which 'simulated emission' occurred for the first time, presented with unforgettable gusto. When, in 1950, I realized that holography could be interesting in light optics if one had a strong source of coherent light, I had the idea of the pulsed laser and proposed it to a student, but he thought it too risky and I agreed with him." It is interesting to note here that Dennis *never* contemplated doing the experimental research himself; although, as has been here recorded, he gave much time and attention to the planning of an experiment he was not a real experimenter, Ivor Williams on B.T.H. and my Aldermaston staff did the experiments: Dennis was, as he told me in 1934, an "ideas man". Mr R. H. McMann, a colleague in the C.B.S. Laboratories where Dennis worked for most of the Michaelmas and Lent terms, contributes a similar assessment: 'Dennis did all his own typing by the hunt and peck method and consequently, since he was a very prolific writer, he seemed to spend all his time out of the laboratory in front of a typewriter. He did all his own drawings despite the fact that we had a large Drafting Department at his disposal! He was a good experimentalist but he liked others to actually do the experiments. He would go into the lab., observe what was going on for 10 minutes, make a few suggestions, then go back to his office where he would hunt and peck several pages of ideas on how to properly conduct the experiment.' Mr McMann also provided another nice touch: 'Dennis was a very hard worker but he worked strictly at his own pace.' (The reader might remember that at this time Dennis had retired, supposedly.) 'If he felt like taking a nap at 2 in the afternoon, and he frequently did, he would simply sit at his desk and go to sleep. At one time he shared an office with a colleague and their desks faced each other. Every afternoon the two of them could be found facing each other, quietly asleep. But of course they could also be found in the laboratory on almost every Saturday and Sunday and if need be, on any day of the week till midnight.'

Laser speckle "is a direct consequence of the high coherence of laser light and has long been recognized as Enemy Number One," wrote Dennis as he worked in America in 1969, (Gabor, 1970e). The reconstructed image is marred by interference fringes set up by every particle of dust, etc. He distinguished between objective speckle, due to uneven illumination of the object, and subjective speckle caused by imperfections in the optical reproduction. He then showed that the former could be eliminated by a special type of illumination of the hologram through two phase-plates of great complexity "The realization of these may require considerable experimental skill," the subjective speckle presented even greater difficulties which were not resolved in this paper.

In April 1970 he spoke at the Engineering Conference at Chicago, (Gabor, 1970b), on the 'Scientific and technological trends for future television'. First he referred to the start of British t.v. when the picture was scanned at 405 lines and now with 625 lines; he suggested that the Americans would jump to the front with 750 or even 1000 lines, though high-quality pictures in the homes might be better provided at 1000 lines from video cassettes. Looking further ahead he thought the large screen, say 3 ft × 4 ft (0.9 m × 1.2 m) might be served with 1000 lines; this should suffice for such screens, as viewers would not be sitting very close to them. He dismissed the 'solid-state' t.v. screen; little progress has been made in 20 years. As for three-dimensional t.v. which could even then actually be achieved, using perhaps holographic techniques, the complications at the receiving end would be stupendous and he did not think "to my regret, that full 3-D. t.v. will arrive except in the remote and rather unlikely future post-economic society."

Throughout the year 1970 he was immersed in writing *The mature society*, (Gabor, 1972a), which he sent to the printers early in 1971. It was a sequel to his earlier (1963) book, (Gabor, 1964b); "in the intervening 8 years more and more thinking people have realized that our free industrial civilization is not likely to survive another generation without fundamental institutional changes." We had seen Soviet expansion in the occupations of Czechoslovakia and interference in the Middle East, and we have witnessed moral erosion manifested by rebellious university youth. Added to this, growth of the National Product which had brought such affluence had slowed down, and though there would be ups and downs of prosperity, Gabor thought that a crisis of saturation was almost upon us. He argued for improved scientific forecasting of socio-economic matters (especially in a lecture 'Cybernetics and the future of our industrial civilization', (Gabor, 1971c)) and set out to sketch a mature society which accepts zero growth in materialistic matters while preserving growth in the quality of life. The retreat from material growth is going to be difficult. Hundreds of government and other bodies exist to make forecasts and long-term plans and many of these concern themselves with material possibilities, but Gabor suggested that it is the psychological reactions which will be the most threatening to the change which has to occur, and he considered that a new science *The new anthropology* will be needed. He advocated, as before, a change in state financing, a different approach to education, first at an early age and then at university level where there could be two branches, one for the gifted minority, the other to prepare the less-gifted majority for entrance into the permissive society. He advocated that everyone should have a chance to change occupation in middle life, and that everyone ought to spend some time in some service occupation. "The programme for the future must not become a platform for political parties," and he offered suggestions for the way it might be put across to the people. With two American colleagues he gave a general review of holography to the American Association for the Advancement of Science, (Gabor et al., 1971).

Gabor had been on the Nobel Prize list for some time. Indeed he had always recognized that the original invention of holography might one day be so rewarded; he and I hoped in Aldermaston years that if a resolution of 1–2 Å were ever achieved, he and the team producing the electron hologram and the optical reconstruction might share a Prize. After the explosion of holography in the 1960s the recognition of his basic work and of the brilliant inventions which followed appeared to be well justified. However, the atomists seemed to win the day and in 1968 in a mood of depression he said that the days of awards for inventions were passed. Now, in December 1971, 'the Winter of his discontent' was ended by the award of the Nobel Prize in Physics and his delight was shared by his friends and the many scientists for whom holography had become a way of life. Several thousand papers had now been written on holography and its applications.

His Nobel Lecture, (Gabor, 1971f), embraced the foundations of wave theory and the invention of holography, and then he reviewed the wonderful achievements of holographic scientists, concluding with a brief mention of two of his favorite holographic brainchildren, Panoramic Holography or Holographic Art, and three-dimensional holography without viewing aids (the old basic concept of 1932 mentioned by Oscar Deutsch). So far, holograms could extend to a depth of a few meters; if they could be extended to infinity a 'picture' on a wall could look like a window through which a distant vista could be seen. Work on both these concepts was proceeding but "I am not sure whether the difficulties will be overcome in this century. Ambitious schemes, for which I have a congenital inclination, take a long time for their realization. I shall be lucky if I shall be able to see in my life-time the realization of holographic electron microscopy on which I started 24 years ago"—we shall see soon the sequel to this reflection—"but I am one of the few lucky physicists who could see an idea of theirs grow into a sizeable chapter of physics. I am deeply aware that this has been achieved by an army of young, talented and enthusiastic researchers and I want to express my heartfelt thanks to them, for having helped me, by their work, to this greatest of scientific honors." To a German friend he wrote "Ich schäme mich beinahe, das ich für eine so einfache Erfindung den Nobel-Preis enhalten habe".

The City University had invited Gabor to give the second Edwards Memorial Lecture early in 1972 and for this he chose to give the Nobel Lecture, (Gabor, 1972c); this was followed in December of that year by an Hon.D.Sc. of the University conferred in Guildhall by the Chancellor, the Lord Mayor of London, this being the first occasion on which a Nobel Prizeman had received a degree in this famous Hall in which, over the centuries, so many distinguished men had been honored. The Orator on this occasion, Professor C. W. Miller, spoke of Gabor's mixture of invention and discovery; 'he has never let himself be restricted by existing technology, and his far-reaching, perhaps futuristic thinking has been directed to fundamental reasoning; some of his contributions at present dormant, will, no doubt, make their full impact at a future date'. Surrey University and the Engineering College, Bridgeport, Conn., likewise honored him, and at his own university, London, he received the LL.D. from the hands of H.M. The Queen Mother in 1973.

The two years 1972, 1973 were greatly occupied with giving lectures, the Tykociner Memorial at the University of Illinois, (Gabor, 1972e), the eighth Annual Science Policy Foundation, (Gabor, 1972f), one to a Government and Public Affairs group at U.C.L.A., (Gabor, 1972h), two in Japan, the Fawley lecture in Southampton, (Gabor, 1972i), one in Budapest, one in 1972 at the Institute for Medical Optics in the university in Munich where a memorial plaque to the naming of the research laboratory after him was unveiled, the street leading to the laboratory being Die Gaborstrasse (there is a Nobel Avenue in Nottingham with one side-street named after Gabor), and a second visit to Munich to the Siemens Foundation, [146.]

Some of these lectures were on holography, others on variants of *The mature society*, (Gabor, 1972a), but all are so different and so brilliantly expounded. He gave a great amount of time also to the Club of Rome: its first report on 'The limits to growth' published in 1972 had been criticized on the grounds that it had neglected the help which science and technology might be able to give in overcoming threatened scarcity of world resources. Early in 1973 Dennis, with Sr Umberto Colombo, proposed that a Study group should be asked to identify those areas of scientific endeavor which could increase man's capacity to exploit and regenerate natural resources. They co-chaired this Working Party, meeting in Rome, Tokyo, and Milan and completed their task in 1976, publishing *Beyond the age of waste*, (Gabor et al., 1978), in 1978.

In the Spring of 1973 the Gabors were guests of the Young Presidents' Organization, an organization comprising only men who, before they are 40 years of age, have become Presidents of their respective Companies, and they must leave the Organization before their 50th birthday. They hold an annual event called the 'International University' and in 1973 took over the Queen Elizabeth II for a 6-day cruise from Southampton into the Mediterranean and invited academics and people prominent in public life. He gave two formal lectures, as a futurist; they were well received as were his rejoinders to queries from his audience. Mr J. F. Engelberger writes 'Dennis was confronted with an assemblage of aggressive and self-confident middle-aged executives from the United States, Europe and Japan. He revelled in this student body, so different from his students in electron physics where classes were heavily loaded with Hindus and Pakistanis whose educational motives distressed him; he deplored their fascination with scientific exotica when their home countries so dearly needed basic engineering contributions.'

He was elected, for four years, a Trustee of the International Federation of Institutes for Advanced Studies. This Federation had been studying problems of the type mentioned in Gabor's books and the October 1979 Seminar, 'From vision to action' will be in memory of him. The Director, Dr S. Nilsson, recalls Dennis saying "The so-called Western civilization is built on extraordinary successful technology, but spiritually on practically nothing".

He prepared several papers for conferences in 1974 but was unable to present them in person as he suffered a severe cerebral haemorrhage when in Lavinio in the summer. To the International Meeting of Communications and Transports (Geneva) he dealt with the energy crises precipitated by the fourfold rise in the price of oil, (Gabor, 1974e), and at the International Institute for Peace, meeting in Vienna, he dealt with the impact of this crisis on power politics, (Gabor, 1974d). He sent two papers to the International Conference on Electron Microscopy meeting in Canberra, (Gabor, 1974b, 1974c): one was an extremely detailed review of the history of the electron microscope, one of the best ever written: in it he repeated more accurately the story Professor Le Poole has told of the 1928 Szilard/Gabor conversation in the Cafe Wien, Berlin. Gabor had replied to Szilard "But one cannot put living matter into a vacuum, and everything would burn anyway to a cinder under the electron beam". It would, of course, but, as he wrote, "who would have dared to believe that the cinder would preserve not only the structure of microscopic bodies but even the shape of organic molecules!" Then he looked ahead; he was still waiting for his 1948 electron holograms, but in the intervening 25 years alternate approaches had been very successful, "holography entering into electron microscopy through the back door", he stated the immediate goal without giving detailed advice, "I do not want to spoil the pleasure of investigators as to how to achieve it."

Dennis had been invited to be the Chairman of this Conference, paper (Gabor, 1974b) was to have been his opening address, and his inability to attend had been a bitter blow. His old friend of Aldermaston days, Professor Tom Mulvey, read the Address and also spoke about the content of the second paper, (Gabor, 1974c). This had been hastily prepared and was not typed in the Conference standard format so it was not published, and indeed has never been published. However, this second paper concerned such an important milestone that its origin must be given here, however briefly. Professor Bartell of the University of Michigan had written to Dennis in April about having achieved direct views of electron clouds in atoms by means of the two-stage holographic microscope based on the original 1948 principle; with his collaborator, Dr Ritz, he had achieved a theoretical resolving power of 0.08 Å. Professor Bartell has kindly presented me with the correspondence which followed. Dennis was delighted with the news (10 April 1974) but thought he saw a flaw: he asked for a slide of the atom picture to show at Canberra where he was to give the opening address, (Gabor, 1974b). Professor Bartell was able to satisfy his doubts and Dennis replied (22 April) that he could "see a good chance that your idea will revolutionise electron microscopy; it is not difficult to build a few heavy atoms into complicated molecules such as DNA and these ought to give *enough coherent background to make visible light atoms nearby*". He indicated how this should be done and wrote a paper, (Gabor, 1974c), in which he suggested that heavy atoms should be evaporated as a widespread lattice onto the object; his calculations indicated that with this technique carbon atoms may be made visible by the ordinary process of electron microscopy. To Bartell he wrote (31 May) "your work has started me thinking and I have worked out an approximate theory of heavy atom holography: it is complicated but it can be done". In his next letter, written by an amanuensis, he had to report his severe illness, but it is so gratifying to note from the above correspondence that he lived to see the realization of his early dream and was able to write about it, and calculate afresh the outline of a new chapter in microscopy.

His final paper, in three parts, (Gabor, 1975), dealt with the transmission of information by coherent light, as in the experiments done by Goss 20 years earlier, (Gabor & Goss, 1966), and with suggested ways to reduce speckle, as previously studied, (Gabor, 1970e).

An account of Gabor's scientific work would not be complete without some reference, however brief, to his many patents. Mention has already been made to a few of the early patents but Gabor did not compile a list and I regret I have not had time to trace them all. Those of later years were mostly assigned to the C.B.S. and I am grateful to the C.B.S. Counsel for sending me copies of the American versions; all of them are erudite in the extreme. Two concern the magnetization of plastic filaments and tapes of the kind used in sound reproducing machines; these date back to 1957, ten years before retirement. The first (U.S.A. 2 911 317) describes a novel way of aligning magnetic particles so that their direction of easy magnetization lies at 45° to the axis of the plastic filament and thus there is less interference magnetically between adjacent turns. The second (U.S.A. 3 064 087) concerns a very clever way to magnetize a tape from both sides. Two years later he invented (U.S.A. 3 108 383) for the C.B.S., apparatus to decode into pictorial form information encoded in a t.v. video tape; it was based on the diffraction patterns produced when monochromatic light traversed the tape and then fell on a mask having as many graded transparent lines in it as the scanning lines of the t.v. signal; the emergent light, focused, produced a recognizable image good enough for the purpose of editing a video film strip.

Then Gabor returned to his early love of 1939–40, a three-dimensional projection for a cinema (a partial version of some new thinking had been written up in 1960, (Gabor, 1960b)), but now he had two aids not available in 1940, holography and laser beams. The essence of the 1966 patent (U.S.A. 3 479 111) was the creation of a 'deep hologram' on a cinema screen, (for theory of the deep hologram, see (Gabor & Stroke, 1968) published two years later). The screen, or segments of it in turn, was to be coated on the reverse side with a photosensitive medium and then processed to produce the deep hologram, the process starting with a laser beam split into two halves, one half being reflected onto the reverse of the emulsion, the other onto the front surface of the emulsion, the direction of the two beams coinciding with the direction of light from one cine projector, say the left-eye view, and with the direction of light to the left eye of the viewer. Both these light sources had then to be made to transverse the whole of the emulsion and then the process repeated for beams of light from the right-eye projector to the viewer's right eye; finally, for color cine films the whole process had to be repeated with lasers of two more different colors. At the time of writing I am not aware that this patent has been developed.

In the previous year, 1965, he had invented an ultrasonic camera (assigned to C.B.S. May 1966, U.S.A. 3 869 904) employing holographic principles based on laser beam illumination; sound waves in a fluid, modified by the object to be 'photographed', fell on a thin silvered membrane and thus distorted it, its position was registered by being illuminated by part of a laser beam, the other part of which bypassed the vessel and interfered with the reflected wave and thus produced a hologram of the object. Some years later he developed this idea further (1971, U.S.A. 3 745 814), replacing the ultrasonic generator by a short sharp pressure wave; prior to this wave the membrane had been illuminated with a laser and a hologram taken of it, thus recording any imperfections it may have had, and then the hologram was combined with the reflection from the membrane distorted by the pulsed wave modified by the object.

Another patent taken out before retirement and made over to C.B.S. (U.S.A. 3 506 952) concerns a sonar system employing holography; it is referred to in 1969, (Gabor, 1969f). A narrow acoustic wave scans an object and each one of an array of small hydrophones collects reflected waves from all parts of the object field. Each hydrophone is also fed with a modulating signal of the same frequency as the acoustic wave but 30° out of phase, this constitutes the 'reference wave', as used in the generation of optical holograms. The combined joint intensity signals are then converted into modified light signals by an ingenious set of micro-shutters which are moved by the electrical signals and they then fall on a photographic film. Also early in 1967 he laid down the principles of holographic imaging (U.S.A. 3 561 838) and later in the year the principles of picture making (U.S.A. 3 545 836) both of which were beautifully portrayed in his Friday Evening Discourse at the Royal Institution in February 1969, (Gabor, 1969c).

Mr McMann also informed me that Dennis analyzed the ideas of other members of the staff, at the request of the Director, and frequently contributed ideas of his own; he was never caught up in the 'not invented here' syndrome and would help with someone else's invention as much as his own. One electrical printing system had been devised in the Laboratories of C.B.S. and Dennis did a great amount of work analyzing the electron optics of the image dissector pick-up tube and c.t.r. display. Even today, the performance of this system is far above anything else that has ever been done with electron beams. All American patents are printed by this system.

The succeeding years must have been most frustrating for Dennis; he could neither read nor write, and later almost lost the power of speech.

Fortunately his keen intellect and hearing were unimpaired so that he could enjoy company and conversation, and the tape recordings of lectures and scientific papers sent to him by his friends. His physique remained good and with Marjorie he was able to travel far and wide, thus unconsciously inspiring many with his grit and determination to overcome the human physical difficulties just as, all his life, he had overcome the difficulties of science.

He observed with great pleasure the invasion of holography into the field of art and of advertising; his friend, Salvador Dali, was most impressed with the possibilities of 3-D since holography provides the third dimension far surpassing any of the conventional tricks based on illusions. Instead of lasers, ordinary lamps, even candles, could now be used to illuminate the hologram and the art museums were quick to seize upon this new medium. The Gabors in March 1977 visited the newly created Museum of Holography in New York of which Dennis's Membership was no. 1; in the winter of that year they were honored at the holography exhibition staged at the Royal Academy and in the following March they paid their second visit to the New York Museum and Dennis was invited to become the Honorary Chairman of the Board of Trustees. Here he sat for his holographic portrait by Hart Perry, a form of art destined to become very popular.

At his home in Italy Dennis swam frequently and they entertained numerous friends Indeed his friends will always treasure memories of the Gabors' happy life and generous hospitality; for 25 years we 'let the New Year in' with a party at which Dennis was the leading spirit. Though not blessed with children he was always fond of young people and little ones, who responded with affection.

After a happy summer in Italy, in 1978 he became bed-ridden and died peacefully in a nursing home. At the funeral service, the Reverend Ian Robson echoed all our thoughts: 'We do indeed give thanks for Dennis's interests in the small things of life, his love of children, of people, of music and of subjects which deeply concern mankind.' We knew him as an extremely genial, warmhearted man who knew how to listen especially when talking to younger folk whom he encouraged by his manifest interest in their work. As Professor McGee wrote 'He walked with kings, yet kept the common touch'.

Pericles provides me with my last thought; 'Famous men have the whole Earth as their memorial; it is not only the inscription on their graves that mark them out; no, in foreign lands and in peoples' hearts their memory abides and grows'. So will it be with Dennis Gabor.

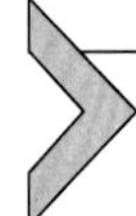

8. Honours

Commander of the Order of the British Empire, 1970
Honorary membership of the Hungarian Academy of Sciences, 1964
The Cristoforo Colombo Prize of Genoa, International Institute of Communications, 1967
The Thomas Young Medal of the Physical Society, 1967
The Michelson Medal of the Franklin Institute, 1968
The Rumford Medal of the Royal Society, 1968
The Medal of Honor of the I.E.E.E., 1970
The Semmelweiss Medal, American Hungarian Medical Association, 1970
Holweck Prize of the French Physical Society, 1971
Nobel Prize in Physics, 1971
Foreign Associate Nat. Acad, of Sciences, U.S.A., 1973
George Washington Award, Amer-Hungarian Studies Foundation, 1973
Hon.D.Sc. University of Southampton, 1970
Hon.D.Sc. Delft University of Technology, 1971
Hon.D.Sc. University of Surrey, 1972
Hon.D.Sc. The City University, 1972
Hon.D.Sc. Engineering College Bridgeport, 1972
Hon.D.Sc. University of Columbia, N.Y., 1975
Hon.LL.D. London University, 1973

My thanks are due to the following who responded to my request for details of their connection with Dennis Gabor.

a. In early years in Budapest, Dr A. Budincsevits, Dr G. Hrehuss, Mr E. Redl, Professor I. P. Valkó, Professor I. Tarján and Dr Millner Tivadar; in Berlin, Professor G. Franke, Dr H. Müller, Dr Z. A. Medvedev and Mrs Reiter.

b. In B.T.H. and A.E.I. Professor G. C. Carter, Dr J. Dyson, Mr C. E. Fenwick, Dr M. E. Haine, Professor C. J. Milner, Professor T. Mulvey, Mr R. L. Rouse, Mr L. Rushforth, Mr K. J. R. Wilkinson and Mr W. I. Williams.

c. In Imperial College years Dr M. J. Albert, Professor E. A. Ash, Professor C. Cherry, Professor W. A. Gambling, Professor M. A. Jaswon, Dr P. Kalman, Professor N. Kurti, Mrs E. Moffatt, Professor J. D. McGee, Professor C. W. Miller, Mrs J. Pingree, and Professor J. Brown.

d. In later years or overseas, Professor L. S. Bartell (University of Michigan), Mr L. Beiser (of N.Y.), Professor H. Boersch (Hempelsberg), Dr J. F. Engelberger (of Unimation), Dr R. Galli (Club of Rome), Professor A. Lohmann (of University of Nürenberg), Mr R. H. McMann (C.B.S.) Mr A. J. Molnar (American Hungarian Foundation), Dr S. Nilsson (IFIAS, Stockholm), Mr K. Ohira (Japan), Professor J. B. le

Poole (Delft), Professor G. W. Stroke (State University N.Y.), Dr A. Tonomura (Hitachi, Japan), Professor J. Tsujiuchi (Tokyo Institute of Technology), Professor W. Waidelich (University of Munich), Dr H. Wheeler (Santa Barbara), and above all to Mr André Gabor and to Marjorie for very many details.

I am grateful to the City University for library and stenographic help (Mrs H. Milloy), and to my wife for reading the memoir and making so many constructive suggestions.

The photograph reproduced was taken by Novotta Ferenc, Budapest, Hungary, about 1970.

List of publications by Dennis Gabor

German Period, 1925–33

Gabor, D. (1925). Berechnung der Kapazität von Sammelschienenanlagen. *Archiv für Elektrotechnik, 14*, 247–258.

Gabor, D. (1926a). Oszillographieren von Wanderwellen. *Archiv für Elektrotechnik, 16*, 296–302.

Gabor, D. (1926b). *Einige Untersuchungen mit dem Kathodenoszillographen zur Aufklärung von Überspannungserscheinungen. Verh. d 34 Hauptversammlung Verein d. Elektrizitatswerke* (p. 52).

Gabor, D. (1927a). Fortschritte im Oszillographieren von Wanderwellen. *Archiv für Elektrotechnik, 18*, 48–55.

Gabor, D. (1927b). Oszillographieren von Wanderwellen mit dem Kathodenoszillographen (Berlin Thesis). With an added article on investigations on surge arresters, ForschHft. Stud-Ges. HochstspannAnl. no. 1 (77 pages). Berlin: Julius Springer.

Gabor, D. (1928a). Ultraviolette Strahlen und Zellteilung (In collaboration with T. Reiter). *Strahlentherapie, 28*.

Gabor, D. (1928b). A Kátodoszcillográf és a vándorhullamok oszillografálásá. *Elektrotschnika, XXI*. 7–8 szam.

Gabor, D. (1930). Ein neuer Gleichrichtertransformator. *Wissenschaftliche Veröffentlichungen aus Den Siemens-Werken, 9*, 144–156.

Gabor, D. (1932). Zur Theorie des Lichtbogenplasmas. *Zeit für Technische Physik, 13*, 560–563.

Gabor, D. (1933a). Die Ausbildung der Maxwellverteilung im Langmuirschen Plasma. *Physikalische Zeitschrift, 34*, 38–45.

Gabor, D. (1933b). Elektrostatische Theorie des Plasmas. *Zeitschrift für Physik, 84*, 474–508.

Gabor, D., & Reiter, T. (1928). *Zellteilung und Strahlung*. Berlin: Julius Springer.

In Britain, 1934–48 at B.T.H. Co.

Gabor, D. (1937a). Mechanical tracing of electron trajectories. *Nature (London), 139*, 373.

Gabor, D. (1937b). Electron optics. *British Electrical and Allied Manufacturers' Association Journal, 41*, 80–86. Also at B.T.H. Report no. 125 and Rugby Eng. Soc.

Gabor, D. (1938). What price peace? *Rugby Observer*.

Gabor, D. (1939). Electric discharges at low gas pressures (Survey). *Electronics Television, 641*, 710. Also as B.T.H. Report no. 135 and Rugby Eng. Soc., 12 January 1938.

Gabor, D. (1942). Electron lenses. *Nature (London), 150*, 650.

Gabor, D. (1942–43). Electron optics (Survey). *Electronics Engineering, 15*, 295. See also pp. 328, 372.
Gabor, D. (1944a). Energy conversion in electronic devices. *Journal of the Institution of Electrical Engineers, 91, III*, 128–141.
Gabor, D. (1944b). Power loss in deflecting condensers. *Wireless Engineer, 21*, 115–116.
Gabor, D. (1945a). Dynamics of electron beams. *Proceedings of the Institute of Radio Engineers, 33*, 792–805.
Gabor, D. (1945b). Stationary electron swarms in electromagnetic fields. *Proceedings of the Royal Society of London. Series A, 183*, 436–443.
Gabor, D. (1946a). A zonally corrected electron lens. *Nature (London), 158*, 198.
Gabor, D. (1946b). Theory of communication. *Journal of the Institution of Electrical Engineers, 93, III*, 429–457.
Gabor, D. (1947a). Acoustical quanta and the theory of hearing. *Nature (London), 159*, 591–602.
Gabor, D. (1947b). Stabilizing linear particle accelerators by means of grid lenses. *Nature (London), 159*, 303.
Gabor, D. (1947c). A space charge lens for the focusing of ion beams. *Nature (London), 160*, 89.
Gabor, D. (1947d). New possibilities in speech transmission. *Journal of the Institution of Electrical Engineers, 94, III*, 369–387.
Gabor, D. (1948a). *The electron microscope*. (1st British ed.) (1945). London: Hulton Press. (2nd British ed.) (1948). London: Electronic Engineering (104 pages). (U.S. ed.) (1948). New York: Chemical Publications Inc. (164 pages).
Gabor, D. (1948b). Recherches sur quelques problémes de Telecommunication et d'Acoustique. *L'Onde Electrique, 28*, 433–439.
Gabor, D. (1948c). A new microscopic principle. *Nature (London), 161*, 777.
Gabor, D. (1949). Microscopy by reconstructed wavefronts. *Proceedings of the Royal Society of London. Series A, 197*, 454–487.

At Imperial College of Science and Technology, 1940–67

Gabor, D. (1950a). Electron optics at high frequencies and at relativistic velocities. In *Reunion d'Opticiens*. Paris: Ed. Rev. d'Optique.
Gabor, D. (1950b). Problems and prospects of electron diffraction microscopy. In *Proc. Electron Microsc. Conf., Delft*.
Gabor, D. (1950c). Photons and waves. *Nature (London), 166*, 724–726.
Gabor, D. (1950d). Communication theory and physics. *Philosophical Magazine*, 7(41), 1161–1187 (In French). In the volume *La Cybernetique*. Paris: Ed. Rev. d'Opt. (1951).
Gabor, D. (1951a). Electron optical systems with helical axis. *Proceedings of the Physical Society. Section B, 64*, 244–255.
Gabor, D. (1951b). Diffraction microscopy (Popular survey). *Research, 4*, 107–112.
Gabor, D. (1951c). Microscopy by reconstructed wavefronts II. *Proceedings of the Physical Society. Section B, 64*, 449–469.
Gabor, D. (1951d). Plasma oscillations (Survey, but mostly original). *British Journal of Applied Physics, 2*, 209–218.
Gabor, D. (1952a). Wave theory of plasma. *Proceedings of the Royal Society of London. Series A, 213*, 73–86.
Gabor, D. (1952b). Louis de Broglie et les limites du monde visible. In M. Albin (Ed.), *Louis de Broglie, Physicien et Penseur* (pp. 241–252). Paris.
Gabor, D. (1952c). *A summary of communication theory*. Reprinted from 'Communication theory,' 1953. Published by Butterworths Scientific Publications. Papers read at a Symposium on 'Applications of Communication Theory held at the I.E.E., London, September 1952.

Gabor, D. (1953a). Electron optical systems with curved axes. In *Comptes Rendus du ler Congres Int. de Microscopie Electronique* (pp. 57–62). Paris: Rev. d'Opt.
Gabor, D. (1953b). Generalized schemes of diffraction microscopy. In *Comptes Rendus du ler Congres Int. de Microscopie Electronique* (pp. 128–137). Paris: Rev. d'Opt.
Gabor, D. (1953c). Der Nachrichtengehalt eines elektromagnetischen Signals. *Archiv für Elektronik und Übertragungstechnik*, *7*, 95–99.
Gabor, D. (1954a). La meilleure utilization des plaques photographiques pour l'enregistrement des quanta de lumiére. *L'Onde Electrique*, 800–803.
Gabor, D. (1954b). Communication theory and cybernetics. In *Proceedings of the Institute of Radio Engineers, Prof. Group on Circuit Theory, CT-1* (pp. 19–31).
Gabor, D. (1954c). Progress in microscopy by reconstructed wavefronts. In *Electron Physics, U.S. Nat. Bur. of Standards, Circ. 527* (pp. 237–246).
Gabor, D. (1954d). Velocity of electron pulses (In collaboration with L.L. Whyte & D.L. Richards). *Nature (London)*, *174*, 398.
Gabor, D. (1955a). *Models in cybernetics.* Estratto del Vol. II degli Atti del Convegno di Venezia (1–4 Ottobre 1955) su 'I Modelli nella Technica'.
Gabor, D. (1955b). What have we learned so far from Cybernetics? (A talk in the Third Programme of the R.A.I.) (Italian Radio).
Gabor, D. (1955c). Electronics in the Imperial College of Science & Technology, London. *Nature (London)*, *175*, 885.
Gabor, D. (1956a). Collective oscillations and characteristic energy losses of electrons in solids. *Philosophical Magazine*, *1*(8), 1–18.
Gabor, D. (1956b). Light and information. In *Proc. Conf. on Astronomical Optics, Manchester, 19–22 April 1955*. North Holland Publ. Co.
Gabor, D. (1956c). Collecting information on partially known objects. In *Proc. Conf. on Astronomical Optics, Manchester, 19–22 April 1955*. North Holland Publ. Co.
Gabor, D. (1956d). Optical transmission. In *3rd London Symposium on Information Theory, 12–16 September*. London: Butterworth.
Gabor, D. (1956e). Plasma oscillations. In *Proc. Symp. on Electromagnetic Wave Theory, Ann Arbor, Michigan, June 1955* (pp. 526–530). I.R.E. New York.
Gabor, D. (1956f). Theory of electron interference experiments. *Reviews of Modern Physics M*, *28*, 260–276.
Gabor, D. (1956g). A new television tube. *Journal of the Television Society*, *8*, 142–145.
Gabor, D. (1957). Die Entwicklungsgeschichte des Elektronenmikroskops. *ETZ. Elektrotechnische Zeitschrift. Ausgabe A*, *78*, 522–530.
Gabor, D. (1958a). A flat television tube. *Journal of the Institution of Electrical Engineers (New Series)*, *4*, 609–612.
Gabor, D. (1958b). A new cathode-ray tube for monochrome and colour television (In collaboration with P.R. Stuart and P.G. Kalman). *Proceedings of the Institution of Electrical Engineers, 105, part B*(24).
Gabor, D. (1959a). *Electronic inventions and their impact on civilization.* Inaugural Lecture, 3 March 1959. Modern Book Co.
Gabor, D. (1959b). Inventing the future. *Encounter*, *1960*, 3–16.
Gabor, D. (1959c). 'Learning' filters, predictors and recognizers. *Supplemento al Nuovo Cimento. Serie X, N.2*, *13*, 455–466.
Gabor, D. (1959d). Television compression by 'countour interpolation'. *Supplemento al Nuovo Cimento. Serie X, N.2*, *13*, 467–473.
Gabor, D. (1960a). The next ten years in engineering and technology. *New Scientist*, 7(165).
Gabor, D. (1960b). Three-dimensional cinema. *New Scientist*, *8*(191), 141–145.
Gabor, D. (1960c). A self-optimising non-linear filter, predictor and stimulator. In E. C. Cherry (Ed.), *Information Theory, 4th London Symposium, 1960* (pp. 348–352). London: Butterworths.

Gabor, D. (1961a). A universal non-linear filter, predictor and stimulator which optimizes itself by a learning process (In collaboration with W.P.P. Wilby and R. Woodcock). *Proceedings of the Institute of Electrical Engineers, 108, Part B*(40), 422–438.
Gabor, D. (1961b). The Bertele arc. *Nature (London), 189*, 821–822 (In collaborate with J. Vithayathil).
Gabor, D. (1961c). A new thermionic generator. *Nature (London), 189*, 868–872.
Gabor, D. (1961d). Television band compression by contour interpolation. *Proceedings of the Institution of Electrical Engineers, 108*, 303–315 (In collaboration with P.C.J. Hill).
Gabor, D. (1961e). Light and information. *Progress in Optics, 1*, 111–153 (Ed. E. Wolf).
Gabor, D. (1961f). Zivilisation und Erfindung E.N.I. *Merkur, 3*, 203–217.
Gabor, D. (1961–62a). Machines and prediction. In *La Scuola in Azione, San Donato Milanese 1961–62* (Chapter 13).
Gabor, D. (1961–62b). Predicting machines. Cambridge. *Opinion, 27*, 14–17.
Gabor, D. (1961–62c). Colour television. *Endeavour, 21*, 25–34.
Gabor, D. (1963a). Theory of gas discharges with extraneous ion supply (Symposium on Thermionic Power Conversion Colorado Springs, May 1962). *Advanced Energy Cornersion, 3*, 307–314.
Gabor, D. (1963b). Technology, life and leisure (George Thomson Lecture, October). *Nature (London), 200*, 513–518.
Gabor, D. (1964a). *The future of computers (an after-dinner talk).*
Gabor, D. (1964b). *Inventing the future.* London: Seeker & Warburg (1963). New York: A. Knopf (1964) (Translated into 7 languages. German: Menschheu Morgen Scherz, 1965; Inventons le Futur, Plon. 1965; Il Paradiso Artificial della Technology, Tamburini, 1967; also Dutch, Spanish and Japanese translations).
Gabor, D. (1964c). Communications tomorrow. *Discovery, 25*, 59–61.
Gabor, D. (1965a). *Inventing the future.* The Joseph Wunsch Lecture given at Technion, Haifa, 1965.
Gabor, D. (1965b). Unborn instruments. *Discovery, 26*, 38–41.
Gabor, D. (1965c). The dimensions of consciousness. In *The scientist speculates* (pp. 66–70) (Chapter 21).
Gabor, D. (1965d). Die Zukunst der technischen Zivilisation. *Elektrotechnik und Maschinenbau, 82*(6), 291–296.
Gabor, D. (1965e). Eine Vorschau über moderne Methoden der Energieerzeugung. *OZE (Österreichische Zeitschrift für Elektnzitatswirtschaft), 18*(8), 319–324.
Gabor, D. (1965f). *Imaging with coherent light.* Summary of C.B.S. Laboratories Seminar.
Gabor, D. (1965g). The smoothing and filtering of two-dimensional images. *Progress in Biocybernetics, 2*, 1–9.
Gabor, D. (1965h). Information theory in electron microscopy (Proc. symp. on quantitative electron microscopy, Washington, April 1964). *Laboratory Investigation, 14*, 801–807.
Gabor, D. (1965i). Character recognition by holography. *Nature (London), 208*, 422–423.
Gabor, D. (1965j). An essay on Aldous Huxley. In J. Huxley (Ed.), *Aldous Huxley—a memorial volume.* London: Chatto & Windus.
Gabor, D. (1965k). *Inventing education for the future.* Lecture delivered Institute of Government and Public Affairs. University of California, Los Angeles. Paper M R 59. (pp. 1–16).
Gabor, D. (1966a). Holography, or the 'whole picture'. *New Scientist, 29*, 74–78.
Gabor, D. (1966b). Wellenfront-Rekonstruktion oder 'Holographie'. *Physikalische Blätter*, 256–265.
Gabor, D. (1966c). Holography—the reconstruction of wavefronts. *Electronics and Power*, 230–234.
Gabor, D. (1966d). Holography and communications. In *Symp. on Generalized Networks. Polytechnic Institute of Brooklyn, April 12–14: Vol. xvi. Microwave research institute symposia series.* Brooklyn: Polytechnic Press.

Gabor, D. (1966e). Les transformations de l'information en optique. *Optica Acta, 13*, 299–310.
Gabor, D. (1966f). La liberté dans une civilization industrielle avancée. *Analyse el Prévision, II*, 819–822.
Gabor, D. (1966g). Ernst Ruska's grosses Lebenswerk. *Optik, 24*, 369–374.
Gabor, D. (1967a). Holography: Photography without lenses in three dimensions. *Journal of the Royal Society of Arts, CXV*, 246–259.
Gabor, D. (1967b). Character recognition. *Scienta, C11*, 48–55.
Gabor, D. (1967c). *Industry and society: The next 30 years.* Opening address to Section V discussion at the Royal Society 8 May 1967.
Gabor, D. (1968a). Three pressing problems. *Isis, 1553*, 15–16.
Gabor, D. (1968b). Holographic model of temporal recall. *Nature (London), 217*(a), 584–585. *217*(b), 1288.
Gabor, D. (1968c). The outlook for holography. *Optik, 28*, 437–441.
Gabor, D. (1968d). Prévision Technologique et Responsabilité Sociale. *Analyse et Prévision, VI*, 719–733.
Gabor, D. (1968e). The prospects of industrial civilisation. 58th Mav Lecture. *Journal of the Institute of Metals, 96*, 193–197.
Gabor, D. (1968f). The future of industrial civilisation. *Student World, 61*, 149–160.
Gabor, D. (1969a). Open-ended planning. In *Perspectives of planning* (pp. 239–350). Paris: OECD. See also pp. 511–512.
Gabor, D. (1969b). Information processing with coherent light. *Optica Acta, 16*, 519–533.
Gabor, D. (1969c). The hologram. *Proceedings of the Research Institute of Great Britain, 43*, 35–70.
Gabor, D. (1969d). Normative technological forecasting. *Technological Forecasting, 1*, 1–3.
Gabor, D. (1969e). Science of civilisation. *New Scientist*, 184–185.
Gabor, D. (1969f). Summary and directions for future progress. *Acoustical Holography, 1*, 267–275.
Gabor, D. (1969g). Associative holographic memories. *IBM Journal of Research and Development, 13*, 156–159.
Gabor, D. (1969h). Ernáhrung, Energiewirtschaft und industrielle Produktion der Zukunft. In *Menschen im Jahr 2000* (pp. 235–247). Frankfurt: Umschau Verlag.
Gabor, D. (1969i). Emlékézes Aldous Huxleyra 1969. *Hid (Ujvidék)*, 839–843.
Gabor, D. (1969j). Progress in holography. *Reports on Progress in Physics, 32*, 395–404.
Gabor, D. (1969k). The outlook for holography. *Optik, 28*, 437–441.
Gabor, D. (1970a). *Innovations: Scientific, technological and social.* Oxford University Press.
Gabor, D. (1970b). Scientific and technological trends and possibilities for future television. In *Engineering Conf., Chicago, 6 April 1970.*
Gabor, D. (1970c). Prognosen des technischen Fortschritts. *Systems, 69*, 21–29.
Gabor, D. (1970d). Material civilisation, hopes and fears. In *History of the 20th century: Vol. 8* (pp. 2653–2656) (Chapter 95). Published by Purnell for B.P.C. Publications, London.
Gabor, D. (1970e). Laser speckle and its elimination. *IBM Journal of Research and Development, 14*, 509–514.
Gabor, D. (1971a). Reply to the Hon. Degree given in University of Delft Feb 1971. *De Ingenieur, 83*(5), 80–81.
Gabor, D. (1971b). In *Proceedings of the NATO Advanced Study Institute, Milan, 24/June 4.* New York: Plenum Press. (a) *The principle of wavefront reconstruction 1948.* (pp. 9–14). (b) *Wavefront reconstruction.* (pp. 15–21). (c) *Information theory in holography.* (pp. 23–40).
Gabor, D. (1971c). Cybernetics and the future of our industrial civilisation. *Journal of Cybernetics, 1*, 1–4.
Gabor, D. (1971d). *Bergedorfer Gesprächskreis zu Fragen der freien industriellen Gesellschaft 1971. Protokoll Nr 40. October 11. Hamburg, Körber und Blanck.*

Gabor, D. (1971e). *L'Holographie. Prix Holweck*. Société Française de Physique.
Gabor, D. (1971f). *Holography, 1948–71* (Nobel Lecture, 11 December 1971). In *Le Prix Nobel en 1971, Stockholm, 1972* (pp. 169–201).
Gabor, D. (1972a). *The mature society*. London: Seeker & Warburg.
Gabor, D. (1972b). Demokratie und Technologie. *Physikalische Blätter*, *12*, 529–535.
Gabor, D. (1972c). Holography 1948–71 is also reprinted as the Edwards Memorial Lecture 1972, City University, 17 February 1972. Also as *Proc IEEE,* 1972, an extract pp. 655–668.
Gabor, D. (1972d). Can we survive our future? Interview conversation. *Notes and Topics*, 54–58.
Gabor, D. (1972e). *The scientist in the new society*. Tykociner Memorial Lecture. University of Illinois.
Gabor, D. (1972f). The new responsibilities of science. *Science Policy*.
Gabor, D. (1972g). Technology autonomous. *New Scientist*, 448–449.
Gabor, D. (1972h). *Thoughts on the future*. Inst. of Government and Public Affairs, Technology Autonomous, UCLA (pp. 448–449).
Gabor, D. (1972i). *The Fawley Lecture. The proper priorities of science and technology*. University of Southampton.
Gabor, D. (1973a). *Social control through communication* In *Understanding the new cultural revolution*. John Wiley & Son (Chapter 7).
Gabor, D. (1973b). Der Bertele-Bogen. *Elektrotechnik und Maschinenbau*, 7, 315.
Gabor, D. (1973c). *Holography 1973*. Lecture given at the Siemens Foundation, München-Nymphenburg.
Gabor, D. (1973d). Informationstheorie in der Optik. *Optik*, *39*, 86–92.
Gabor, D. (1973e). Technology and society in the last quarter of the 20th century. *Omega. International Journal of Management Science*, *1*, 231–233.
Gabor, D. (1973f). Environment versus energy 1973. *International Journal of Environmental Studies*, *5*, 160–161.
Gabor, D. (1973g). *Thoughts on the future* (Bowra Memorial Lecture, Cheltenham College 6 October 1973).
Gabor, D. (1974a). In *Contribution to management symposium, Davos*.
Gabor, D. (1974b). History of the electron microscope, from ideas to achievements. In *8th Int. Congr. Electron Microsc., Canberra* (pp. 6–12).
Gabor, D. (1974c). Heavy atom holography in electron microscopy. In *8th Int. Congr. Electron Microsc., Canberra*. N.B. this reference was not published in the Canberra Conference papers but will be presented by T. E. A. at Haifa on the occasion of a Holography Conference dedicated to the memory of Gabor, 25 March 1980, and will be printed by the *Israel Journal of Technology*.
Gabor, D. (1974d). *Wissenschaft und Frieden; Der Friede und die Erdölkrise Inst. für der Frieden*. Wien (pp. 73–76).
Gabor, D. (1974e). Development of communications in the frame of energetic and environmental constraints. In *International Meeting of Communications and Transport, Geneva, 7–12 October 1974. Open-ended planning*.
Gabor, D. (1974f). The history of holography. *Fizikai Szemle*, *24*, 289–303.
Gabor, D. (1975). Transmission of information by coherent light. Part I: Classical theory. *Journal of Physics. E*, *8*, 73–78. Part II: Transmission in the presence of noise. *8*, 161–163. Part III: Speckle noise. *8*, 253–255.
Gabor, D., & Ash, E. A. (1955). Experimental investigations on electron interaction. *Proceedings of the Royal Society of London. Series A*, *228*, 477–490.
Gabor, D., & Gabor, A. (1954). An essay on the mathematical theory of freedom. *Journal of the Royal Statistical Society. Series A*, *117*, 31–59.
Gabor, D., & Gabor, A. (1958). L'Entropie comme Mésure de la Liberté Économique et Sociale. *Cahiers de Science Économique Appliquée, Paris. Serie N.2*, 13–26.

Gabor, D., & Goss, W. P. (1966). Interference microscope with total wavefront reconstruction. *Journal of the Optical Society of America, 56*, 849–858.
Gabor, D., & Hampton, B. (1957). A Wilson cloud chamber with time-marking of particle tracks. *Nature (London), 180*, 746–749.
Gabor, D., & Hartley, H. (1970). T.R. Merton. *Biographical Memoirs of Fellows of the Royal Society (London), 16*, 421–428.
Gabor, D., & Jull, G. W. (1955). Experimental evidence for the collective nature of the characteristic energy losses of electrons in solids. *Nature (London), 175*, 718–720.
Gabor, D., & Nelson, C. V. (1954). Determination of the resultant dipole of the heart from measurements on the body surface. *Journal of Applied Physics, 25*, 413–416.
Gabor, D., & Sims, G. D. (1954). On the theory of the magnetron. In *U.S. Nat. Bureau of Standards, Circular 527* (pp. 253–255).
Gabor, D., & Sims, G. D. (1956). Theory of the preoscillating magnetron. *Journal of Electronics, 1*. I, 29–34. II, 231–262 (1955). III, 449–452 (1956).
Gabor, D., & Stroke, G. W. (1968). The theory of deep holograms. *Proceedings of the Royal Society of London. Series A, 304*, 275–289.
Gabor, D., & Stroke, G. W. (1969). Holography and its applications. *Endeavour, 28*, 40–47.
Gabor, D., Ash, E. A., & Dracott, D. (1955). Langmuir's paradox. *Nature (London), 176*, 916–919.
Gabor, D., Stroke, G. W., Restrick, R., Funkhouser, A., & Brumm, D. (1965). Optical image synthesis (complex amplitude addition and subtraction) by holographic Fourier transformation. *Physics Letters, 18*(2), 116–118.
Gabor, D., Stroke, G. W., Brumm, D., Funkhouser, A., & Labeyrie, A. A. (1965). Reconstruction of phase objects by holography. *Nature (London), 208*, 1159–1162.
Gabor, D., Albert, M. J., & Atta, M. A. (1967). A new type of composite all-metal electron emitter for valves and electron-optical devices. *British Journal of Applied Physics, 18*, 627–633.
Gabor, D., Tophill, H. A. W., & Smith-Whittington, J. E. (1968). A fully electrostatic flat, thin television tube. *Proceedings of the Institution of Electrical Engineers, 115*, 467–478.
Gabor, D., Koch, W. E., & Stroke, G. W. (1971). Holography. *Science, 173*, 11–23.
Gabor, D., Colombo, V., King, A., & Galli, R. (1978). *Beyond the age of waste*. Oxford: Pergamon.

CHAPTER ELEVEN

The French electrostatic electron microscope (1941–1952)

P. Grivet[a] and Peter Hawkes (Afterword)[b,*]
[a]Formerly Institut d'Electronique Fondamentale, Université de Paris-Sud, Orsay, France
[b]CEMES-CNRS, Toulouse, France
*Corresponding author. e-mail address: hawkes@cemes.fr

Contents

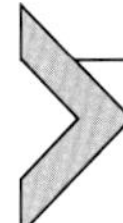

1. Introduction

1.1 Electron optics, a proper domain of "electronics"

A vigorous movement in favor of the history of science is presently gathering momentum in all the developed countries, an encouragement for the specialized International Scientific Union. Naturally, it is dealing neatly with "electronics," a part of physics deeply engaged in the social evolution of mankind, with such works as the 50th volume of *Advances in Electronics and Electron Physics* (Grivet, 1980) and a special issue of the journal *Elec-*

Advances in Imaging and Electron Physics, Volume 220
ISSN 1076-5670
https://doi.org/10.1016/bs.aiep.2021.08.011

tronics (*An Age of Innovation (The World of Electronics, 1930-2000)*, 1980). The roots of "electron optics" have not been neglected and recently two important books have been dedicated to electron diffraction (Goodman, 1981), and to "The Early Development of Electron Lenses and Electron Microscopy" (Ruska, 1979). These works already show that electron optics developed early as a three-pronged domain: one branch directly linked with the discovery of de Broglie's waves in 1924, and their most immediate application, diffraction spectroscopy; one branch linked with electron lenses with an early subdivision into electrostatic and magnetic lenses; and one branch on electron "super" microscopy with two variants, electrostatic and magnetic microscopes. In the first case, research remained active until 1945, the magnetic route appearing afterward as the most efficient for reaching high resolving power and high penetration of the electron beam.

1.2 The magnetic microscope

The saga of the *magnetic* electron supermicroscope is very ably and vividly reported by Ruska (1979) up to about 1940 and shows that this achievement, essentially due to Ernst Ruska, was at the limit of the possible, given the structure of science and technology in an advanced country, at that time: it appears also as a multidisciplinary enterprise, the role of the younger brother of Ernst, Helmut Ruska, a medical student in 1931, clearly being very important in sustaining the courage of the team in a critical period during the 1930s and bringing the moral help of distinguished biologists for maintaining hope in the final goal, supermicroscopic resolving power. Ruska's relation is in agreement with other historical pictures, more limited in scope, such as L. Marton's (1968) and W. Glaser's (1952), or more encyclopedic, such as that of Bodo von Borries (1949). Finally, the fundamental value of Ruska's contribution is evident in the difficulty still encountered in improving on his results; a prospective review of the present possibilities in this direction is published in the Acta of a recent ad hoc symposium (Parsons, 1978).

1.3 The electrostatic microscope

Surprisingly, electrostatic and magnetic lenses were first described nearly at the same time and in the same place, Max Knoll's laboratory, circa 1928 at the Technische Hochschule in Charlottenburg, near Berlin, by E. Ruska and Max Knoll (Knoll, 1929; Ruska, 1930). This innovation emerged from the pioneer work of physicists on the dynamics of electrons in the corresponding fields and the application of these theories to the early form of cathode-ray oscillographs by Braun (1897), Dufour (1914, 1920), and their followers, prominent among whom was Hans Busch (Ruska, 1979, p. 15). Nevertheless, during the 1930s the investigators seemed to divide progressively into two families, depending on the personal choice of a final goal, such as E. Ruska, thinking in terms of *supermicroscopy* and dreaming already of biological discoveries by overcoming the optical limits imposed by Abbe's formula in light microscopy. Conspicuous among this group was L. Marton, with his first experiments at Brussels in 1934 (Marton, 1934, 1935, 1936) (five communications to the Académie Royale de Belgique between 1934 and 1937). All were extremely preoccupied by the interaction of the beam with the specimen and also conscious of the numerous technological difficulties, especially with high voltage (it was early clear that 50 kV was necessary). They chose the magnetic solution: the lenses could be put outside the vacuum envelope; the region submitted to high voltage was reduced to the electron gun, and a high vacuum was easier to maintain, largely eliminating spurious discharges and voltage fluctuations, as well as any deleterious effect by oil vapors on the delicate object. Once this choice, first promoted by E. Ruska, had been made as a first step, it remained unaltered due to the necessity of concentrating efforts in such a difficult task.

To other investigators, the field of future applications of electron optics appeared both broader and brighter. Microscopy was certainly an important topic, but only one among many, such as sealed-tube oscillographs already popular in the United States in the late 1920s (Zworykin & Morton, 1940) and television tubes, first black and white, with color being introduced in the late 1930s by inventors in England (Baird, 1938) and France (Barthélémy & de France, 1932) who made use of the sequential solution in the trichrome version. In the United States, V. K. Zworykin at RCA was very active with both kinescopes and iconoscopes [see Ref. (Zworykin & Morton, 1954) for more details]. In this context, E. Brüche, the leader of an important team of scientists at the laboratory of the Allgemeine Elektrizität Gesellschaft (AEG) in Berlin, adopted this more flexible attitude as regards the future of electron optics. This is already clear in the publication of the first treatise on electron optics by E. Brüche and O. Scherzer, the well-known theoretician of the group (Brüche & Scherzer, 1934). This tendency is even more strongly expressed in their last book of the prewar period, "Elektronen Geräte" ("Electronic Instrumentation"), published in 1941 (Brüche & Recknagel, 1941). For E. Brüche, during the early 1930s, research on electron optics was first a preferred tool for studying directly electron sources (Brüche, 1932), an attitude also favored by Davisson and Calbick in the United States (Davisson & Calbick, 1931). Nevertheless, later on the progress of E. Ruska working at Siemens after 1936 (Ruska, 1979, p. 66) persuaded Brüche's concurrent team to concentrate more on electrostatic "Übermikroskopie" (supermicroscopy). Proceeding from a rather elaborate study of an electrostatic converging lens already appropriate for transmission microscopy (Johannson & Scherzer, 1933), H. Mahl succeeded in 1939 in building a classical transmission supermicroscope (Mahl, 1939a, 1939b), H. Boersch proposing simultaneously a new type of arrangement, the shadow microscope, which was not studied further [see Section 1.5 and (Boersch, 1939)]. All this development was more widely publicized and documented a year later by C. Ramsauer, the older scientific superviser of the AEG, in a booklet of 127 pages, brilliantly illustrated (Ramsauer, 1940) and in Ramsauer (1941).

1.4 Impact of the German pioneer work outside the frontiers

The diffusion of electron microscopy around the world was very slow in the prewar period, much slower than could be inferred today judging from the present-day speed in the spreading of technical innovations and scientific discoveries, even when of much lesser importance. The British were the most active, with the spontaneous effort of Louis Claude Martin, who in 1937 interested the Royal Society in the matter and with its help developed a magnetic microscope (Martin et al., 1937) at the Metropolitan Vickers Electrical Company, the seed of further action in Great Britain after the war, and the stimulus for a book on electron optics (Myers, 1939) and for a similar effort in Canada (Prebus & Hillier, 1939). The obstacles hampering diffusion are acutely shown by the numerous difficulties encountered by Marton: escaping from the Nazi terror, he found a refuge in Brussels (1932), where he constructed a first magnetic microscope (1933) and a second (1936) with high resolving power (slightly better than 1 μm) and founded the techniques of observation for biological objects (Marton, 1934, 1937). The approach of war sent him to the United States, where he was able to interest V. K. Zworykin in supermicroscopy for biology and constructed at RCA a third magnetic microscope (Marton, 1936, 1940). His numerous previous publications in French [see, e.g. (Marton, 1934, 1935, 1936, 1937)] aroused some academic interest in France but not much action: two experimenters, F. Holweck at the Paris Laboratory of Madame Curie and J.-J. Trillat with his pupil R. Fritz (1936) at the University of Besançon, constructed lenses and demonstrated their imaging properties by obtaining enlarged images of an oxide cathode; more theoretically minded professors reviewed the basis of electron optics (Bricout, 1938; Henriot, 1935). Two original and very valuable publications appeared later: the general theory of thin pencils of electrons by M. Cotte (1938), a thesis initiated by Louis de Broglie; and the use of strong "grid" lenses by Louis Cartan (1937) in order to increase strongly the luminosity of a mass spectrograph, doctoral work done in the laboratory of Maurice de Broglie, the older brother of Louis by 17 years.

These two publications were the first to awaken interest in university circles, e.g., professors and research students in physics. The present author was one of these at the time and Cartan's thesis, very clear and well balanced, exerted a strong influence on his thinking about possible postdoctorate scientific activity. He became eager to understand more fully the mysteries of "grid" and other electron lenses, but—let it be known—the first step was the reading of the very accessible American book by I. G. Maloff and D. W. Epstein, "Electron Optics in Television" (Maloff & Epstein, 1938).

Indeed, the American way of life—in science and technology—began to be strongly influential on younger generations in France: the success of broadcasting, the development of television and of short-wave radio communications, and long-distance flights even over the Atlantic Ocean were deeply impressing, and their appeal was a kind of early answer to the disquieting rearmaments east of the Rhine, even orienting some individuals toward science. The American approach to the delicate technologies of television photoelectric devices, luminescent screens, and pickup tubes was already capable of diverting the attention of mature experts in electron optics such as Max Knoll in Germany (Ruska, 1979, p. 37) and was still more prevalent in France. Naturally, television and aids to navigation and flight were also monopolizing efforts in the United States, as shown in detail in (Grivet, 1980) and (*An Age of Innovation (The World of Electronics, 1930-2000)*, 1980) and as appears in the slow evolution of the mind of the American television pioneer (Marton, 1968) V. K. Zworykin toward electron microscopy under the influence of L. Marton during the years 1939–1940; the list of publications from the RCA Laboratory in 1941 is a witness of this change, as shown by Marton (1941a, 1941b) and Morton (1941).

1.5 Supermicroscopy and diversity of microscopes

Finally, it is also of general historical interest to note that a high rate of progress in the main branch of supermicroscopy, so ably described by E. Ruska (1979), does not absolutely guarantee final success, because the task is very complex and each step forward is made at the limit of the possibilities of the day: quick progress in overcoming many difficulties generates parallel avenues of investigation, sometimes very promising as further evolution shows, and encourages a kind of dispersion of efforts which is not an effect of fancy or change of mood but is inevitably a natural source of weakness of the mainstream. The only remedy is the skill and ability of the leader of the team to drive it. One may illustrate this remark by the following examples: (1) The discovery of the field emission microscope, born in the United States from hot-cathode research by R. P. Johnson and W. Shockley (1936), then developed by E. W. Muller at Siemens for cold field emission (Muller, 1937) [see (Ruska, 1979, p. 96), for historical details]. (2) The shadow microscope of H. Boersch (1939), soon transformed into an X-ray shadow microscope by von Ardenne (1939). Manfred von Ardenne was a brilliant advocate of supermicroscopy, a domain in which his outstanding talents as organizer and engineer and his gift for understanding the needs of users, mostly biologists and chemists at this time, were recognized by Siemens. With the help of this company [cf. (Ruska, 1979, p. 100)], his outstanding engineering work (von Ardenne, 1940a, 1940b) would lead him a year later to publish a comprehensive book (von Ardenne, 1940c) on a universal electron microscope "for operation in a bright field or dark field and stereo-imaging." This treatise is striking in its clarity, concision, and efficient architecture, qualities which are reminiscent of the contemporary achievement of V. K. Zworykin and his team in television (Zworykin & Morton, 1940) in the United States, a new and brilliant style for scientific books. This bold takeoff would be brought to halt by World War II, but for 5 years only, and V. K. Zworykin achieved a bright comeback in 1945, with a new team, with "Electron Optics and the Electron Microscope" (Zworykin et al., 1945).

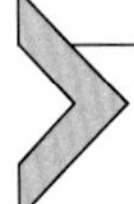

2. Electron optics in occupied France

2.1 The academic year 1940–1941 in France

In 1940, Manfred von Ardenne dedicated his treatise on electron supermicroscopy to his brother Gothilo, who fell in the field during the war in

Poland. Soon, hundreds of thousands of French losses were added to the Polish ones in the battle of France. The armistice of June, 1940, also put an end to liberty in the northern half of the French territory, and particularly to the freedom of scientific study. The German rule in the occupied half of the country precluded even any teaching on radio science. Consequently, after July, 1940, the author found it difficult to finish the writing of his doctoral thesis (in the field of what is now called plasma physics) and to earn his living: his prewar function as assistant at the Ecole Normale Supérieure was a temporary one (a generous term of 4 years for obtaining a doctor diploma, after which the stipend was passed on to another fellow of the school), and after a period as chief of metrology at the Conservatoire National des Arts et Métiers, he felt it wise to abandon a career as professor because civil servants were severely screened for prejudice against the new "racial laws" of the Vichy Government. He turned to research in industry, at the Laboratory of the Compagnie Générale de TSF (CSF) in Levallois near Paris. There, Dr. Maurice Ponte, the dynamic French pioneer in electron diffraction during the 1920s (Goodman, 1981, p. 59) tried to maintain a decent standard of research in electronics; with great courage and a sense of national responsibility he disregarded any consideration for the new regulations. There the official problems on hand were new oscillograph tubes, especially minitubes for control, and the technology of powerful emission tubes. Dr. Ponte also managed to start an exploratory side activity in television and electron optics, with an eye on electron diffraction and electron microscopy, first for materials analysis and cathode studies and soon as a possible mark of prestige for the laboratory. The team in charge of this program first consisted of three new young engineers, just graduated from the Ecole Supérieure de Physique et Chimie, later joined by two members of University Laboratories, D. Charles, concerned with tube technology, and still later H. Brück, who in the first months of the year 1942 began to study strongly converging lenses. With this in mind, he remained for a time in his laboratory at the Ecole Normale Supérieure, until electron optics could be properly treated in a quiet house at Levallois, near the tube plant and adjacent to the CSF laboratory. The latter designed, and the plant produced, powerful shortwave and microwave emission tubes; it was not at all suited for delicate, metrological work, as electron microscopy is in fact. Charles and Brück were also refugees happy to find a permissive administration as regards racial rules,[1] the case of Brück being particularly delicate

[1] This was also the case for E. Regenstreif later on, and this was the reason for delaying any publication from the laboratory until the end of the war. For this reason, too, this

as he was an Alsatian, son of a French mother and a German father, having chosen French nationality in 1933 as permitted by a tiny paragraph of the Peace Treaty of Versailles in 1921. On the other hand, Brück was a well-trained and brilliant physicist formed in the laboratory of Professor Stern in Hamburg and an expert in molecular beams and high-resolution light spectroscopy; to the author, he was also primarily a laboratory friend since 1934 and a knowledgeable, imaginative, and rigorous physicist.

During May, 1940, the CSF laboratory and tube plant had been removed to a little town, Cholet, in the southwest of France, but afterward it settled again in Levallois during the end of 1940 and the beginning of 1941. When the new team started its activity in summer, 1941, the plant was already working, but with many difficulties: modern machines had been confiscated by the occupying forces, and many uncommon materials were lacking. Nevertheless, it was a valuable asset to work in symbiosis with the plant, because scarcity for a plant may sometimes be a temporary and welcome abundance for a laboratory. This was the case, for example, for refractory metals, tungsten, and molybdenum, for mercury, pure solvents, and chemicals, for magnetic shields, fluorescent powders, and so on, and also for services of various kinds; vacuum equipment was plentiful as well as liquid nitrogen, all of which had disappeared months before from university laboratories. On the other hand, the morale of scientists and technicians was outstanding, as well as their patriotic coherence; many engineers, particularly those aged between 25 and 40 years, were wanted, many being still prisoners of war; for many others it had been necessary to change their speciality and therefore great needs for teaching arose.

For these reasons a kind of small technical and specialized university was created in the plant, with 1-hour courses at the end of the working day ($\frac{1}{2}$ hour paid, $\frac{1}{2}$ hour unpaid). Teaching was part of the duty of research fellows, even in a nascent team. The need for a higher level of scientific knowledge was largely recognized as desirable, in the ambiance of bitter changes of responsibilities among the renewed staff of such a plant, and

group was also engaged in various activities, such as producing tungsten wire when this material became scarce, or powders for fluorescent lamps or iron carbide when tungsten steel for tools disappeared. Another flux of workers and a clever foreman, Gladieux, arrived later in 1943, when the departure of slave labor for Germany was organized by Minister Sauckel in 1943 (the so-called Sauckel Army). Most of these people used false names and faked identity papers. Even a distinguished young secondary school professor, F. Bertein, enrolled at that time, because he had no more pupils, all sucked away by Sauckel action (STO in French: Service du Travail Obligatoire).

also at the university: P. Grivet lectured on thermionic and photoelectric emission at Levallois and H. Brück on quantum mechanics at the Ecole Normale Supérieure; D. Charles prepared documentation on the preparation of fluorescent powders, ceramics, and photoelectric films, and planned the construction of an electrolytic tank for the mapping of the electrostatic fields in possible future lenses. Indeed, in late 1941 all materials became scarce, a danger to the bread and butter daily activity, but happily in the summer of 1941 the aerial Battle of Britain appeared to have been lost by the Germans and a faint hope arose that the West could finally resist, if not win. Scientific dreams are not science fiction, and decisions had then to be made between magnetic and electric lenses for the future. P. Grivet, during the last quarter of 1941, started a meticulous report on electron optics, a work[2] which was multigraphed at Easter, 1942, and could then circulate among the group. Enough data could be gained on the prewar German work to make a decision: notwithstanding the fact that the work of the AEG team was less well documented in this period (the 1939 issues of the technical journals *Zeitschrift für technische Physik* and *Zeitschrift für angewandte Physik* were not available in 1941), the electrostatic technique appeared to be more accessible in those difficult times; it was then very difficult to obtain good alloys for magnetic polepieces, or even accurately regulated and efficient voltage sources for the electron gun or current regulators for magnetic lenses, and one could not even dream of performing fine engineering, such as was described in von Ardenne's book (von Ardenne, 1940c), an impressive achievement indeed even today.

2.2 Spartan times ahead in 1942: choice of an electrostatic instrument

The decision was finally made in the spring of 1942, with the help of a first vivid experiment by H. Brück. He had succeeded in building what we now would call a simple emission microscope of some 0.75 m in length (Fig. 1a), with an underheated oxide cathode and a main glass body, con-

[2] The first half, relating to electrostatic lenses, was printed in the same year, 1942, in the November issue of the *Revue Générale d'Electricité* (Grivet, 1942); the second half was delayed until 1945 (*Optique électronique*, Grivet, 1945); the internal report included a comparison of the existing microscopes which was not included in the articles. The report was the root of an article in English for Volume 2 of *Advances in Electronics* (Grivet, 1950) and of a book published later by the team, originally in French (Grivet et al., 1955); the book was later published in two successive and enlarged English editions (Grivet et al., 1965).

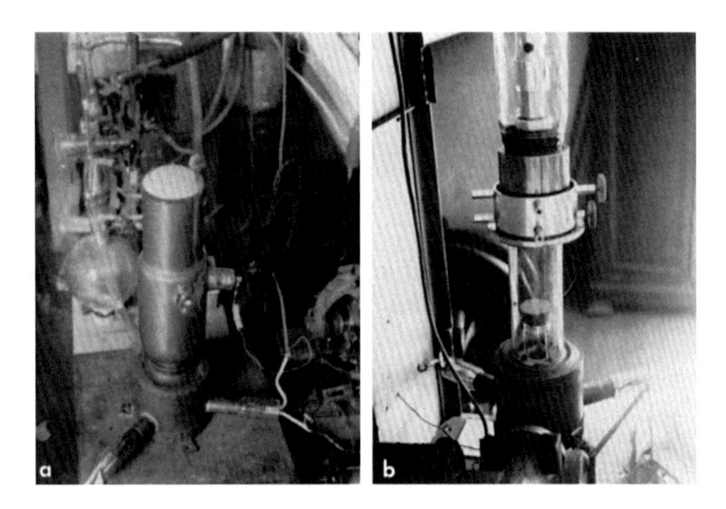

Figure 1 H. Brück's first electron-optical bench: (a) for an emissive object, (b) for a transparent object.

taining a fluorescent screen and a stand for a second lens of any reasonable structure. This electron-optical bench was fed with a 30-kV X-ray source; one could use it for a single lens (Fig. 1b). A careful study of this experiment showed to both P. Grivet and H. Brück the outstanding attractions of the fundamental simplicity and reliability of this electrostatic device, just when a very difficult period loomed ahead. Indeed, the extension of the war (June 20, 1941) to Soviet Russia, disquieting in itself, had also distressing local consequences: a good part of the Wehrmacht withdrew from France, but giving place and importance to the Gestapo brute beasts. Two of the earlier advocates of electron optics in France, Fernand Holweck in 1941 and later Louis Cartan,[3] paid with their lives for their engagement in the underground movement. In the last quarter of 1942, numerous workers were deported to Germany to work in armaments plants. The big battle on the Dniepr was in full swing, but very soon the defeat of Stalingrad would end the last blitzkrieg operation and destroy the last illusions of the civilians in Germany regarding the issue of this World War.

The Levallois group then delayed any secondary investment: for instance, instead of building a proper electrolytic tank, Brück organized a collaboration with former French specialists in applied aerodynamics, and with the help of one of them, L. Romani, worked with equipment put

[3] Deported to the German Camp in Lubeck and executed. Maurice de Broglie wrote a biography of his pupil Louis Cartan in the July, 1953, issue of the *Cahiers de Physique*.

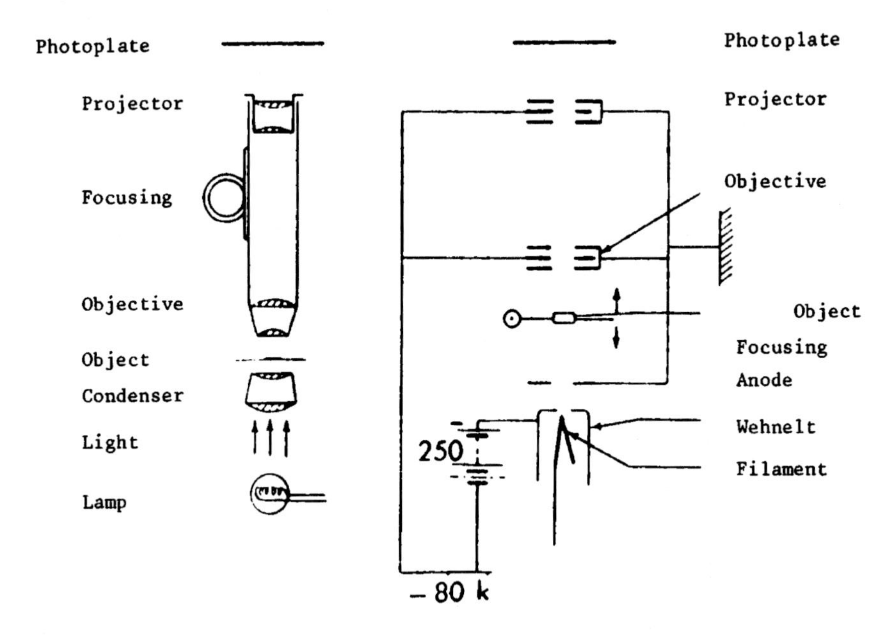

Figure 2 Supplying an electrostatic EM with only one voltage.

at their disposal by the Laboratoire d'Analogie Rhéographique, Sorbonne (Directors L. Malavard and Professor J. Pérès); the aim was to determine the dimensions of an optimal electrostatic objective lens and also of a projective lens by calculating Gaussian rays and spherical and chromatic aberrations for various dimensions of holes and distances between electrodes, keeping in mind two constraints: (1) the use of a high working voltage of the order of 60 kV, for a good transparency of the prospective objects, and (2) the necessity to keep the focus of the lens outside the entrance electrode, to protect the focusing field against any perturbation and the delicate object against electric forces.

These constraints are merely the optical expression of the simplicity conditions as applied to the voltage source. If the optics is to be made insensitive to potential fluctuations to first order, the most simple and efficient solution is to use *one* voltage only for the electron gun, the objective lens, and the projector (which is of course much less sensitive), as shown in Fig. 2; the focusing is then obtained by mechanical displacement of the object, the angular aperture of the beam being very small ($2\alpha \approx 10^{-3}$ rad) and consequently the depth of focus very high as compared to the conditions

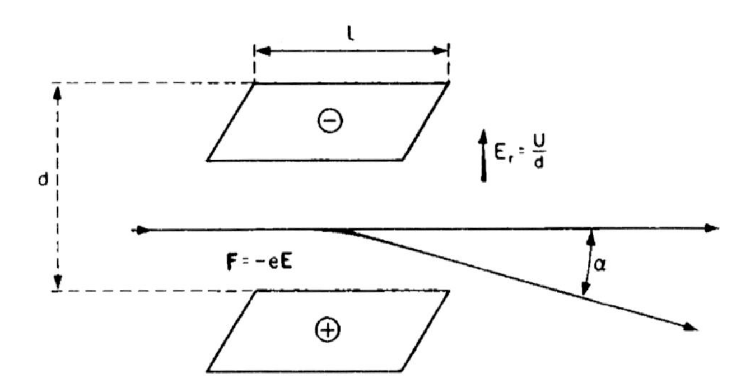

Figure 3 Electrostatic deviation of an electron beam.

prevalent in a light microscope ($\alpha \approx 90°$).[4] A potentiometer arrangement would have been more accurate and convenient, but mains hum and accidental fluctuations must then be rigorously filtered out, which could not be achieved easily at the time. A family of lenses was simulated in the electrolytic tank and the inside field was measured with an impressively high relative accuracy of some 10^{-3}, as was later shown by comparison with electron-optical measurements. A selection of these results is given in two articles by H. Brück and L. Romani (1944) and H. Brück and P. Grivet (1950). The optical arrangement obtained is very close to that of a light microscope apart from the absence of any condenser lens and the success is dependent on a true linearity in the dynamical laws of electron movement. There is no doubt about the linearity of the electric potential under these conditions, but relativity introduces nonlinear correcting terms into the fundamental law of dynamics. They were first evaluated from the simple formula, for angular deviation α by a plane condenser held at a voltage u, with plates parallel to a beam of energy eU (Fig. 3).

The classical formula is

$$\alpha_0 = \tfrac{1}{2}(u/U)(l/d)$$

where l is the length of plates, d is the distance of plates, and u is the deviation voltage.

Relativistic conditions are

$$\alpha = \alpha_0\left(1 + \tfrac{3}{4}\tfrac{U}{V_e}\right)$$

[4] With an "immersion" objective.

and

$$V_e = m_0c^2/e \approx 512 \text{ kV}$$

is the voltage equivalent to the rest energy of the electron.

J. Laplume made a detailed study of the correction to the electron-optical rays themselves and to the Gaussian canonical parameters of the lens (Laplume, 1947). He showed, for example, that the correction would be of the order of 1% at $U = 100$ kV and that a safety limit for the amplitude of fluctuations in U would be 1000 V or $\delta U/U \approx \frac{1}{100}$, if a beam aperture of 5×10^{-3} rad could be reached in the future. This was highly desirable for luminosity, but appeared then to be accessible only in the distant future. The conditions for fluctuations would have been much more severe for the voltage (4×10^{-5}) and the current (2×10^{-5}) in a magnetic instrument, and seemed wholly unattainable under the 1942 conditions.

2.3 Designing the elements of a first prototype (1942)

2.3.1 Electron gun

In the available literature, the most obscure point was the breakdown limit in the electron gun and the electric lenses. Apparently, the gun was the easiest to build, because there were no stringent limits on the room available. Moreover, as it was a common organ to all microscopes but also to X-ray tubes, a large body of empirical knowledge was available in the literature. von Ardenne in particular was rather explicit in his descriptions; he had also paid great attention to the problem in his attempt to use 200 kV in a special prototype (von Ardenne, 1941). Thus P. Grivet and H. Brück, after some preliminary measurements, sketched a first rudimentary form (Fig. 4). Apart from the fine mechanical adjustments for positioning and orientation of the beam, it was the same as the gun shown in Fig. 5; it was constructed with glass insulation, and tests showed that it was still reliable at 80 kV (measured with a high-voltage Abraham electrostatic voltmeter) provided that it was assembled with much care taken over cleanliness and "formed" by a process of trial and error in which the voltage was slowly increased under an attentive oscillographic control of the tiny concomitant discharges, appearing in the current as small spikes. The choice of the damping serial resistance necessary as a passive and ultimate protection against the hazard of any destructive avalanche proved also to be possible by shear empiricism.

Figure 4 First sketch of a French electrostatic EM.

2.3.2 The objective lens

The problems raised by lenses and their supplies were of a much wider scope, but the experience gained with the gun already gave some indications concerning its easiest aspect, the order of magnitude of the minimal distance between the central electrode and the two side screening plates, because the geometry is similar to two parallel planes. Nevertheless, the nature of the electrodes is important as well as their degree of polishing; the only high-quality material available was amagnetic stainless steel (18% Cr, 8% Ni), and its mode of production could not be influenced in any way; only the polishing could be perfected, the control of perfection being ultimately achieved by breathing patterns. Brück conducted a rigorous series of trials and finally reached 80 kV over 4 mm steadily, but this was a real tour de force and could not be reproduced, even by very skilled and conscientious workers. So the team was satisfied to adopt a more conservative limit of 60 kV over 4 mm, accessible to a good, meticulous worker accurately following a sequence of operations in a clean room. In 1943, reports

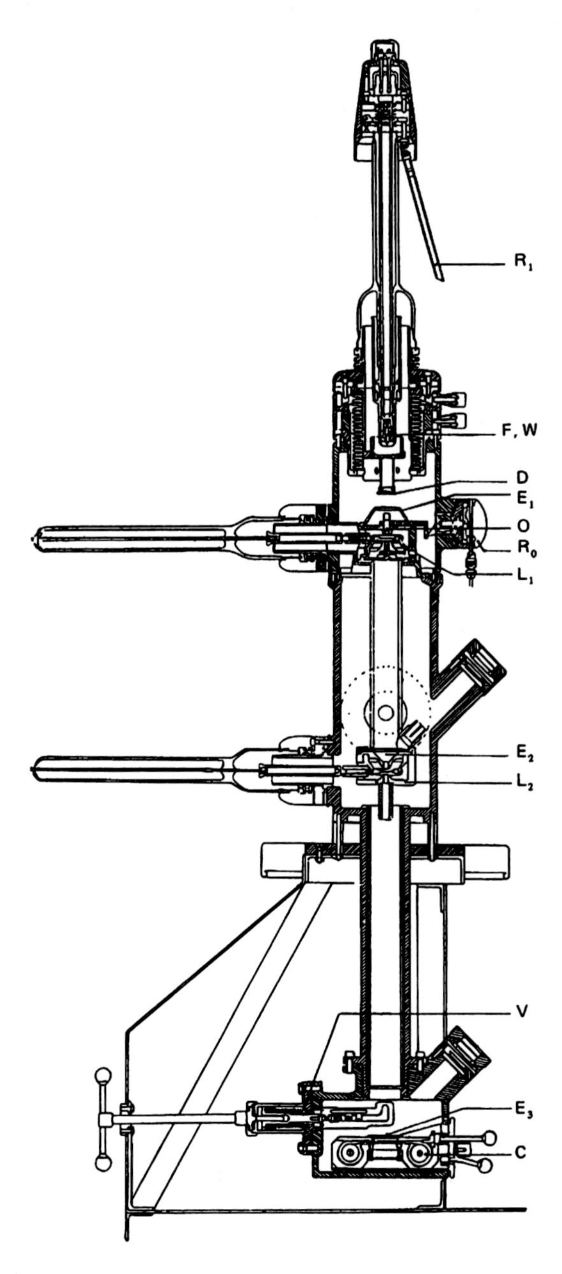

Figure 5 The first prototype with glass insulation of the feeder. General layout of the instrument: R_1, adjusting the filament–Wehnelt gap; F, filament; W, Wehnelt; D, diaphragm limiting the illumination aperture; E_1, fluorescent screen for illumination control; O, object, R_0, exploration of the object; L_1, objective; E_2, fluorescent screen for observing the intermediate image; L_2, projective; V, lock-gate; E_3, final fluorescent screen and photo shutter; and C, film.

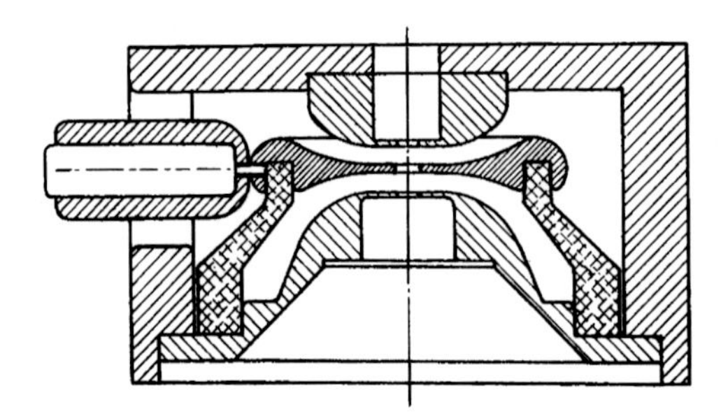

Figure 6 Electrostatic lens with "ebonite" insulation.

(Brüche, 1943; Gölz, 1940; Ramsauer, 1943) from the AEG group became available in Paris and a study by E. Gölz on breakdown revealed that the same difficulties had been encountered in Germany and surmounted in a similar way. Ten years later, a deeper and more systematic study was made by R. Arnal (1953, 1954a, 1954b, 1955, 1956), who analyzed more clearly and rigorously the useful phase of "prebreakdown"; he revealed some important factors and also found some means of increasing the active life of lenses between periodic cleanings.

Renewed research during the following decades revealed the intricacies of the problem, where the degree of vacuum, the state and cleanliness of surfaces, and their secondary emission properties all play important roles: the most efficient remedy to the hazard of discharge, found in recent years, was the introduction of very high vacuum (10^{-9} mm Hg instead of 10^{-6}, the current limit with mercury pump and liquid-nitrogen traps), with new ion pumps after a preliminary backing of the entire instrument, which became possible with the availability of new materials for the joints (Capton and so on).

Returning to the wartime lenses, it has to be noted that the insulator of the central electrode is a sensitive part as regards breakdown; mechanical accuracy is also important for centering of the hole and ensuring electrode parallelism: the first material in the prototype was hard ebonite, an excellent dielectric in static conditions, but difficult to clean and of poor mechanical reliability (Fig. 6); Lucite was better behaved and was used in the next generation of lenses, being produced in France just at the right time for the needs of medical optics. The form of the central insulator (Fig. 7) was changed at the same time, in order to put the two junctions between metal and insulator in a recessed region of low field and to maintain the body of the insulator in a constant field. The lens supplies, at first insulated with

Figure 7 Electrostatic lens with Lucite insulation.

glass, were soon enclosed in Lucite too, the central conductor being then easily protected and the alkali ions of the glass being eliminated as a source of electron multiplication inducing breakdown. The optical properties of the objective were those of type VIII in the electrolytic tank measurements (Brück & Romani, 1944); its focal lens was 5.4 mm and the focus was just outside the first electrode of the lens, but still inside the magnetic screen as shown in Figs. 6 and 7. At the time, the quality of the objective was $K_s = 12$, where $C_s = K_s f$ and f denotes the focal length; this reduced value was some seven times larger than that of Ruska and von Ardenne's magnetic objective. This was not too disquieting because the resolving power resulting from the optimization of conditions was proportional to the fourth root of K_s. In fact the situation was not so simple, as was later proved by the group and as explained in Section 2.4.3 concerning the "ellipticity" defect. At this point a discussion of the aperture diaphragm would be in order but for the same reasons it will be delayed to a later section.

In the first tests of the whole instrument the variety of specimens was very limited: some fine powders such as ZnO and MgO, already in common use in electron diffraction, and a few thin films of polymers such as collodion and Formvar. The diameter of the contrast diaphragm was not critical and $d_a = 20$ μm was adopted as the smallest dimension conveniently produced by pressure of a fine point of tungsten on a thin disk of tantalum.

2.3.3 Projector lens

The projector lens was also of the type VIII characterized by H. Brück, but with a slightly smaller minimal focal length, $f = 4.7$ mm, as the intermediate image given by the objective may be immersed in the field of the projector. The magnification could be varied by changing the potential of the central electrode of the projector; this adjustment was initially done by a potentiometer made of some 200 high-precision radio resistors which, associated with a microammeter, controlled the high voltage. Later it was noted that for setting the magnification, a stick of common polymer, especially a kind of bakelite (high-frequency bakelite), could be used instead of the chain of resistors, with the advantage of using a sliding metal contact and hence adjusting continuously the magnification between 1000 and 6600× with a negligible current consumption ($i \leq 10$ μA). The distance between the two lenses being 320 mm, the magnification at the intermediate image was 60×; it could be magnified 10× optically for a first control of focusing. It was enlarged by the projector 110×, obtaining a total direct magnification of 6600× and some 26,000× when observing the final screen with a magnifying glass under optimum conditions of luminosity and resolving power for fine focusing. This structure had the advantage of giving a rather large field, which was convenient before the advent of the grid holders for the object, so long as one was using a circular disk with a small hole: the field diameter was $D = 5$ μm for a final film 24 × 36 mm, 10 μm for 6 × 6-cm film.

A description of this first prototype was only given at the end of the war on the occasion of the first "national" symposium presided over by Louis de Broglie on June 12, 1945, in a reunited France, and was published in 1946 (Grivet, 1946; cf. Bertein et al., 1946).

2.3.4 Magnetic screening and general equipment

Magnetic screening was provided by placing a common cylindrical very soft magnetic case around both lenses and the beam in the intermediate space, but Permalloy was no longer available, except for some remnants of thin band and tubing for common oscillograph tubes. So ferronickel ("Anhyster," the French equivalent of Hyperm) was used in greater thickness, some 10 mm, which also ensured a better mechanical stability: a Permalloy band was also tried as internal lining. The mechanical layout was carefully devised by the designer of the group, H. Blattman, who was eager to compete with von Ardenne's apparent excellence, while the research scientists were more impressed by Dr. Boersch's striking taste for simplicity and ruggedness, very probably helped by a keen physical insight.

2.3.5 Friendly users' help

Fig. 8 shows the general layout of the instrument in its second stage of development (1943) when it spread to a few research laboratories such

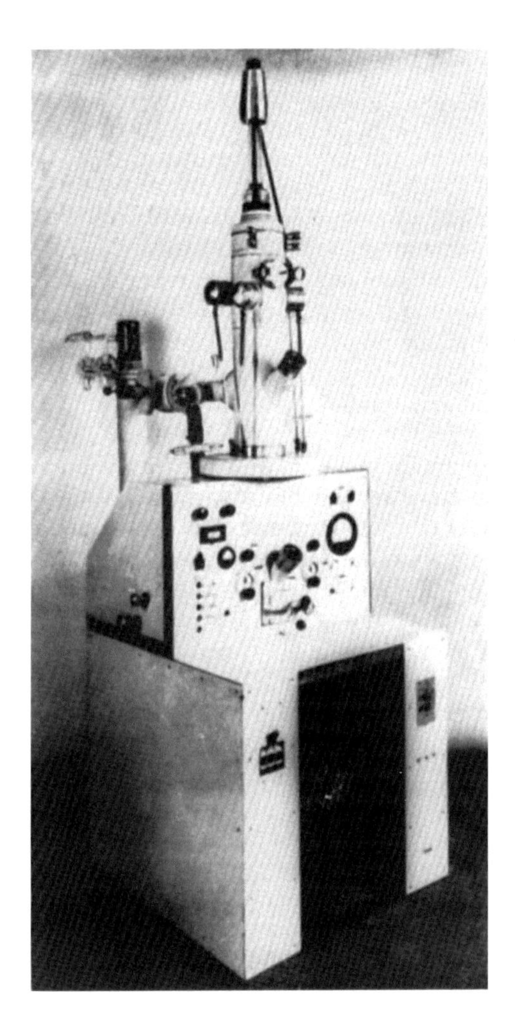

Figure 8 Second stage of development (1943).

as the Institut Pasteur, Paris (Professor P. Lépine) (Lepine, 1944, 1946), and Michelin, Clermont-Ferrand (Professor Travers). This extremely small-scale "production" was initially supported by a grant from the French CNRS, which at that time had an applied science division (Dr. Rivière) and a geologist as director, Professor Jacob; a generous grant from Mrs. Deutsch de la Meurthe, sponsor of the Institut Pasteur, started the movement and was a great encouragement for the Levallois team.

During 1943, the deterioration of the general situation in France made it imperative to remedy the frequent and large fluctuations in the mains voltage, still more conspicuous in the industrial environment of Levallois: the simple X-ray rectifier became insufficient; a stabilizer composed of a small group made of a synchronous motor and a 600-Hz alternator linked the high dc voltage to the frequency of the mains but not to its amplitude, restoring a notably better stability; this arrangement was later pushed to a high degree of stability of some 10^{-4} by a clever young engineer, J. Vastel, introducing a feedback loop into the system. On the other hand, the use of the instrument outside its birthplace extended the domain of applications, and a few young technicians were recruited to better the preparation of specimens by methods such as shadowing powders on films, by sputtering obliquely a metal on the surface (Muller, 1942) or by taking an oxide or organic film replica of the relief of the etched surface of a metal (Mahl, 1940a, 1940b, 1942).

By a feedback effect this furnished the Levallois team with a choice of better images, all these processes increasing the contrast; thanks to their flexibility, they permitted the scattering properties of objects to be more efficiently adapted to the rather rigid structure of a microscope deprived of condenser lenses: the adjustment of the negative polarization of the Wehnelt in the electron gun was the only means of matching the aperture of the illuminating beam to that of the objective; but, simultaneously, the luminosity of the image was modified, and choosing transparent and contrasted objects was imperative for obtaining the best resolution. The mechanical setting was at first completed by a variable polarization of the objective lens through a dry battery and potentiometer with $\delta V = 500$ V, $\delta f = 0.1$ mm. The resolving power was increased, but unfortunately a plateau was soon reached and the main staff was obliged to concentrate all their efforts on explaining and eventually bridging the gap between these first encouraging results and those given by the basic theory. At the end of 1943, a critical review of the whole construction process and of the instrument's structure was started with the help of a new engineer, E. Regenstreif. He was initially regarded as representing faithfully the "advanced users" of the instrument and expected to synthesize usefully their remarks, criticism, and desires, but soon he was doing much more as a devoted and gifted electron optician and microscopist.

2.4 A first industrial two-stage electron microscope

2.4.1 Brück's gun

H. Brück first studied and weighed the pros and cons of the simple electron gun, and tried to design and build with M. Bricka a three-electrode gun

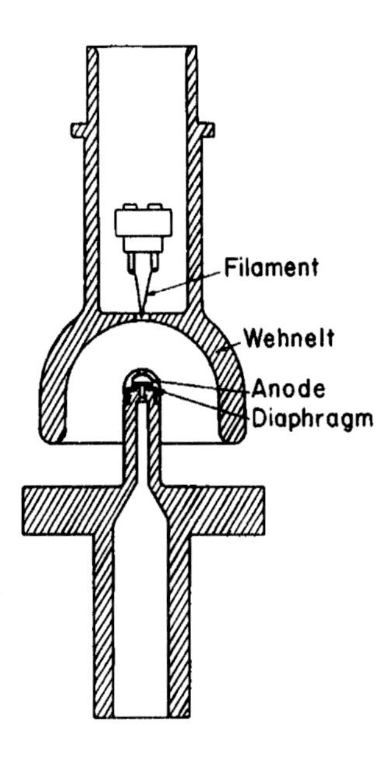

Figure 9 Structure of Brück's three-electrode gun.

which could compete more successfully with the classical arrangement (gun plus condenser lens). Soon (Bricka & Brück, 1948; Brück & Bricka, 1945), inspired by the Pierce gun of sealed oscillograph tubes (Pierce, 1940), they succeeded in obtaining a marked converging effect at the exit of the triode (Figs. 9–11); a similar arrangement was later called a "fernfocus" gun by Steigerwald (1949) of the AEG group. Brück's gun increased considerably the image luminosity for a great range of polarization of the Wehnelt electrode; consequently, by adjusting this voltage one could cover efficiently a wide range of aperture angles with a convenient margin in luminosity. In other words, one could choose a suitable aperture by means of the Wehnelt polarization and adjust almost independently the luminosity by the direct magnification (2000–14,000×) and the filament temperature (*Bettering of the focussing devices for electron microscopes*, 1946a).

2.4.2 Industrial instrument

This instrument (Figs. 12 and 13) may be called "industrial" because it was constructed outside the laboratory, by a production branch of CSF in its section for "models" under the responsibility of two distinguished engineers, R. Bonne and Clerc. They introduced many refinements such as an important miniaturization of the voltage supplies, which were contained in

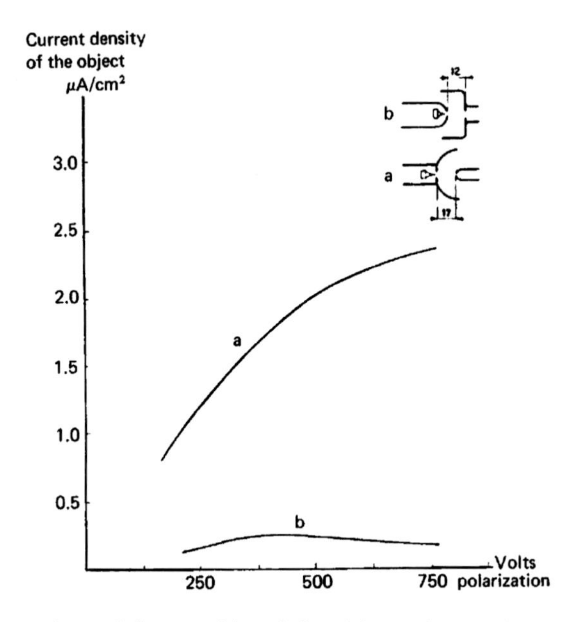

Figure 10 Comparison of the new (a) and the old gun (b). Anode object distance = 150 mm; filament diameter $D = 89$ μm; heating current = 1.32 Å; lifetime, >100 h.

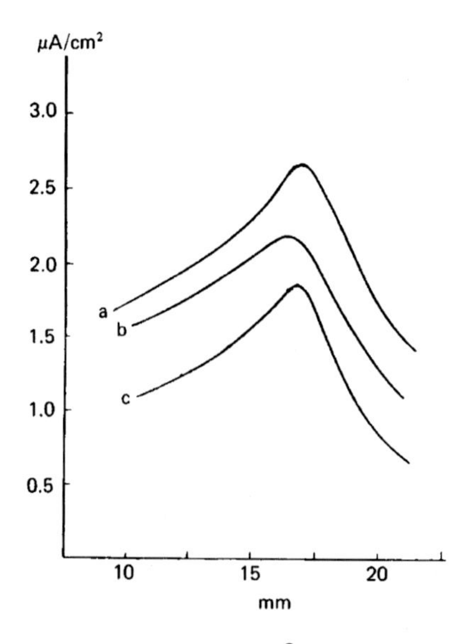

Figure 11 Current density at the object (μA/cm^2) versus Wehnelt anode distance (millimeters) for various polarizations of the Wehnelt electrode. Polarization: (a) −720 V; (b) −450 V; (c) −270 V.

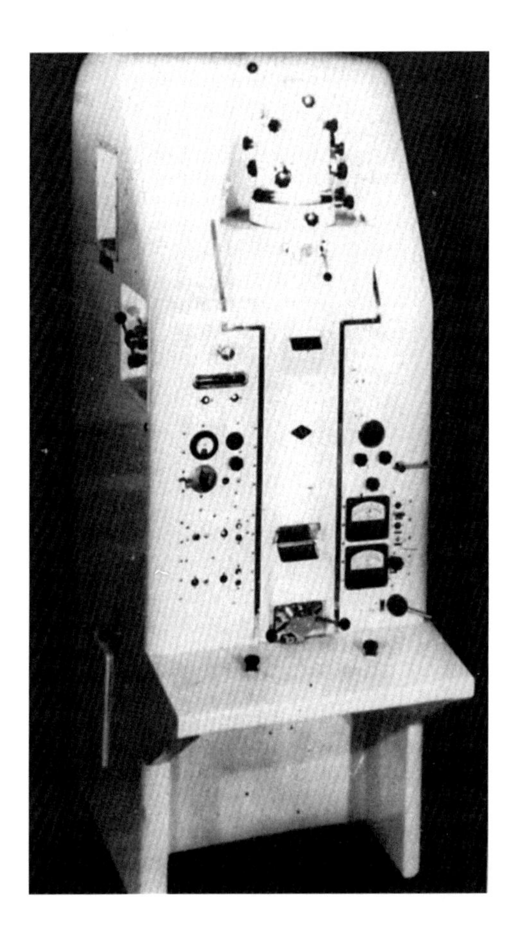

Figure 12 The CSF first commercial EM: general aspect.

a 20-in. cubicle (Fig. 14), efficiently screened so that stray magnetic fields were not induced, but nevertheless containing a saturated diode to limit unwanted discharge currents to a safe value of a few hundred microamps (*Bettering of the HT sources for electron diffraction and electron microscopy*, 1946b). The pumping was provided by an oil diffusion pump of the Burch type (not silicone oils) made by the CGR (Compagnie Générale de Radiologie), which proved to preserve satisfactorily the cleanliness of lenses and diaphragms, but with increased pumping speed and less vibration. Manipulations and service were made much more handy, and the use of forged Duralumin for the external casing both eliminated small leaks (the earlier vacuum envelope was *cast* aluminum bronze, showing episodically some porosity) and permitted binocular observations through large windows. All

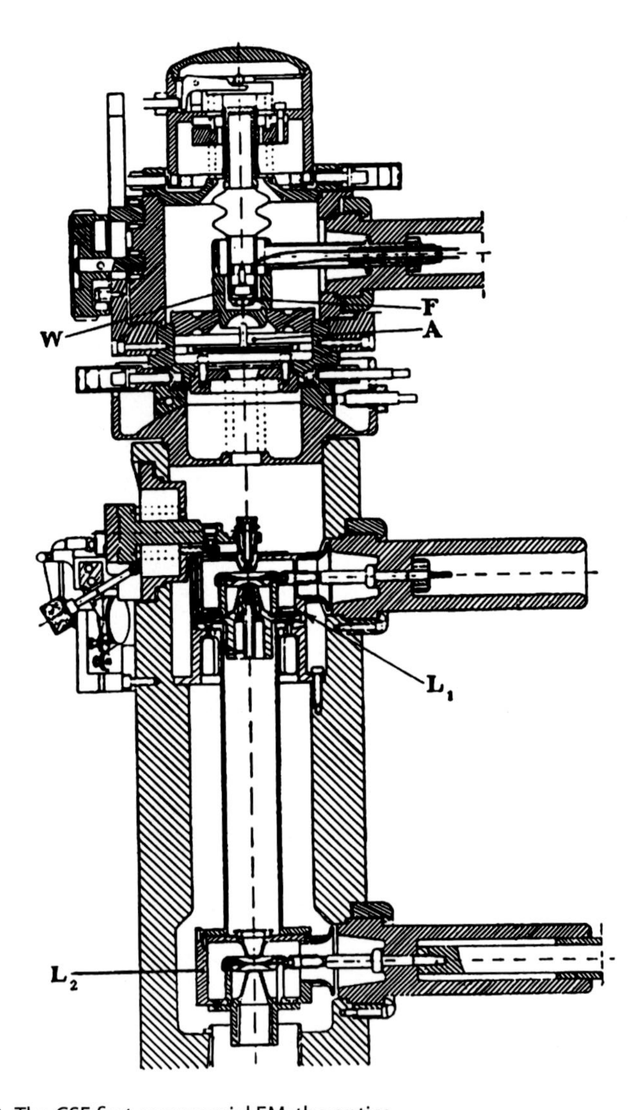

Figure 13 The CSF first commercial EM: the optics.

these developments were well received by the laboratory, for which the work was eased, and, more important, by the public: this product was launched in July, 1946, and more than 60 such instruments were sold during the following 5 years. This success was a great encouragement to the laboratory, which remained in charge of solving various special problems of the customers (Robillard, 1949a) and of teaching electron microscopy at large, with a new influx of young staff.

Figure 14 A screened, stabilized, current-limited source of high tension for the French EM: under the HT lead is the diode limiter.

During this period the only competitor was the magnetic instrument, especially the "simplified" type such as the EMC of RCA, sold for slightly less than half the price of the EMU; for various other products, see the reviews published in 1949; for example, that by von Borries (1949, pp. 83–115) and that by J. Robillard (1949b). The AEG microscope was no longer available due to the destruction of the war, although scientific research was vigorously recovering in a new laboratory at Mosbach on the Neckar. On the other hand, an early American effort on the electrostatic instrument had stopped in 1943 (Bachman, 1943; Bachman & Ramo, 1943); the development in Switzerland by G. Induni (1945) of a very complete instrument with an original gas cathode (Induni, 1947), providing electrons of low-speed dispersion, was more lucky, but its high price outside Switzerland due to the high value of the Swiss franc limited its success.

The preceding short description of the CSF microscope is largely based on its presentation at the Sixth British Symposium on Electron Microscopy at Oxford, March, 1947 (Brück & Grivet, 1947a).

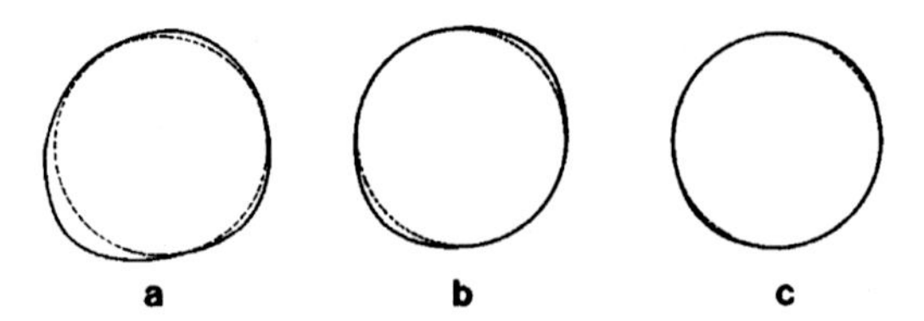

Figure 15 Hole profiles in the central electrode of an objective for various finishing techniques. (a) $\Delta r = 2.7\ \mu m$; (b) $\Delta r = 1.3\ \mu m$; (c) $\Delta r = 0.3\ \mu m$.

2.4.3 "Ellipticity" aberration and the "natural" resolving power of EM

As often remarked, one can materialize in an industrial construction only a part of the scientific results acquired at the same time in the research laboratory. These were presented in a second communication (Bertein et al., 1947) by F. Bertein, H. Brück, and P. Grivet at the Sixth British Symposium on EM in 1947.

The efforts of the whole research team had begun to open up new directions for research when in 1944 mechanical measurements on central electrodes, prepared for the purpose by different procedures and polished by various means, showed a lack of circular symmetry normally amounting to a fraction of a micrometer and to some 3 μm in the worst case (Fig. 15a-c). Soon the mechanical accuracy of the measurement had been pushed to its limits and the present author was led to look at the optical consequences of these defects and to scrutinize many pictures obtained by various objectives under diverse conditions: for example, at different direct magnifications or focusing but at the same total magnification for simple objects, fine crystalline powders, or organic or mineral films, shadowed with metals or in their natural state. This quest finally showed a definite trend to "oriented fuzziness," varying in direction with focusing. On the other hand, H. Brück and E. Regenstreif, who were in charge of the experiments, checked directly the existence of this effect on Fresnel fringes (Boersch, 1940a, 1940b, 1940c, 1943; Hillier, 1940) on the edges of ZnO and MgO crystalline powders, and noted also that the best pictures were always obtained with the smallest gun aperture angles, clearly smaller than the theoretical optimum of the basic theory. These observations were all reminiscent of the theory of "thin pencils," classic in light optics, which was rigorously extended to the continuous refraction of electron optics by M. Cotte (1938). Indeed the illumination is properly a thin pencil in the object space, and still more in the image space [$\alpha \approx 10^{-3}$, $\alpha' = \alpha/G$ (G being the objective magnification)]. But how could one reconcile this idea with axial symmetry? Moreover, at first glance, it was rather surprising that the variety of forms of the mechanical defects so conspicuous in Fig. 15 was apparently not influencing in any way the most careful electron optical observations. E. Regenstreif shared this impression, and after studying a theory of W. Glaser (1942) on a disturbance of definite elliptic form, he succeeded in realizing [cf. (Bertein et al., 1947)] an elliptical hole with a difference between the semiaxes a_1 and a_2, $a_2 - a_1 = 50$ μm. He thus obtained a first quantitative check of the theory, as the images showed clearly the existence of two focal lines offering two possibilities of focusing, each giving an elongated image of a point, at 90° to each other; the sharpest image of a point, e.g., a small hole in a film, was at an intermediate adjustment, on the spot of "least confusion" (Fig. 16).

The experiment in itself gave no direct explanation of the insensitivity of the optical effect to the form of the hole. But it gave a clear statement of the problem and interested F. Bertein in its solution, and so it became unnecessary to proceed further with experimentation; it was indeed rather difficult at this high degree of mechanical accuracy to provide holes with ternary or higher dissymmetries. Fortunately, Bertein was a skilled and deep theoretician, always finding his own way, with a complete disregard for previous theories as soon as he could be convinced that he was facing a really good problem. This then appeared to be the case, and during the year 1945 he succeeded in building an original theory, first published in a few notes to the Academy of Sciences (Bertein, 1947a, 1949; Bertein & Regenstreif, 1947), and later in a thesis submitted November 28, 1947, and printed in the *Annales de Radioélectricité* (Bertein, 1947b), the scientific journal of the Compagnie Générale de Télégraphie Sans Fil (see also Bertein, 1947c, 1948a, 1948b). This field theory gives a complete solution for any perturbation to rotational symmetry: the first term of the development of the optical effect is always an astigmatism which may be conveniently attributed to a "representative" ellipticity in the case, for example, of a hole of irregular form; the theory defines quantitatively this first-order equivalence, which is not always geometrically evident. More important, it shows that if the electron pencil meets two independent perturbations in succession (independence meaning that the perturbing field may be split into two regions without noticeable interaction) the two may cancel. The residual astigmatism of a strongly convergent objective can be corrected by completing it with a feeble lens of strong astigmatism: just the process used by the optician to cure an astigmatic malformation of the eye. The precious minus sign may be obtained by various means, the simplest being the crossing of

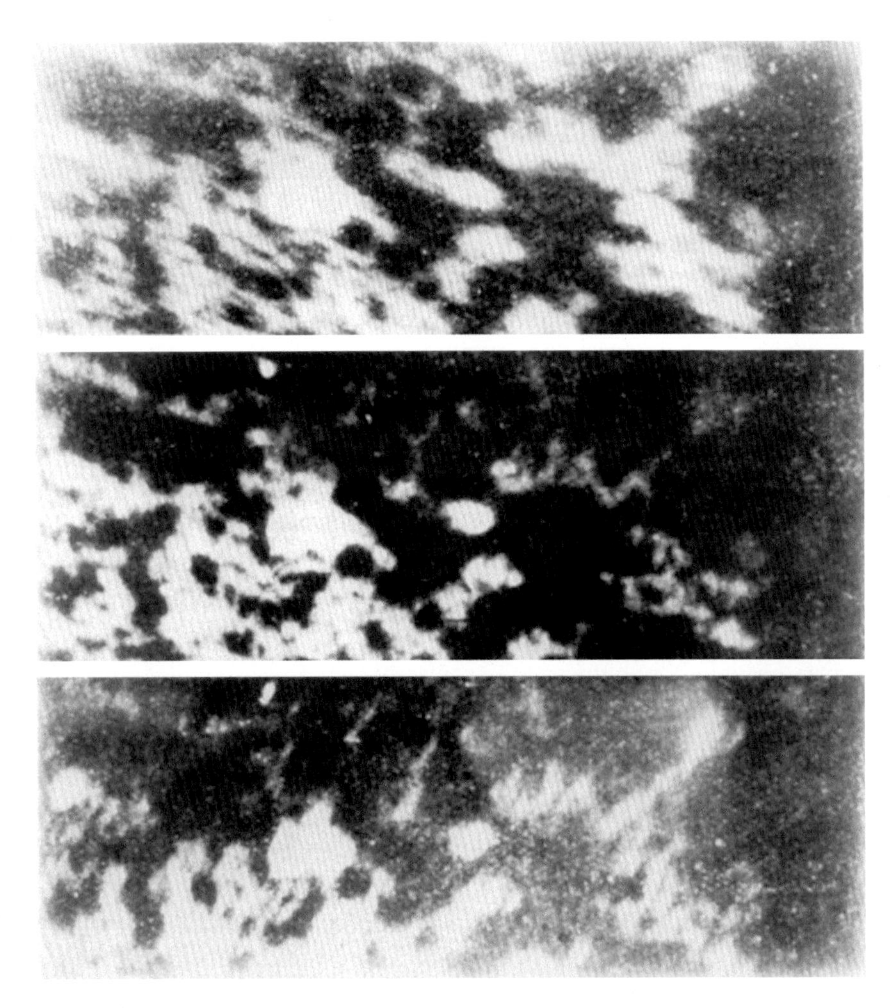

Figure 16 Pictures obtained for different focusing adjustment for a strongly elliptical hole: $a_2 - a_1 = 50\ \mu m$.

the axis of the cylindrical lenses (Fig. 17), which may represent faithfully the first-order action of the defect and of the corrector.

A peculiar form of corrector *in situ* was independently proposed by Hillier and Ramberg for the magnetic objective with a set of screws put directly in the active gap of the lens (Hillier, 1946, 1947; Hillier & Ramberg, 1947) but Bertein was the first to give a general and rigorous theory of the correction which could work in every case, and to demonstrate the efficiency of this innovation by shaping an electrostatic corrector, which with four (Fig. 18) or eight (Fig. 19) sectors and potentials of the order of

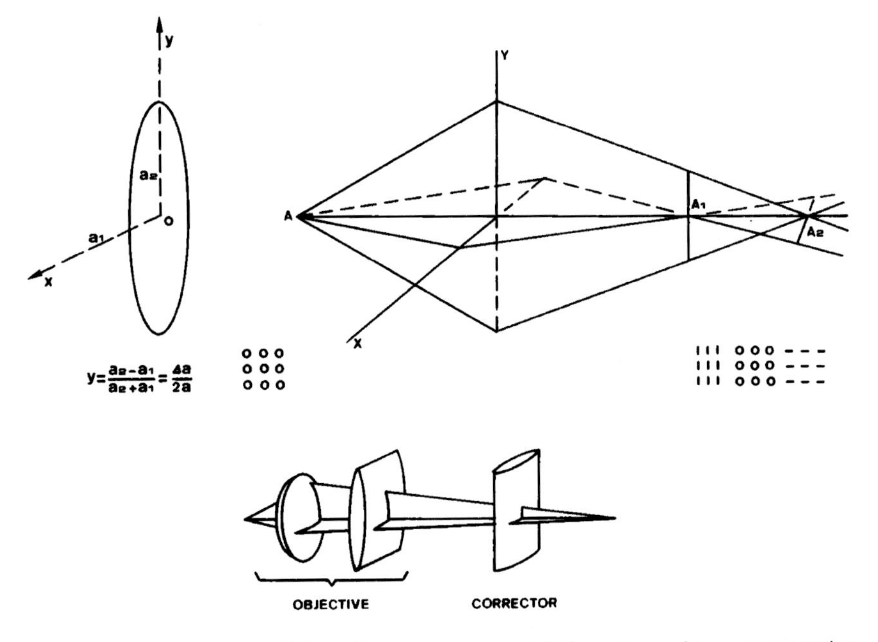

Figure 17 Astigmatism, a defect of transverse vectorial character and its representation by a cylindrical glass lens.

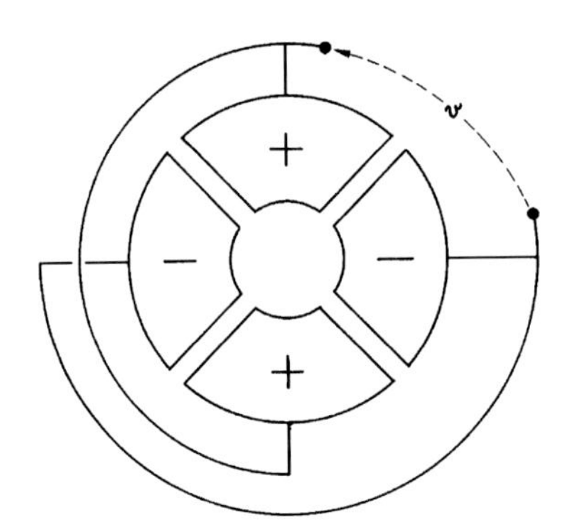

Figure 18 Bertein's corrector: with four sectors it corrects an astigmatism of known transverse direction.

500 V were able to correct the asymmetry of good objectives, whatever its origin. E. Regenstreif studied this device in detail; four electrodes constitute the equivalent of a cylindrical lens, but with a fixed direction of the axis; using the vectorial character of the corrector, eight electrodes allow one to rotate the axis of the equivalent cylinder electrically, which is both

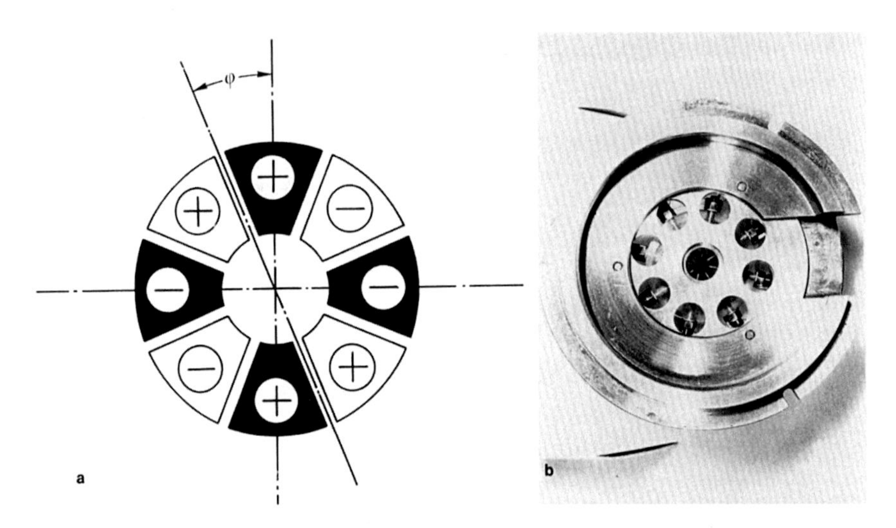

Figure 19 Bertein's corrector: with eight sectors it becomes possible to adjust electrically the transverse direction of the correction. (a) Scheme, (b) realization.

efficient and convenient for maintaining the vacuum in the electrostatic case (it abolishes any mechanical transmission of movement through the vacuum envelope; on the CSF microscope it permitted us to double the resolving power without any delicate adjustment). This kind of corrector was later thoroughly studied at Mosbach by O. Rang[5] (Rang, 1949; Scherzer, 1946a, 1946b, 1947), who coined the term "stigmator," a welcome correction of an etymological barbarism introduced long ago by photographers, who for "stigmatic" used "anastigmatic." H. Brück and P. Grivet were very early convinced of the validity of the explanation by a first-order "ellipticity," and Brück, working critically on Glaser's formalism, could improve its conclusion by taking into account the already deep analysis of such light opticians as A. Maréchal in France (Maréchal, 1944, 1947, 1948): using essentially Lord Rayleigh's simpler theory of diffraction, he showed that the circle of least confusion is halfway between the two focal lines, and not in the plane of the Gaussian image. He could also evaluate its radius (Brück, 1947a, 1947b), first linked to the dissymmetry of the field ϵ and then to that of its source, e.g., the hole η ellipticity. If a_1 and a_2 are the two semiaxes

[5] The possibility of the correction probably occurred to H. Mahl independently. He told H. Brück that he formulated in this sense a patent proposal which could not succeed due to the turmoils of the war; see (Rang, 1949, p. 521) for details and, for another independent proposition, see Scherzer (1946a, 1946b).

of the ellipse and a their mean value, he obtained

$$\epsilon = (E_1 - E_2)/(E_1 + E_2) \quad \eta = (a_1 - a_2)/(a_1 + a_2) = \Delta a/2a$$

and then proceeded by heuristic reasoning checked by measurement in the electrolytic tank; he also discussed the problem with Cotte, who later published a summary of his thinking (Cotte, 1949); the result of this semiempirical work was a precious formulation by Brück and Grivet (1947b) of the ultimate factor in resolving power δ, and of the optimal aperture of the illuminating pencil u, distinguishing carefully between two cases. In this way, the following formulas showing a square-root dependence in the ellipticity aberration C_e were established; now they pertain to the objective in its "natural" state of deficient rotational symmetry, whereas the older formulas in the fourth root of the spherical aberration are applicable to the new case in which the ellipticity is eliminated by the corrector (leaving aside the still hypothetical case in which cleanliness against any spoiling dielectric, perfection of machining, or homogeneity of any magnetic polepieces would guarantee a rigorous circular symmetry). These last conditions may be accurately defined in the general theory by

$$C_e < 2\lambda(C_s/\lambda)^{1/2}$$

C_e being the ellipticity aberration, C_s the spherical aberration, and λ the wavelength of the electron at accelerating voltage V. In the common "natural" case on the other hand,

$$C_e > 2\lambda(C_s/\lambda)^{1/2}$$

the best resolving power $\delta_{\min}$ and optimum aperture u_{opt} are given by

$$\delta_{\min} = 1.1\lambda(C_e/\lambda)^{1/2} \quad u_{\text{opt}} = 0.78(\lambda/C_e)^{1/2}$$

On the other hand, in the case of practical objectives, the following approximations are provided by the field theory and by Cotte's and Brück's analysis of astigmatism (Brück, 1947a; Cotte, 1949)

$$\epsilon \approx \eta \quad C_e \approx l \approx \Delta a$$

where l is the longitudinal distance between the focal lines, $\Delta a = a_2 - a_1$.

The results shown in Table 1 are obtained (Bertein et al., 1947; Brück & Grivet, 1947b), the typical characteristics of the objective being $U = 50$ kV,

Table 1 Ellipticity and resolving power δ.

Parameter	Values				
$C_e = \Delta a$ (μm)	0	0.32	1	4	16
$\epsilon = \eta$	0	4×10^{-5}	1.25×10^{-4}	5×10^{-4}	2×10^{-3}
δ_{min} (nm)	0.79	1.13	2.0	4.0	8.0
$10^3\ u_{opt}$	3.4	2.9	1.7	0.82	0.41

$f = 6.5$ mm, $C_s = 75$ mm. Thus it was realistic to hope for a "natural" resolving power of 4 nm at 50 kV, but with a small aperture slightly less than 10^{-3} rad. The use of the corrector could practically easily lower the limit to less than 2 nm, i.e., divide it by more than 2. The classical theoretical limit was still significantly lower, but notable progress in this direction proved to be delicate, demanding an exceptionally careful adjustment both of the corrector and of the illumination aperture.

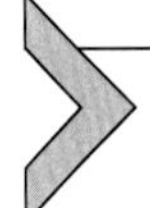

3. The liberation of the North of France and the transition to peace

3.1 Transition period

The research related in the preceding paragraph was carried on with as much continuity as possible during the transition years 1946–1948. Nevertheless, the hopes of 5 years fulfilled, the return to peacetime conditions brought great changes in economic conditions and a general reorientation of the activities of the CSF Company, in a great effort to cope with the technical progress achieved 20 miles away over the Channel and, still more extensive, some 3000 miles over the Atlantic Ocean. This tremendous change was encouraged by the Allied Powers, who generously opened their files during the golden years of the Marshall Plan. It was of great concern for the laboratories, which had to look at many important innovations such as television or radar with their new electronic tools for the present, but without neglecting for the future any opening toward the revolution in avionics and the nascent nuclear physics and nuclear energy; all these problems were felt to be fundamental for the company and the nation. The first massive impact on the electron optics group in 1945 was the displacement of their activity to a new laboratory in the town, near an old place renamed Place de Stalingrad, in an old plant refitted for the general purpose of research. Part of the wartime activity, such as tungsten metallurgy or fluorescent powder fabrication, was transferred to specialized plants elsewhere and electron optics and electron microscopy were retained as a valuable activity together with television, cathode-ray oscillographs, and soon some aspects of nuclear electronics such as mass spectrography. As regards microscopy, the first move was to think of a new apparatus embodying all the progress achieved in the laboratory. Notwithstanding the ambient public enthusiasm, the first step was to check all the useful new results and especially to measure properly the astigmatism in an effort to master it completely and to popularize the use of the ellipticity corrector. The new instrument was also conceived realistically as a complete system in the modern sense of the term. Practical formulas for expressing the useful properties of lenses and the general properties of the optics were urgently needed. Indeed, the greater stability made desirable by a notable progress in resolving power and the need for a large magnification and for convenience and accuracy in providing diffraction diagrams from a well-defined spot in the specimen were determining factors in adopting a set of *three* lenses: objective, first projector, and second projector for the future model MIX. E. Regenstreif was elected chief of the project and in fact devoted all his time for a few years to its success.

3.2 An elaborate study of asymmetry

The use of an artificially enlarged ellipticity such as that shown in Fig. 15 associated with the linear law summarized in the formula $C_e \approx \Delta a$ already gave an estimate of the defect with a 10% accuracy. Nevertheless, it remained desirable to find a new, more accurate method than the determination of the focusing voltages of the two focal lines as a measure *in situ* of the astigmatism. This would mean that the right focus could be determined exactly, and the precise correction of ellipticity would be eased. A partial answer to these needs was given by the use of transmission rays with a rather large aperture u in the illumination beam and simultaneously of a voltage nearly closing the lens as shown in the shaded zone of Fig. 20; under these conditions, working only with the objective in the absence of the projector at a 0.5-m distance, diagrams such as those of Fig. 21 could be recorded, with four cusps, looking like a square if the focusing is sharp (Fig. 21A), and distorted in the opposite case (Fig. 21B)—a very sensitive test; moreover, the half side of the square $2S$ divided by L (the distance between lens and screen) gives η directly: $\eta = S/L$ (Bertein & Regenstreif, 1949; Grivet & Regenstreif, 1948; Grivet et al., 1949). The evolution of the figures with the choice of aperture confirms the general theory of Bertein; third-order defects appear as a threefold symmetry, first

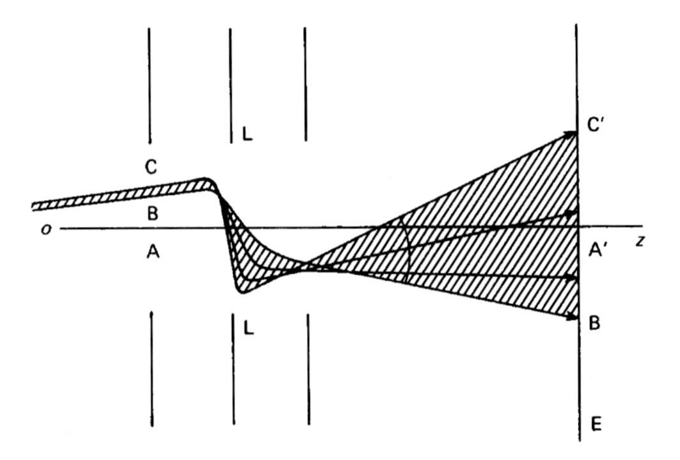

Figure 20 The domain of "trans-Gaussian" rays.

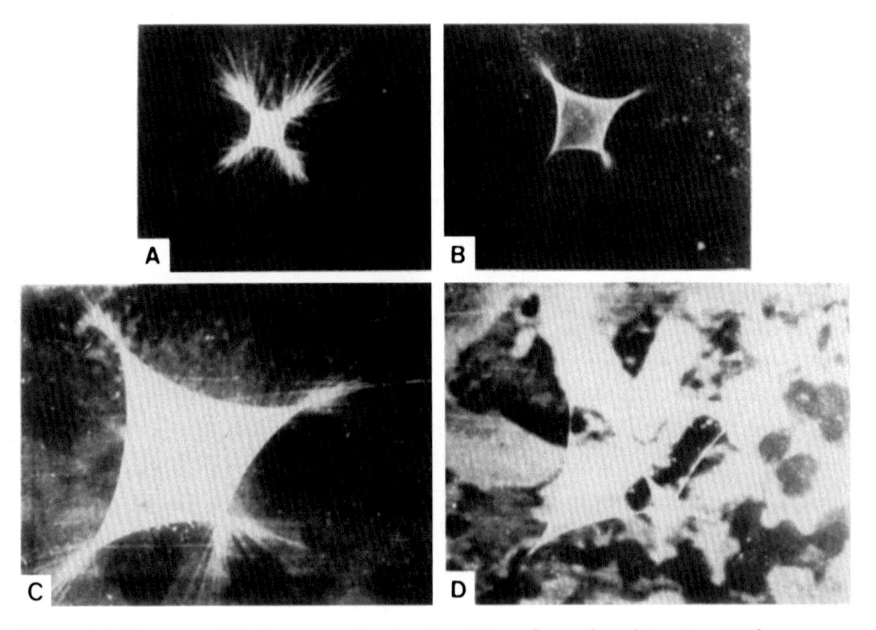

Figure 21 Regenstreif's diagrams obtained at 50 cm from the objective: (A) four cusps corresponding to sharp focusing and slight ellipticity, (B) out of focus and slight ellipticity, (C and D) sharp focus when third-order defects are showing and prevailing by rays grazing nearer to the rim.

slightly (Fig. 21C), then more neatly when the rays graze the central electrode rim (Fig. 21D). This sensitive method was used to determine and to minimize the role of the electrical lead as an observable cause of ellipticity (Bertein, 1949). Simultaneously, E. Regenstreif was able to describe electrostatic lenses by introducing a convenient parameter $x = \varphi(0)/\varphi(z_0)$, $\varphi(z)$ being the potential on the axis with $z = 0$ at the center of the central electrode, z_0 measuring the gap between the hole center and the external electrodes; x is a synthetic parameter depending on the voltage applied to the central electrode and on the dimension of the holes and gaps via a relatively simple formula. Thus, more clarity and universality are obtained for the curves representing the results of the original and accurate theory, developed on this occasion and confirmed by comparison with numerous measurements, including those of German research groups (Heise & Rang, 1949). We refer on this point to the digest given in (Grivet et al., 1965, 2nd Ed., pp. 211–224) and to the article devoted to the model EM objective (Regenstreif, 1951).

3.3 A last prototype

On this basis the following structure (Grivet & Regenstreif, 1952) was adopted:

1. A sturdy construction, shown in Figs. 22 and 23, with a Brück gun, without any condenser or mechanical orientation adjustment, and with three lenses forming a solid block (Fig. 24a). A preliminary study by H. Brück (1947) showed that a mechanical accuracy of $\frac{1}{100}$ mm in centering and in assuring parallelism of flanges offered a sufficiently safe security margin for clean optical conditions. The length of the microscope proper was reduced to 0.75 m, 1.40 m comprising the entire length for the whole apparatus.
2. The optics: objective $f = 6$ mm, $G_1 = 20$; first projector $f = 2.5$–50 mm, $G_{2_{max}} = 20$; second projector $f = 2.5$ mm, $G_3 = 100$ (fixed for EM).
3. The total magnification could be chosen between 500 and 40,000×. A retractable contrast diaphragm at the image focus of the objective provided a wide field for diffraction, and for this purpose G_3 could also be varied electrically. Focusing was also obtained electrically by varying the polarization of the central electrode of the objective between 0 and 500 V (Fig. 24b).
4. Astigmatism could be corrected in two ways: first, by choosing a peculiar objective, with a hole of larger diameter than usual, the focal length of 6 mm being restored by an additional negative voltage of one-third of the accelerating voltage, highly stabilized—an idea of E. Regenstreif (1950a, 1950b) based on the failure of the equality $\epsilon = \eta$ valid in the unipotential arrangement and discussed in a separate paper at the Paris

Figure 22 General layout of the M IX instrument (E. Regenstreif).

Conference on EM (1950), where the apparatus shown in Figs. 22 and 23 was fully described; second, by a Bertein corrector.

Only two examples of this apparatus were built: one of them was tested with success by Regenstreif in a quiet basement at the Ecole Normale Supérieure. But no industrial production followed: all energies in the CSF Laboratory had to be fully dedicated to reconstruction and reconversion tasks. The staff concerned with the electron microscope had inexorably vanished as many opportunities had opened in various fields. P. Grivet returned to the ENS Laboratory as Professor in 1948, soon followed by F. Bertein. H. Brück found a new home in Saclay at the CEA, studying Van de Graaff's and the associated accelerator, as well as circular machines such as "Saturne" later; H. Blattman followed the same path; J. Vastel remained at CSF and designed and built many of the mass spectrographs essential for the nascent French nuclear industry. E. Regenstreif was the last to abandon electron microscopy, but the first to enter a European agency, the CERN[6]

[6] Centre Européen de Recherches Nucléaires.

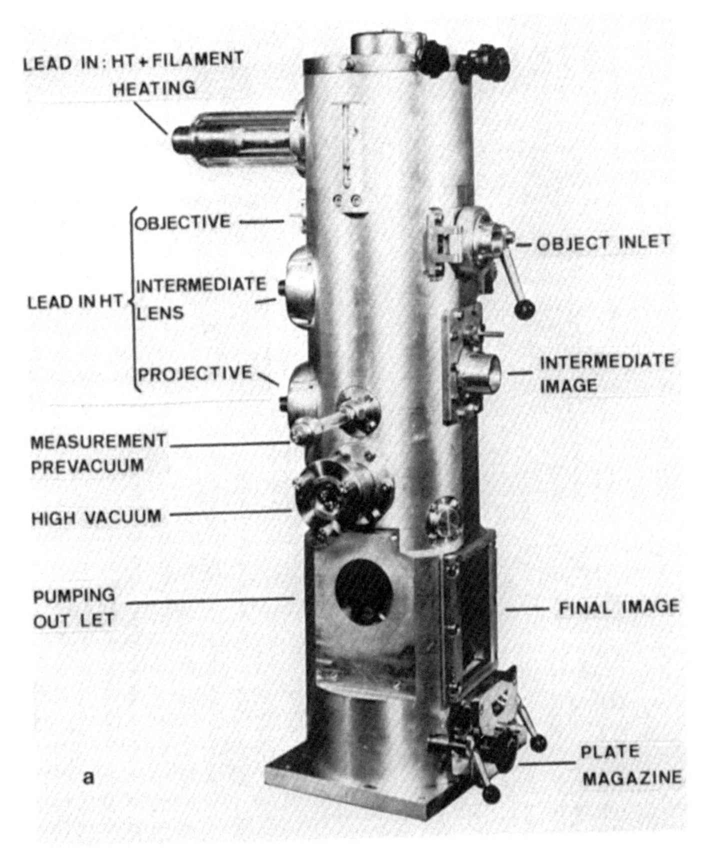

Figure 23 M IX: (a) forged Duralumin casing, (b) internal optical block: gun anode plus three lenses.

in Geneva: invited by Professors P. Auger and F. Perrin, two of the founding fathers of CERN, he undertook first a wide-ranging journey in the United States to collect advice and data on circular accelerators with classical or strong focusing and took a major part in the conception of the large circular accelerator at CERN near Geneva and actively supported the new focusing scheme.

3.4 The ill fate of electrostatic lenses in supermicroscopy and their present revival in ion microprobe analyzers

Nowadays the prime mover of scientific progress is not only embodied in outstanding individuals, but also, for the most part, in sustained teamwork. The decline in electrostatic electron supermicroscopy on the eve of a brilliant peacetime period was not peculiar to France and Germany, but was

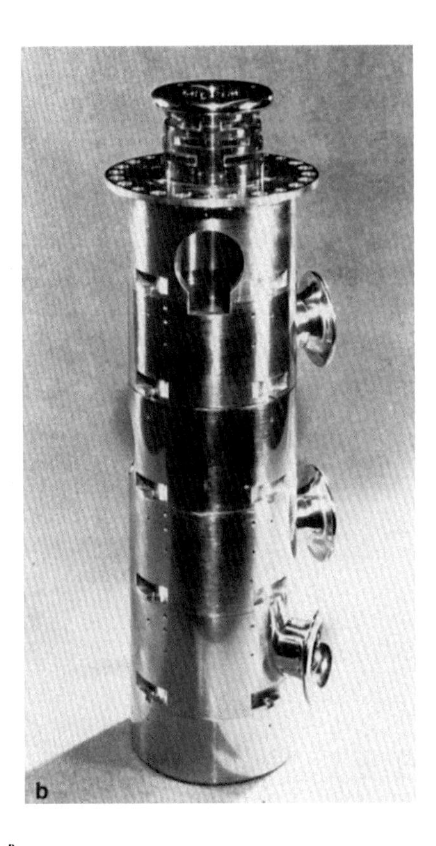

Figure 23 (*continued*)

recognized worldwide during a decade. The later development of magnetic supermicroscopy revealed that electrostatic lenses suffered in two respects: the converging effect is essentially a differential action, the "einzellens" (or single lens) being really a composite of two divergent parts with a stronger converging element in between [cf. (Brüche & Scherzer, 1934; Grivet, 1950; Ruska, 1979) or (Grivet et al., 1965)]; these parts add as regards spherical aberration. On the contrary, the total convergence is the result of the positive contribution of the center and of the negative one of the two external side lenses; therefore C_s/f is at least seven times larger than for a uniformly acting magnetic lens. Moreover, the balance inclines in the same direction when weighing the "practical" advantages of high tension, which appeared so favorable to the magnetic supermicroscope (see the relevant contributions to this volume). Nor was it possible to take advantage directly in an electrostatic EM of the very short de Broglie wavelength of

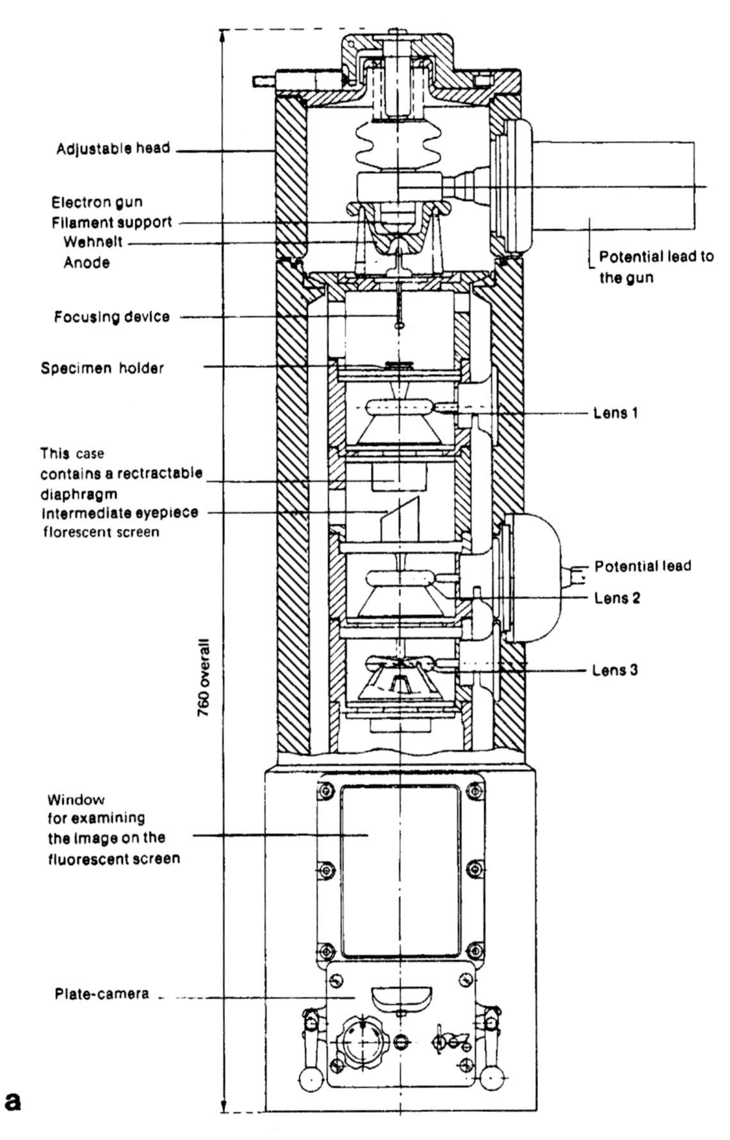

Figure 24 (a) Optical structure of the three-stage M IX, (b) electrical focusing and adjustment of magnification.

ions, due to their large mass for a given voltage ($\lambda = h/mv$): the low intensity of ion sources (Couchet, 1951, 1954; Gauzit, 1951, 1954) is a great difficulty and this heavy mass also limits the transparency of the thinnest films. The history of ion microscopy retraced in (Grivet et al., 1965) was very disappointing in this respect.

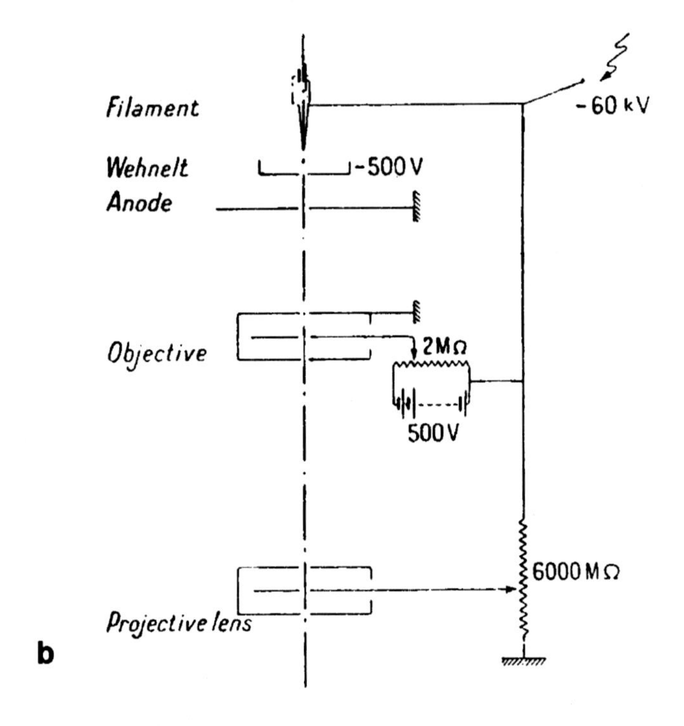

Figure 24 (*continued*)

It should not, however, be hastily concluded that the electrostatic solution should be consigned to domains in which extreme sharpness of images is of secondary importance, such as oscillographs, image tubes, image converters, and so on. New devices combining elementary qualities in a new way were conquering new provinces in electronics—for example the ion[7] microanalyzer. After 1945, the most enthusiastic defenders of the EM were the metallurgists, probably because theoreticians explained most of the characteristics of *real* metals and alloys, e.g., the important properties making them strongly different from the crude *ideal* models, by introducing many submicroscopic elements such as slip lines, various forms of dislocations and of impurity clusters, which at first no one could observe with light microscopes. Among these interesting hypothetical objects were the small platelike precipitates introduced by A. Guinier (1938a, 1938b) and D. G. Preston (1938a, 1938b) to explain "age hardening" of certain alloys,

[7] The differential equation governing the movement of charged particles in an electrostatic field contains coefficients independent of the mass of the particle.

for example, aluminum alloyed with 4% copper. This concept was indeed born from an interpretation of their X-ray diagram, obtained by long time exposures of several hours to a few days in specialized high-luminosity X-ray spectrographs. This is of course a volume effect, but one could hope to detect it as well on the surface by replica techniques. In fact it first escaped the sagacity of electron microscopists; when the aging is conducted at low temperatures, the platelets are very thin and their diameter is small (5 nm), so the replica techniques made it possible to visualize only precipitates obtained at slightly higher temperatures (Castaing, 1949). In 1948, that problem of precipitation in alloys oriented the reflections of a laboratory friend of the author, Professor A. Guinier, a distinguished metallurgist and well-known expert in X rays. He began to look at a new technique making use of X rays for point analysis; he proposed (Castaing & Guinier, 1949a, 1949b, 1949c, 1950) reversing the role of magnification of an ordinary EM: "demagnification" in two stages producing a very small spot much less than 1 μm in diameter (Fig. 25). One had only to detect the X rays produced by the impact of electrons in the spot on the specimen considered as an anticathode. By analyzing spectroscopically the X rays, a fine local determination of the alloy chemical composition in a small region under the spot would then be obtained. A. Guinier proposed this problem as a goal to a young thesis candidate he had just recruited, R. Castaing, a brilliant student, both imaginative and rigorous, skilled and efficient. He worked in a well-equipped laboratory of the recently developed state agency ONERA.[8] He soon recognized that a good tactic would be to study and modify an existing EM for the purpose and he went to the nearest place, the Paris EM group, working hard on their own problems as well as his in the new experiment. The last lens was the sensitive one, the equivalent of the objective in the classical scheme, but it was subjected to the new constraint of allowing the X rays to escape, so that its minimal image focal length was of the order of 20 mm, and Castaing checked first the astigmatism of such a lens. He conceived a new and very accurate experiment (Castaing, 1950a, 1950b, 1950c) for the purpose, using a very fine wire of Lucite as object at the first lens focus in a shadow experiment; under these conditions astigmatism produces a strong distortion in the image, as shown in Fig. 26; the form of the image is extremely sensitive both to the focusing and to the value of astigmatism, its accurate measure being derived (Fig. 27). In this new case Castaing checked the early theories

[8] Office National d'Etudes et de Recherches Aéronautiques (National Office for Aeronautical Studies and Research).

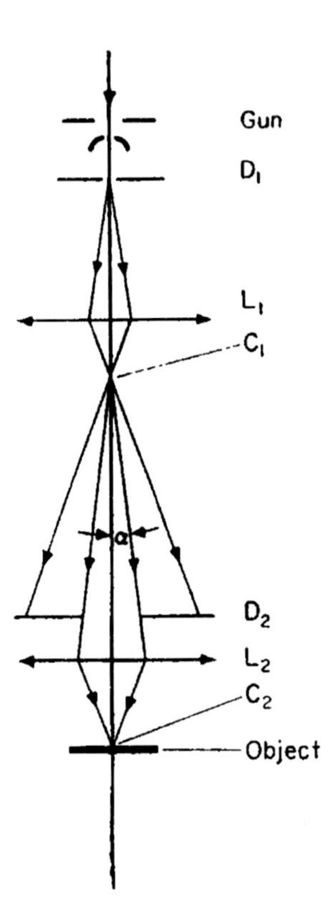

Figure 25 "Demagnification" in the Castaing microprobe analyzer: L_1, L_2, lenses; D_1, D_2, diaphragms; C_1, first crossover; and C_2, second crossover.

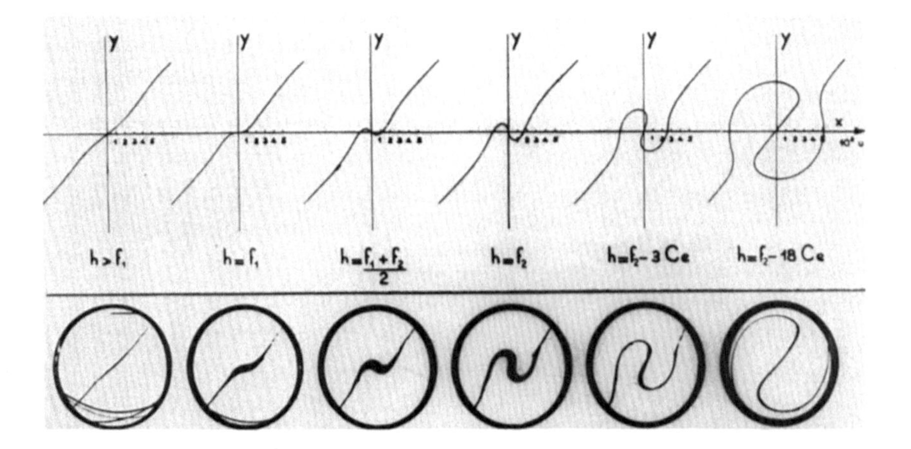

Figure 26 Image of a very fine Lucite wire for different degrees of focusing.

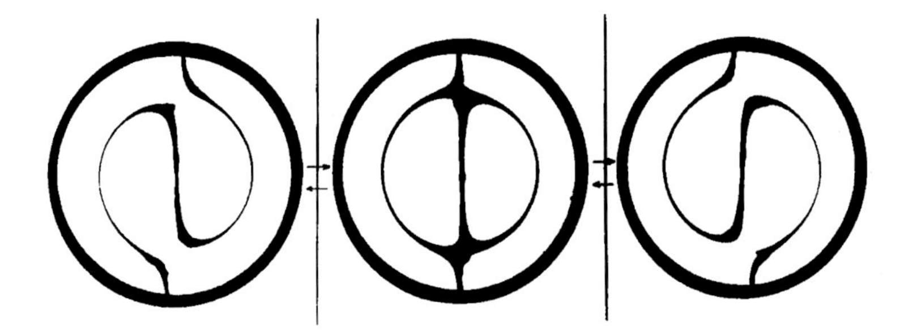

Figure 27 Left and right: astigmatic lens. Center: lens corrected for astigmatism.

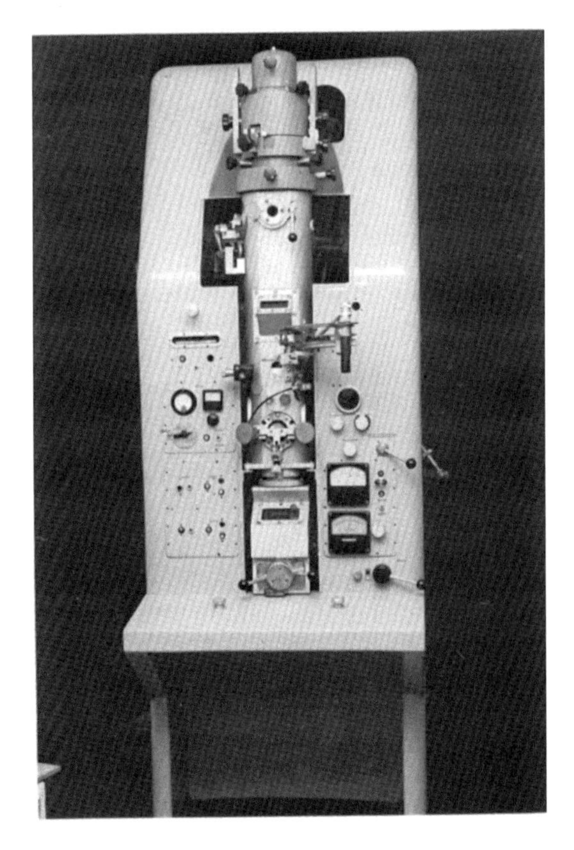

Figure 28 Castaing's prototype of an "electrostatic" microprobe analyzer.

and experimental results of the EM group, and, more important, he obtained favorable conditions for his own experiment. Soon the construction of an *electrostatic* prototype (Fig. 28) demonstrated clearly the feasibility of

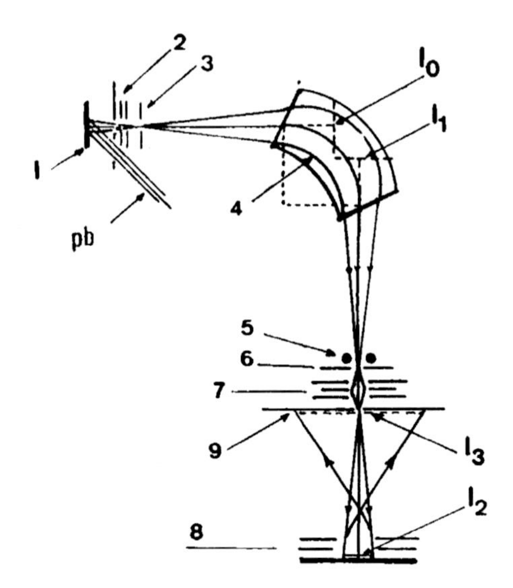

Figure 29 Optical scheme for the "direct-imaging ionic microanalyzer" of R. Castaing and G. Slodzian (Castaing & Slodzian, 1962; Morrison & Slodzian, 1975): pb, primary-ion beam; I_0, nonselected image; I_1, mass-selected image given by the magnetic prism; I_2, final ionic image; 3, diaphragm; 4, magnetic prism; 5, stigmator; 6, mass-selecting slit; 7, projective lens; 8, image converter; and 9, fluorescent screen.

the scheme, and striking results were obtained in 1951 (Castaing, 1951). This so-called "electron microprobe analyzer" (Castaing, 1960) was a great success, as was proved by its tremendous development (Grivet et al., 1965, pp. 752–762). But the appetite of research workers for more and more data is inexhaustible, and to extend substantially the domain explored by the electron microprobe, R. Castaing and G. Slodzian proposed and built in 1962 a "direct imaging microanalyzer" (Castaing & Slodzian, 1962; Morrison & Slodzian, 1975). In this new instrument (Fig. 29) the specimen is bombarded by a primary-heavy-ion beam; the atoms located at the surface of the object are thrown out, a substantial part as ions which are focused by a lens, then filtered in mass and received on an ion–electron converter; afterward they feed an electron lens producing a final image on a fluorescent screen. One kind of ion or the other can be selected obtaining an analysis of the specimen not only chemical but also isotopic (through the mass of the ions used). Moreover, the primary-ion bombardment scours the surface and explores progressively a small depth of the specimen; the final image shows details of the order of 0.2 μm over a field of 250 μm. Parsons (1978) offers further details (pp. 168–172) on this very successful and promising technique.

An old hand cannot refrain from feeling gratified by such a clever arrangement of so numerous electrostatic lenses and their stigmators; he was also moved by the first encouraging results obtained in the biology field by modern proton scanning supermicroscopes from the American group of W. H. Eskowitz, Y. R. Fox, and R. Levi-Setti at the Enrico Fermi Institute of Chicago (Grivet et al., 1965, pp. 183–202). This all looks like a modern illustration of the old phoenix legend of Greece. This old tradition leads us to hope that all the work done in various countries on electrostatic particle optics may receive valuable new rewards in future inventions.

4. Afterword by Peter Hawkes

Pierre Grivet was born on 14 November 1911 in Lyon where his father was a German master at the Lycée du Parc, which no doubt explains his mastery of that language. Accepted by both the Ecole Polytechnique and the Ecole Normale Supérieure (ENS), he chose the latter (1931–1935). In July 1934, he married Thérèse Meyer, whose doctoral thesis was defended, like that of her husband, in 1941. In view of his later activity, it is interesting to note that the subject of her "Deuxième Thèse' (propositions données par la Faculté) was 'Microscope Electronique'.

Work on Pierre Grivet's doctoral thesis on improvement of methods of measuring the speed of light began in 1936, in the ENS laboratory. But with the declaration of war in 1939, Grivet was called up and despatched to the Maginot Line as an artillery lieutenant. Soon though, he was recalled, worked briefly on radar and was entrusted with the task of redirecting research at the ENS laboratory towards the detection of electromagnetic waves. When Paris was invaded by the German army, Grivet moved briefly to the Conservatoire National des Arts et Metiers before joining the CSF as he describes here.

In later years, Grivet directed the Institut d'Electronique Fondamentale at Orsay after a spell as Director of the Laboratoire de Radioélectricité in the EDF building in Fontenay-aux-Roses., then an attractive village in the Paris suburbs, where petanque was played on the village square and wisteria spilled over the surrounding walls. Among many other projects, he worked on accelerator optics, leading to long studies with Albert Septier on quadrupole lens properties, and on transmission lines, on which he wrote a two-volume treatise, later translated into English. He was elected member of the Académie des Sciences in 1972.

Grivet died on 1 June 1992. An obituary by Grivet's close colleague Albert Septier can be found in *Microscopy Microanalysis Microstructures* **3** (1992) iii–vi and a long Notice on La Vie et l'Œuvre de Pierre Grivet by André Blanc-Lapierre appeared in *Comptes Rendus de l'Académie des Sciences* **10** (1993) No. 5, 502–508. Etienne de Chambost has written at length about Grivet, see siteedc.edechambost.net/CSF/Grivet.html.

References

An age of innovation (the world of electronics, 1930-2000). New York: McGraw-Hill, (1980). Initially *Electronics*, *53*(9) (1980).

Bettering of the focussing devices for electron microscopes. French Patent 562,548. (1946).

Bettering of the HT sources for electron diffraction and electron microscopy. French Patent 524,177. (1946).

Arnal, R. (1953). *Comptes Rendus Hebdomadaires des Séances de l'Académie des Sciences*, *237*, 308.

Arnal, R. (1954a). *Comptes Rendus Hebdomadaires des Séances de l'Académie des Sciences*, *238*, 2061.

Arnal, R. (1954b). *Comptes Rendus Hebdomadaires des Séances de l'Académie des Sciences*, *238*, 2402.

Arnal, R. (1955). *Comptes Rendus Hebdomadaires des Séances de l'Académie des Sciences*, *240*, 610.

Arnal, R. (1956). Electric "microdécharges" in a dynamic vacuum. *Annales de Physique (Paris)*, *10*, 830–955.

Bachman, C. H. (1943). Simplified electron microscopy. *Electronics*, *16*, 78–81, See also 195–200.

Bachman, C. H., & Ramo, S. (1943). Electrostatic electron microscopy. *Journal of Applied Physics*, *14*, 8–18, See also 69–76, 155–160.

Baird, J. L. (1938). *2 color sequential system.* London. *3 color sequential system.* London (1938).

Barthélémy, R., & de France, H. (1932). *Early demonstrations.* Paris.

Bertein, F. (1947a). On shape's imperfection in electron optical devices. *Comptes Rendus Hebdomadaires des Séances de l'Académie des Sciences*, *224*, 106–107.

Bertein, F. (1947b). France patent application 531,287.

Bertein, F. (1947c). On some defects in electron optical instruments and on their correction. *Annales de Radioélectricité*, *2*(10).

Bertein, F. (1948a). On some defects in electron optical instruments and on their correction. *Annales de Radioélectricité*, *3*(11).

Bertein, F. (1948b). Influence of electrode deformations in electron optics. *Journal de Physique et le Radium*, *9*, 104–112.

Bertein, F. (1949). On the perturbing potential originating in the ovalisation of the holes in electrostatic lenses. *Comptes Rendus Hebdomadaires des Séances de l'Académie des Sciences*, *224*, 560–562.

Bertein, F., & Regenstreif, E. (1947). On the ellipticity aberration in electrostatic lenses. *Comptes Rendus Hebdomadaires des Séances de l'Académie des Sciences*, *224*, 737–739.

Bertein, F., & Regenstreif, E. (1949). Use of marginal ("transgaussian") rays in the study of asymmetries in electrostatic lenses. *Comptes Rendus Hebdomadaires des Séances de l'Académie des Sciences*, *228*, 1854–1856.

Bertein, F., Brück, H., & Grivet, P. (1946). The electron microscope and its applications in metrology [sic]. *Métaux & Corrosion*, *21*, 1–10. As cited by Grivet. See Grivet et al. (1946) for the correct reference.
Bertein, F., Brück, H., & Grivet, P. (1947). Influence of mechanical defect of the objectives on the resolving power of the electrostatic microscope. Communication to the 6th British Symposium on E.M. (March 1947) reprinted in *Annales de Radioélectricité*, *2*(9).
Boersch, H. (1939). The shadow microscope. *Naturwissenschaften*, *27*, 418. *Z. Tech. Phys.*, *20*, 346-350 (1939).
Boersch, H. (1940a). Fresnel's fringes for electrons. *Naturwissenschaften*, *28*, 709–711.
Boersch, H. (1940b). Fresnel's diffraction in a Supermicroscope. *Naturwissenschaften*, *28*, 711–712.
Boersch, H. (1940c). Diffraction experiments with a very thin electron pencil. *Zeitschrift für Physik*, *116*, 469–479.
Boersch, H. (1943). Fresnel diffraction in the electron microscope. *Physikalische Zeitschrift*, *44*, 202–211.
Braun, F. (1897). *Annalen der Physik und Chemie*, *60*, 552–559.
Bricka, M., & Brück, H. (1948). Sur un nouveau canon électronique pour tubes à haute tension. *Annales de Radioélectricité*, *3*, 339–343.
Bricout, P. (1938). Optique des charges électriques. *Revue Générale de l'Electricité*, *44*, 405–430 and 439–450.
Brüche, E. (1932). Elektronen Mikroskop. *Naturwissenschaften*, *20*, 49.
Brüche, E. (1943). The development of the electron microscope. *Physikalische Zeitschrift*, *44*, 176–180.
Brüche, E., & Recknagel, A. (1941). *Elektronen-Geräte*. Berlin and New York: Springer-Verlag.
Brüche, E., & Scherzer, O. (1934). *Geometrical electron optics*. Berlin: Springer-Verlag.
Brück, H. (1947a). On the limit of resolution of the electron microscope with (perfectly) round lenses. *Comptes Rendus Hebdomadaires des Séances de l'Académie des Sciences*, *224*, 1553–1555.
Brück, H. (1947b). On the limit of resolution of the electron microscope with asymmetric lenses. *Comptes Rendus Hebdomadaires des Séances de l'Académie des Sciences*, *224*, 1628–1629.
Brück, H. (1947). Resolving power of an electrostatic objective with a hole out of center. *Comptes Rendus Hebdomadaires des Séances de l'Académie des Sciences*, *224*, 1818–1820.
Brück, H., & Bricka, M. (1945). *Electron gun for fine spots and 3 electrodes*. France patent application 513,906.
Brück, H., & Grivet, P. (1947a). Improvements in the electrostatic E.M. Communication to the 6th British Symposium on E.M. (March 1947) reprinted in *Annales de Radioélectricité*, *2*(9).
Brück, H., & Grivet, P. (1947b). On the limit of resolution of an electrostatic objective with an ovalized central electrode. *Comptes Rendus Hebdomadaires des Séances de l'Académie des Sciences*, *224*, 1768–1769.
Brück, H., & Grivet, P. (1950). Sur la lentille électrostatique. *Revue d'Optique, Théorique et Instrumentale* , *29*, 164–170.
Brück, H., & Romani, L. (1944). Sur les propriétés de quelques lentilles indépendantes. *Cahiers de Physique*, *24*, 15–28.
Cartan, L. (1937). Une nouvelle méthode de focalisation d'un faisceau d'ions positifs rapides. Application à la spectrographie de masse. *Journal de Physique et le Radium*, *7*, 111–120.
Castaing, R. (1949). Recherches au microscope électronique sur les précipitations dans les alliages d'aluminium. *Comptes Rendus Hebdomadaires des Séances de l'Académie des Sciences*, *228*, 1341–1343.
Castaing, R. (1950a). Une méthode de détection et de mesure de l'astigmatisme d'ellipticité. *Comptes Rendus Hebdomadaires des Séances de l'Académie des Sciences*, *231*, 835–837.

Castaing, R. (1950b). Détection et mesure directede l'astigmatisme d'ellipticité d'une lentille électronique. *Comptes Rendus Hebdomadaires des Séances de l'Académie des Sciences*, *231*, 994–996.
Castaing, R. (1950c). Lentille corrigée de l'astigmatisme et son utilisation pour l'obtention de sondes de grande brillance. ICEM, *Paris*. (pp. 148–154).
Castaing, R. (1951). *Application des sondes électroniques à une méthode d'analyse ponctuelle chimique et cristallographique*. (Publ. O.N.E.R.A. No. 55). (Ph.D. Thesis). Paris.
Castaing, R. (1960). Electron probe microanalysis. *Advances in Electronics and Electron Physics*, *13*, 317–386.
Castaing, R., & Guinier, A. (1949a). Sur les images au microscope électronique des alliages d'aluminium-cuivre durcis. *Comptes Rendus Hebdomadaires des Séances de l'Académie des Sciences*, *228*, 2033–2035.
Castaing, R., & Guinier, A. (1949b). Etude au microscope électronique du vieillissement des alliages aluminium-magnésium–silicium. *Comptes Rendus Hebdomadaires des Séances de l'Académie des Sciences*, *229*, 1146–1148.
Castaing, R., & Guinier, A. (1949c). *Application des sondes électroniques à l'analyse métallographique*. ICEM, *Delft*. (pp. 60–63).
Castaing, R., & Guinier, A. (1950). *Exploration et analyse élémentaire de l'échantillon par une sonde électronique*. ICEM, *Paris*. (pp. 391–397).
Castaing, R., & Slodzian, G. (1962). Microanalyse par émission ionique secondaire. *Journal de Microscopie (Paris)*, *1*, 395–410.
Cotte, M. (1938). Some researches on electron optics. *Annales de Physique (Paris)*, *10*(11), 333–406.
Cotte, M. (1949). Potential and field around the elliptic hole in a plane electrode. *Comptes Rendus Hebdomadaires des Séances de l'Académie des Sciences*, *228*, 377–379.
Couchet, G. (1951). A study of the positive ions emission in the family of salts like Li_20, Al_20_3, $4Si0_2$. *Comptes Rendus Hebdomadaires des Séances de l'Académie des Sciences*, *233*, 1013–1016.
Couchet, G. (1954). A study of ion's solid sources. Their application to secondary electron emission and to the ion emission microscope. *Annales de Physique (Paris)*, *9*, 731–781.
Davisson, C. J., & Calbick, C. J. (1931). Electron lenses. *Physical Review*, *38*, 585. *Phys. Rev.*, *42*, 580 (1932); Electron optics, *Phys. Rev.*, *45*, 764 (1934).
Dufour, A. (1914). *Comptes Rendus Hebdomadaires des Séances de l'Académie des Sciences*, *158*, 1939–1941. *Phys. Radium*, *1*(6), 147-160 (1920).
Dufour, A. (1920). *Journal de Physique et le Radium*, *1*(6), 147–160.
Fritz, R. (1936). The electron microscope. *Revue Générale des Sciences*, *47*, 338–342.
Gauzit, M. (1951). On ion supermicroscopy by transmission. *Comptes Rendus Hebdomadaires des Séances de l'Académie des Sciences*, *233*, 1586–1589.
Gauzit, M. (1954). Microscope corpusculaire l̀'aide d'ions lithium. *Annales de Physique (Paris)*, *12*, 683–700.
Glaser, W. (1942). Electron optical imaging by disturbing rotational symmetry. *Zeitschrift für Physik*, *120*, 1–15.
Glaser, W. (1952). *Grundlagen der Elektronenoptik* [Fundamentals of electron optics]. Berlin and New York: Springer-Verlag.
Gölz, E. (1940). Research on behaviour of metal electrodes in microscope lenses under high tension. *Jahrbuch der AEG-Forschung*, *7*, 57–59.
Goodman, P. (1981). *Fifty years of electron diffraction*. Dordrecht, Netherlands: Riedel Publ.
Grivet, P. (1942). Optique électronique. *Revue Géérale de l'Electricité*, *51*, 473–484.
Grivet, P. (1945). Optique électronique. *Revue Générale de l'Electricité*, *54*, 45–54.
Grivet, P. (1946). Industrial realization of an electrostatic electron microscope. In de BroglieL. (Ed.), *L'Optique Électronique* (pp. 129–160). Paris: Revue d'Optique.
Grivet, P. (1950). Electron lenses. *Advances in Electronics and Electron Physics*, *2*, 47–100.

Grivet, P. (1980). Sixty years of electronics. *Advances in Electronics and Electron Physics, 50*, 89–174.
Grivet, P., & Regenstreif, E. (1948). The resolving power of an electrostatic objective. Ninth British Congress on Electron Microscopy (20.09.1948), Communication No. 22.
Grivet, P., & Regenstreif, E. (1952). On a new electrostatic microscope with three magnifying stages. Paper No. 32, *Paris, 1950*. (pp. 230–236).
Grivet, P., Brück, H., & Bertein, F. (1946). The electron microscope and its applications in metrology. *Métaux & Corrosion, 21*, 1–10.
Grivet, P., Bertein, F., & Regenstreif, E. (1949). The use of marginal rays for the study of asymmetry in electrostatic lenses. ICEM, *Delft*. (pp. 186–189).
Grivet, P., Bernard, M.-Y., Bertein, F., Castaing, R., Gauzit, M., & Septier, A. (1955). *Optique electronique: Vol. I*. Paris: Bordas. Vol. II (1958).
Grivet, P., et al. (1965). *Electron optics* (1st ed.). Oxford: Pergamon (Revised by A. Septier; P. Hawkes, transl.); (2nd ed.) (Revised afresh by A. Septier; P. Hawkes, transl.) (Accompanied by a "popular edition").
Guinier, A. (1938a). A new type of X-ray diagram. *Comptes Rendus Hebdomadaires des Séances de l'Académie des Sciences, 206*, 1641–1643.
Guinier, A. (1938b). Structure of age hardened AlCu alloys. *Nature*, 569–570. With a note of G.D. Preston referring to *Proc. Roy. Soc. London, Ser. A, 166*, 72 (1938).
Heise, F., & Rang, O. (1949). Experimental investigations on electrostatic lenses. *Optik, 5*, 201–216.
Henriot, E. (1935). Optique électronique des systèmes centres. *Revue d'Optique Théorique et Instrumentale, 14*, 146–158.
Hillier, J. (1940). Fresnel diffraction of electrons as contour phenomenon in electron supermicroscope images. *Physical Review, 58*, 842.
Hillier, J. (1946). A study of distortion in EM projector lens. *Journal of Applied Physics, 17*, 411–419.
Hillier, J. (1947). Present status and future possibilities of the electron microscope. *RCA Review, 8*, 29–42.
Hillier, J., & Ramberg, E. G. (1947). The magnetic electron microscope objective: Contour phenomena and the attainment of high resolving power. *Journal of Applied Physics, 18*, 48–71.
Induni, G. (1945). The Swiss E.M.. *Vierteljahresschrift der Naturforschenden Gesellschaft in Zürich, 90*, 181–195.
Induni, G. (1947). On an electron source for electron microscope. *Helvetica Physica Acta, 20*, 463–466.
Johannson, H., & Scherzer, O. (1933). On the electrostatic electron converging lens. *Zeitschrift für Physik, 80*, 183–192.
Johnson, R. P., & Shockley, W. (1936). An electron microscope for filaments. *Physical Review, 49*, 436–440.
Knoll, M. (1929). German Patent 690,809.
Laplume, J. (1947). Les lentilles électroniques en mécanique relativiste. *Cahiers de Physique, 29–30*, 55–66.
Lepine, P. (1944). French electron microscope, first applications to biological research. *Bulletin de l'Académie de Médecine (Paris), 129*, 653–655.
Lepine, P. (1946). First biological images obtained with an electron microscope of French construction. *Annales de l'Institut Pasteur, 72*, 656–657.
Mahl, H. (1939a). On the electrostatic electron microscope of high resolution. *Zeitschrift für Technische Physik, 20*, 316.
Mahl, H. (1939b). Diatomeenaufnahmen mit dem elektrischen Übermikroskop. *Naturwissenschaften, 27*, 417.
Mahl, H. (1940a). Metallurgical researches with the electron microscope. *Zeitschrift für Technische Physik, 21*, 17–18.

Mahl, H. (1940b). Ein plastisches Abdruckverfahren zu übermikroskopischen Untersuchung von Metalloberflächen. *Metattwirtschafi, 19*, 488–491.
Mahl, H. (1942). Observation of surfaces with the replica method. *Naturwissenschaften, 30*, 207–217.
Maloff, J. G., & Epstein, D. W. (1938). *Electron optics in television*. New York: McGraw-Hill.
Maréchal, A. (1944). *Cahiers de Physique, 26*, 1–27.
Maréchal, A. (1947). *Revue d'Optique Théorique et Instrumentale, 26*, 257–277.
Maréchal, A. (1948). *Revue d'Optique Théorique et Instrumentale, 27*, 73–92. and 269–287.
Martin, L. C., Whelpton, R. V., & Parnum, D. H. (1937). A new electron microscope. *Journal of Scientific Instruments, 14*, 14–24.
Marton, L. (1934). La microscopie électronique des objets biologiques. *Bulletin de la Classe des Sciences. Académie Royale de Belgique [5], 20*, 439–446.
Marton, L. (1935). La microscopie électronique et ses applications. *Revue d'Optique, Théeorique et Instrumentale, 14*, 129–145.
Marton, L. (1936). Le microscope électronique. *Revue de Microbiologie Appliquée à l'Agriculture, à l'Hygiène, à l'Industrie, 2*, 117–124.
Marton, L. (1937). Le microscope électronique des objects biologiques. Cinquième communication. *Bulletin de la Classe des Sciences. Académie Royale de Belgique [5], 23*, 672–678.
Marton, L. (1940). A new electron microscope. *Physical Review, 58*, 57–60.
Marton, L. (1941a). The electron microscope. *Journal of Bacteriology, 41*, 397–413.
Marton, L. (1941b). The electron microscope in bacteriology. *Annual Review of Biochemistry, 12*, 587–614.
Marton, L. (1968). *Early history of the electron microscope. History of technology monographs*. San Francisco, California: San Francisco Press.
Morrison, G. H., & Slodzian, G. (1975). Ion microscopy. *Justus Liebigs Annalen der Chemie*, 932A–943A.
Morton, G. A. (1941). Survey of research accomplishments with the RCA electron microscope. *RCA Review, 6*, 131–166.
Muller, E. W. (1937). Electron microscopical observation of field cathodes. *Zeitschrift für Physik, 106*, 541–550.
Muller, H. O. (1942). Die Ausmessung der übermikroskopischer Objekte. *Kolloid-Zeitschrift, 99*, 6–28.
Myers, L. M. (1939). *Electron optics*. London: Chapman & Hall.
Parsons, D. F. (Series Ed.). (1978). *Annals of the New York Academy of Sciences: Vol. 306. Short wavelength microscopy*. New York: N.Y. Acad. Sci. See P. Grivet and A. Septier, *Ion Microscopy: History and Actual Trends*. (pp. 158-182).
Pierce, J. R. (1940). Rectilinear electron flow in beams. *Journal of Applied Physics, 11*, 548–555.
Prebus, A., & Hillier, J. (1939). The construction of a magnetic electron microscope of high resolving power. *Canadian Journal of Research, Section A, 17*, 49–63.
Preston, G. D. (1938a). The diffraction of x-rays by age hardened AlCu alloys. *Proceedings of the Royal Society of London. Series A, 167*, 526–538.
Preston, G. D. (1938b). The diffraction of x-rays by age hardened AlCu alloys. The structure of intermediate phase. *Philosophical Magazine, 26*, 855–871.
Ramsauer, C. (Ed.). (1940). *Das freie Elektron in Physik und Technik* [Free electrons in physics and technics]. Berlin and New York: Springer-Verlag.
Ramsauer, C. (Ed.). (1941). *Ten years of electron microscopy. A self-portrait of the A.E.G. Research Institute*. Berlin: Springer.
Ramsauer, C. (Ed.). (1943). *Elektronenmikroskopie. Report on the Research Work on Electron Microscopy. A.E.G Research Centre, 1930–1942*. Berlin: Springer.
Rang, O. (1949). The electrostatic stigmator, a corrector for astigmatic electron lenses. *Optik, 5*, 518–530.

Regenstreif, E. (1950a). France patent application 590,102.
Regenstreif, E. (1950b). Constantes optiques des lentilles électrostatiques calculées à partir de leurs dimensions. ICEM, *Paris, 1* (pp. 175–190).
Regenstreif, E. (1951). A theory of the "single" electrostatic lens with perturbed rotation symmetry. *Annales de Radioélectricité*, *6*, 244–267.
Robillard, J. (1949a). Le microscope électronique et ses applications en métallurgie. *Mikroskopie*, *1*, 65–116.
Robillard, J. (1949b). Le microscope électronique et ses applications en métallurgie. *Mikroskopie*, *1*(3), 165–182.
Ruska, E. (1930). *Investigation of electrostatic concentration as a substitute for the magnetic concentrating coils in a cathode ray oscillograph* (Diploma Project).
Ruska, E. (1979). *Die Frühe Entwicklung der Elektronenlinsen and der Elektronenmikroskopie. Acta Historica Leopoldina: Vol. 12*. Leipzig: Barth (English transl. by Thomas Mulvey, S. Hirzel, Stuttgart, 1980).
Scherzer, O. (1946a). Zur Korrigierbarkeit von Elektronenlinsen. *Physikalische Blätter*, *2*, 110.
Scherzer, O. (1946b). Germany patent application 13,554D.
Scherzer, O. (1947). *Optik*, 115.
Steigerwald, R. (1949). A new electron gun for electron microscopes. *Optik*, *5*, 469–479.
von Ardenne, M. (1939). On the performance of the electron shadow microscope and on X-ray shadow microscope. *Naturwissenschaften*, *27*, 485–486.
von Ardenne, M. (1940a). On a universal electron microscope for operation in bright field, dark field and stereoimaging. *Zeitschrift für Physik*, *115*, 329–368.
von Ardenne, M. (1940b). Results with a new supermicroscope installation. *Naturwissenschaften*, *28*, 113–127.
von Ardenne, M. (1940c). *Electron supermicroscope*. Berlin and New York: Springer-Verlag.
von Ardenne, M. (1941). A 200 kV universal electron microscope. *Zeitschrift für Physik*, *117*, 657–688.
von Borries, B. (1949). *Die Übermikroskopie*. Berlin: Saenger-Verlag.
Zworykin, V. K., & Morton, G. A. (1940). *Television* (1st ed.). New York: Wiley.
Zworykin, V. K., & Morton, G. A. (1954). *Television* (2nd ed.). New York: Wiley.
Zworykin, V. K., Morton, G. A., Ramberg, E. G., Hillier, J., & Vance, A. W. (1945). *Electron optics and the electron microscope*. New York: Wiley.

Index

F

G

H

Printed in the United States
by Baker & Taylor Publisher Services